W9-BYQ-134

Common Sense Mathematics

Ethan D. Bolker
University of Massachusetts, Boston
and
Maura B. Mast
Fordham University

Published and Distributed by
The Mathematical Association of America

MAA TEXTBOOKS

Bridge to Abstract Mathematics, Ralph W. Oberste-Vorth, Aristides Mouzakitis, and Bonita A. Lawrence

Calculus Deconstructed: A Second Course in First-Year Calculus, Zbigniew H. Nitecki

Calculus for the Life Sciences: A Modeling Approach, James L. Cornette and Ralph A. Ackerman

Combinatorics: A Guided Tour, David R. Mazur

Combinatorics: A Problem Oriented Approach, Daniel A. Marcus

Common Sense Mathematics, Ethan D. Bolker and Maura B. Mast

Complex Numbers and Geometry, Liang-shin Hahn

A Course in Mathematical Modeling, Douglas Mooney and Randall Swift

Cryptological Mathematics, Robert Edward Lewand

Differential Geometry and its Applications, John Oprea

Distilling Ideas: An Introduction to Mathematical Thinking, Brian P. Katz and Michael Starbird

Elementary Cryptanalysis, Abraham Sinkov

Elementary Mathematical Models, Dan Kalman

An Episodic History of Mathematics: Mathematical Culture Through Problem Solving, Steven G. Krantz

Essentials of Mathematics, Margie Hale

Field Theory and its Classical Problems, Charles Hadlock

Fourier Series, Rajendra Bhatia

Game Theory and Strategy, Philip D. Straffin

Geometry Illuminated: An Illustrated Introduction to Euclidean and Hyperbolic Plane Geometry, Matthew Harvey

Geometry Revisited, H. S. M. Coxeter and S. L. Greitzer

Graph Theory: A Problem Oriented Approach, Daniel Marcus

An Invitation to Real Analysis, Luis F. Moreno

Knot Theory, Charles Livingston

Learning Modern Algebra: From Early Attempts to Prove Fermat's Last Theorem, Al Cuoco and Joseph J. Rotman

The Lebesgue Integral for Undergraduates, William Johnston

Lie Groups: A Problem-Oriented Introduction via Matrix Groups, Harriet Pollatsek

Mathematical Connections: A Companion for Teachers and Others, Al Cuoco

Mathematical Interest Theory, Second Edition, Leslie Jane Federer Vaaler and James W. Daniel

Mathematical Modeling in the Environment, Charles Hadlock

Mathematics for Business Decisions Part 1: Probability and Simulation (*electronic textbook*), Richard B. Thompson and Christopher G. Lamoureux

Mathematics for Business Decisions Part 2: Calculus and Optimization (*electronic textbook*), Richard B. Thompson and Christopher G. Lamoureux

Mathematics for Secondary School Teachers, Elizabeth G. Bremigan, Ralph J. Bremigan, and John D. Lorch

The Mathematics of Choice, Ivan Niven

The Mathematics of Games and Gambling, Edward Packel

Math Through the Ages, William Berlinghoff and Fernando Gouvea

Noncommutative Rings, I. N. Herstein
Non-Euclidean Geometry, H. S. M. Coxeter
Number Theory Through Inquiry, David C. Marshall, Edward Odell, and Michael Starbird
Ordinary Differential Equations: from Calculus to Dynamical Systems, V. W. Noonburg
A Primer of Real Functions, Ralph P. Boas
A Radical Approach to Lebesgue's Theory of Integration, David M. Bressoud
A Radical Approach to Real Analysis, 2nd edition, David M. Bressoud
Real Infinite Series, Daniel D. Bonar and Michael Khoury, Jr.
Thinking Geometrically: A Survey of Geometries, Thomas Q. Sibley
Topology Now!, Robert Messer and Philip Straffin
Understanding our Quantitative World, Janet Andersen and Todd Swanson

MAA Service Center
P.O. Box 91112
Washington, DC 20090-1112
1-800-331-1MAA FAX: 1-240-396-5647

Contents

Contents

When you borrow money—on your credit card, for tuition, for a mortgage—you pay it back in installments. Otherwise what you owe would grow exponentially. In this chapter we explore the mathematics that describes paying off your debt.

Pierre de Fermat and Blaise Pascal invented the mathematics of probability to answer gambling questions posed by a French nobleman in the seventeenth century. We follow history by starting this chapter with simple examples involving cards and dice. Then we discuss raffles and lotteries, fair payoffs and the house advantage, insurance and risks where quantitative reasoning doesn't help at all.

Unlikely things happen—just rarely! Here we calculate probabilities for combinations like runs of heads and tails. Then we think about luck and coincidences.

In Chapter 12 we looked at probabilities of independent events—things that had nothing to do with one another. Here we think about probabilities in situations where we expect to see connections, such as in screening tests for diseases or DNA evidence for guilt in a criminal trial.

Preface

Philosophy

One of the most important questions we ask ourselves as teachers is "what do we want our students to remember about this course ten years from now?"

Our answer is sobering. From a ten year perspective most thoughts about the syllabus—"what should be covered"—seem irrelevant. What matters more is our wish to change the way our students' minds work—the way they approach a problem, or, more generally, the way they approach the world. Most people "skip the numbers" in newspapers, magazines, on the web and (more importantly) in financial information. We hope that in ten years our students will follow the news, confident in their ability to make sense of the numbers they find there.

To help them, we built this book around problems suggested by the news of the day as we were writing. We also consider issues that are common—and important—such as student loans. Common sense guides the analysis; we introduce new mathematics only when it's really needed. In particular, you'll find here very few problems invented just to teach particular mathematical techniques.

This preface is meant for students. There's also an instructor's manual that offers more detail about how we carry out our intentions. Students are free to read that too.

Organization

Most quantitative reasoning texts are arranged by topic—the table of contents reads like a list of mathematics to be mastered. Since the mathematics is only a part of what we hope to teach, we've chosen another strategy. Each chapter starts with a real story that can be best understood with careful reading and a little mathematics. The stories involve (among other things):

- Back of an envelope estimation.
- Discounts, inflation and compound interest.
- Income distribution in the United States.
- Reading an electricity bill.
- The graduated income tax.
- Reading a credit card bill.
- Paying off a mortgage or a student loan.

- Lotteries, gambling, insurance and the house advantage.
- False positives and the prosecutor's fallacy.

The best tool for understanding is common sense. We start there. When more is called for we practice "just in time" mathematics. The mathematics you need to understand a question appears when we ask the question. We don't ask you to learn something now because you'll need it later. You always see the mathematics in actual use.

This book differs from many at its level because we focus on how to *consume* numbers more than on how to *produce* them.

Paying attention to the numbers

We hope that when you've finished this course you will routinely look critically at the numbers you encounter every day. Questions like these should occur to you naturally:

- What do the numbers really mean?
- What makes them interesting (or not)?
- Are they consistent? Distorted?
- Do I believe them? Where do they come from?
- How might I check them?
- What conclusions can I draw from them?

To help you answer these kinds of questions we will think about:

- Relative and absolute change.
- Percentages.
- Units: all interesting numbers are numbers *of something*.
- Estimation skills. Counting zeroes: million, billions, trillions and beyond.
- Significant digits, orders of magnitude, quick and dirty mental arithmetic.
- Using a spreadsheet to ask "what-if" questions.
- Using a spreadsheet to display data.
- Models: when simple mathematics can clarify how data may be related.
- Probability and randomness.

Common sense and common knowledge

You can understand the numbers in a paragraph from the newspaper only if you understand something about the subject it addresses. Many of the discussions in the text and the exercises provide opportunities to explore—to learn things you might not know about economics or history or psychology or sociology or science or literature. When words or concepts are unfamiliar, or you're unsure of their meanings, look them up. Explore the ideas before you focus on the numbers.

The exercises

> In guessing a conundrum, or in catching a flea, we do not expect the breathless victor to give us afterwards, in cold blood, a history of the mental or muscular efforts by which he achieved success; but a mathematical calculation is another thing.
>
> Lewis Carroll
> *A Tangled Tale*
> Answers to Knot 4[R1]

Many textbooks give you a head start on the problems because they occur at the end of the section in which the relevant mathematics is taught. You need only look back a few pages to find a sample problem like the one you're working on. Often all you have to do is change the numbers.

There are few like that here. Most of ours call for a more extended solution—at least several sentences, sometimes several paragraphs. You can't simply calculate and then circle the answer on the page. On exams and on homework assignments we frequently remind you of what we expect with boilerplate like this:

> Be sure to write complete sentences. Show how you reached the answer you did. Identify any sources you used. When you refer to a website you should indicate why you think it is a *reliable* source—there's lots posted online that's just plain wrong.

Some exercises have hints at the back of the book. Try not to look at them until you've thought about the problem yourself for a while. If you invent or solve a problem and are particularly pleased with what you've done, send it to us and we'll consider incorporating it in a later edition of the book, with credit to you, of course. If you find an error, please let us know.

When you think you've finished an exercise, read your answer carefully just to see that it makes sense. If you estimate an average lawyer's annual income as $10,000 you have probably made a mistake somewhere. It's better to write "I know this is wrong, but I can't figure out why—please help me" than to submit an answer you know is wrong, hoping no one will notice.

We've annotated some of the exercises. Here's what the tags mean:

- [S]: The solution manual contains an answer.
- [U]: This exercise is untested. We think it might be a good one, but haven't yet tried it out in a class.
- [C]: This exercise is complex, or difficult, or ambiguous. Many problems in the real world are like that, so this book has a few too.
- [W]: This is a worthy exercise. It's particularly instructive, perhaps worth taking up in class.
- [R]: A routine exercise.
- [A]: An exercise with artificial numbers. Sometimes problems like this are good for emphasizing particular points, but we try to avoid them when we can.
- [N]: The idea for an exercise, with no details yet.
- [Goal $x.y$]: Contributes to mastery of Goal y of Chapter x.
- [Section $x.y$]: Depends on or adds to material in Section y of Chapter x.

The world is a messy place

When you're reading the newspaper or come across a web page or see an ad on television you're not told which chapter of the book will help you understand the numbers there. You're on your own.

When we ask open ended questions like those triggered by the news of the day our students are often uncomfortable. Here are comments expressing that discomfort, from two students, part way through the course:

- I still don't understand sometimes how we are given questions that are almost meant to confuse the reader.
- The only improvements I would make would be that we have more structured problems for homework, occasionally they can be broad.

But at the end of the semester students wrote anonymously in answer to the question "What are the strong points of this course?"

- In Quantitative Reasoning we are learning how to look into numbers instead of just looking at them.
- Use math in everyday settings instead of thinking when will I ever use this.
- It covers a real-world perspective of math.
- This course taught material that will be extremely helpful in the future.
- It teaches math that can be used every day, and skills in Excel that are useful and that I will definitely be using down the line.
- This course is very useful for me in the outside world and I feel that I will benefit from the education I received from this class and I will be able to apply my new knowledge to situations outside of the classroom.
- This course has taught me a lot about obtaining information, and using it in ways that I had not before.
- Very applicable subject matter for other areas of academics and professions. The course was something of a blend of refresher mathematics, and a new ways to apply them to everyday life.
- The math was more interesting, relevant. Good examples, news articles employed.
- I think that hardest part about this class was thinking. When you usually enter a math class the only thinking that you have to do is remember equations but in this math class I had to do research and find things on my own to help me out to answer a question.
- For the love of all that is good, why is an English major/poet/musician forced to take math all these years? I am not well-rounded or more comfortable with math, it has just drawn out my college career, costing me time and money that I don't have. I will never use math in my life, the types that I will employ I learned in elementary school. This was the best math class I have ever taken though.

Real and up to date

Our philosophy demands that the examples and exercises in *Common Sense Mathematics* pose real questions of genuine interest. Therefore they usually come from the news of the day at the time we wrote them. You don't have to go to the original sources to answer the questions, but if you're curious you can. You will find the bibliographic details in the References section at the end of the book.

One problem with our philosophy is that the text is out of date as soon as it's printed. Our remedy is to rewrite the course and add to the exercises on the fly each time we teach it. We hope other instructors will do that too. That way nothing is ever stale, which is good. What's less good is that the discussions here may not correspond to what actually happens in the course you are taking.

Common Sense Mathematics on the web

The home page for this text is `commonsensemathematics.net`. There you will find the spreadsheets we refer to, a link to a teaching blog for the years 2008-2014, errata we find (or you tell us about) and other information teachers and students of quantitative reasoning might find useful.

Technology

We wrote this text to help students understand questions where quantitative reasoning plays a part. To that end, we take advantage of any tools that will reduce drudgery and prevent careless errors.

For most applications, an ordinary four function calculator will do—and these are ubiquitous. You probably have one on your cell phone. When more advanced arithmetic is called for you can use a spreadsheet or the calculator on your computer or the internet.

You'll find references to websites in the exercises and elsewhere throughout the book. The references were accurate when we wrote the book, but we know that the web changes. If a link doesn't work, don't give up. Most likely it's moved somewhere else. A broken link isn't an excuse to skip a homework problem. Instead, be resourceful. Look around on the web or email us or your instructor.

We think an educated citizen these days should be able to refer to the internet wisely and effectively and be comfortable using a spreadsheet. In *Common Sense Mathematics* we use Excel, not because we are particularly fond of Microsoft, but because it is the most common spreadsheet in use today. But almost all our spreadsheets can be recreated in any spreadsheet program now or (we imagine) in the near future.

A spreadsheet program is good for data analysis, for asking "what if" questions and for drawing graphs. It also helps make mathematical abstractions like "function" real, rather than formal. We introduce Excel in Chapters 6 and 7. From then on we ask you to create spreadsheets, and to use simple ones we've built for you—more complex than ones you could write, but not too hard to read and understand as well as use.

We regularly refer to searching with Google, because it is the most commonly used search engine. But any other should do; use your favorite.

One search feature turns out to be particularly useful. Both Google and Bing will do arithmetic for you when you type a numerical calculation in the search field.

Old vs. new

If you know one way to do a problem should you learn another? That depends. ("That depends" is the answer to most interesting questions. If the question calls for a straightforward "yes" or "no" or just a number or something you can discover in one step with a web search the question is probably not very interesting.)

If you rarely encounter similar problems it's not worth the effort needed to understand and remember a new way to do them. But if you expect to see many, then it may pay to learn that new method.

For example, if you plan to spend just a day or so in a foreign country, get a phrase book with the common words you'll need to communicate. But if you plan to live there half the year, learn the language.

Here's a second example. When using a computer, there are many things you can do with either the mouse or the keyboard. The mouse is intuitive. You can see just what's happening, and there's nothing to remember. Just pull down the menu and click. But the keyboard is faster. So if you're going to do something just once or twice, use the mouse, but if you're going to do it a lot, learn the keyboard shortcut. In particular, in computer applications these days you often copy text from one place to paste it in another, whether that's from a web page, or in your word processor or spreadsheet. You can do that from the edit menu, or you can use the keyboard shortcuts `control-C` and `control-V`. Learn the shortcuts!

We have tried in this book to teach you new ways to do things when we think those new ways will serve you well in the future. We've resisted the temptation when those new ways are just clever tricks mathematicians are fond of that don't really help you in the long run.

Truth and beauty

We've worked to limit the mathematics we cover to just what you need, along with common sense and common knowledge, to help you deal with the quantitative parts of a complex world. But there is another important reason to study mathematics.

You read not only because it's useful, but because reading can give you access to poetry. You cook not only because you must eat to live, but because there can be pleasure in preparing tasty meals and sitting down in good company to enjoy them. We became mathematicians not only because mathematics is useful, but because (for us and some other people like us) it's beautiful, too.

This passage from Henry Wadsworth Longfellow's 1849 novel *Kavanagh* captures both the truth and the beauty of mathematics (as we hope we have).

> "For my part," [says Mary Churchill] "I do not see how you can make mathematics poetical. There is no poetry in them."
>
> "Ah, that is a very great mistake! There is something divine in the science of numbers. Like God, it holds the sea in the hollow of its hand. It measures the earth; it weighs the stars; it illumines the universe; it is law, it is order, it is beauty. And yet we imagine—that is, most of us—that its highest end and culminating point is book-keeping by double entry. It is our way of teaching it that makes it so prosaic." [R2]

Contact us

We welcome questions, feedback, suggested problems (and solutions) and notes about errors. You may contact us by email at `eb@cs.umb.edu` (Ethan Bolker) or `mmast@fordham.edu` (Maura Mast).

Acknowledgements

We owe much to many for help with *Common Sense Mathematics*.

Many years ago Linda Kime shepherded the first quantitative reasoning requirement at UMass Boston. In 2007 then Mathematics Chair Dennis Wortman allowed us to coteach Math 114 in hopes of reinventing the course. In early years Mark Pawlak pilot tested early versions of this text. He was the first to believe that we were onto a good thing—soon he was scouring the newspaper for examples to use in class, on exams, and in the exercises. His input, drawn from his long involvement in quantitative reasoning and his deep engagement with student learning, has made this a better book. When *Common Sense Mathematics* became the official textbook for quantitative reasoning at UMass Boston Mark was the course administrator, trained tutors, developed new approaches to assessing student learning and recruited instructors: George Collison, Karen Crounse, Dennis DeBay, Monique Fuguet, Matt Lehman, Nancy Levy, John Lutts, Robert Rosenfeld, Jeremiah Russell, Mette Schwartz, Joseph Sheppeck, Mitchell Silver, Karen Terrell, Charles Wibiralske and Michael Theodore Williams.

We benefited from feedback from colleagues at other schools who asked to use the text: Margot Black (Lewis & Clark), Samuel Cook (Wheelock College), Grace Coulombe (Bates College), Mike Cullinane (Keene State College), Timothy Delworth (Purdue University), Richard Eells (Roxbury Community College), Marc Egeth (Pennsylvania Academy of the Fine Arts), Ken Gauvreau (Keene State College), Krisan Geary (Saint Michael's College), David Kung (St. Mary's College of Maryland), Donna LaLonde (Washburn University), Carl Lee (University of Kentucky), Alex Meadows (St. Mary's College of Maryland), Wesley Rich (Saginaw Chippewa Tribal College), Rachel Roe-Dale (Skidmore College), Rob Root (Lafayette College), Barbara Savage (Roxbury Community College), Q. Charles Su (Illinois State University), Joseph Witkowski (Keene State College) and several anonymous reviewers.

Students caught typos, suggested rewordings and provided answers to exercises. We promised to credit them here: Courtney Allen, Matt Anthony, Vladimir Altenor, Theresa Aluise, Selene Bataille, Kelsey Bodor, Quonedell Brown, Katerina Budrys, Candace Carroll, Jillian Christensen, Katie Corey, Molly Cusano, Sam Daitsman, Hella Dijsselbloem-Gron, Michelle DiMenna, Shirley Elliot, Lea Ferone, Solomon Fine, Murray Gudesblat, Frances Harangozo, Irene Hartford, Katilyn Healey, Anna Hodges, Anthony Holt, Laura Keegan, Jennifer Kunze, Kevin Lockwood, Jacob Looney, Ashley McClintock, Edward McConaghy, Nicole McKenna, Amanda Miner, Antonio de las Morenas, Daniel Murano, Hannah Myers, Matt Nickerson, Rodrigo Nunez, Gabby Phillips, Jaqueline Ramirez, Hailey Rector, Taylor Spencer, Jaran Stallbaum, Melinda Stein, Willow Smith, Nick Sullivan, Robert Tagliani, Julia Tran, Marcus Zotter, ... and many others.

Cong Liu worked on the index. Monica Gonzalez and Alissa Pellegrino tested the links to the web. Paul Mason ferreted out the newspaper headlines for the cover.

The National Science Foundation provided support from grant DUE-0942186. Any opinions, findings and conclusions or recommendations expressed in this material are those of the contributors and do not necessarily reflect the views of the National Science Foundation. We hope they approve of what we've done with their generosity.

The Boston Globe graciously gave us blanket permission to reproduce here the quotes we found in our morning paper and brought to class.

Wizards at `tex.stackexchange.com` were always quick to answer TeXnical questions.

Carol Baxter, Stephen Kennedy, Beverly Ruedi and Stanley Seltzer at the Mathematical Association of America were enthusiastic about our book and brought wisdom and competence to design and production.

Ethan: My wife Joan's ongoing contribution began 56 years ago when she asked me how I'd feel if I went to medical school and did no more mathematics. It continued with constant support of all kinds—most of the details would be inappropriate here. I will say that I recommend living with a writing coach to hone writing strategies. I've talked for years with my professor children about mathematics and teaching—Jess and Ben make cameo appearances in the text. I dedicate *Common Sense Mathematics* to the next generation: Solomon Bixby and Eleanor Bolker.

Maura: I owe a greater debt than I could ever express to my husband Jack Reynolds. We met in Boston over 21 years ago (thanks in part to a National Science Foundation grant). Living in Iowa three years later, we saw that UMass Boston wanted to hire a mathematician to work on quantitative reasoning. With Jack's support, I left a tenured position to accept that challenge. I couldn't have done that and my other work at UMass as well as I did without his faith in me and his support. There's more. Jack has been my conscience as well as my partner. He fundamentally believes that each person can make a difference in the world. Because of that, I now see my work in quantitative reasoning, and mathematics, as a way to change the world. I thank my children Brendan, Maeve and Nuala Reynolds for their patience and support, especially for the times when I turned their questions into homework problems, as in the tooth fairy exercise. I dedicate *Common Sense Mathematics* to the memory of my parents, Cecil and Mary Mast, who set high ideals grounded in reality. I miss them dearly.

Newton, MA and The Bronx, NY
October 2015

1

Calculating on the Back of an Envelope

In this first chapter we learn how to think about questions that need only good enough answers. We find those answers with quick estimates that start with reasonable assumptions and information you have at your fingertips. To make the arithmetic easy we round numbers drastically and count zeroes when we have to multiply.

Chapter goals

Goal 1.1. Verify quantities found in the media, by checking calculations and with independent web searches.

Goal 1.2. Estimate quantities using common sense and common knowledge.

Goal 1.3. Learn about the Google calculator (or another internet calculator).

Goal 1.4. Round quantities in order to use just one or two significant digits.

Goal 1.5. Learn when not to use a calculator—become comfortable with quick approximate mental arithmetic.

Goal 1.6. Work with large numbers.

Goal 1.7. Work with (large) metric prefixes

Goal 1.8. Use straightforward but multi-step conversions to solve problems.

1.1 Billions of phone calls?

On September 20, 2007, the United States House of Representatives Permanent Select Committee on Intelligence met to discuss legislation (the Protect America Act) that expanded the government's surveillance powers. The committee was concerned with the balance between protecting the country and preserving civil liberties. They questioned Admiral Michael McConnell, Director of National Intelligence, about governmental monitoring of international telephone calls. In the course of the hearing, Admiral McConnell said that he did not know how many Americans' telephone conversations may have been overheard through US wiretaps on foreign phone lines saying "I don't have the exact number … considering there are billions of transactions every day." [R3]

McConnell knows he's not reporting an exact number. Is his claim that there are "billions of transactions" a genuine estimate, or just a way of saying "lots and lots of transactions?" We can find out using just a little arithmetic and a little common sense.

In 2007 there were about 300 million people in the United States. There are more now, but 300 million is still often a good approximation for back-of-an-envelope calculations.

If everyone in the United States talked daily on a foreign phone call that would come to about 300 million calls. That's 300,000,000: 3 with eight zeroes. McConnell talked about "billions." He didn't say how many, but "billion" means nine zeroes, which is ten times as large. To get to billions of phone calls in a day each person in the country would have to make ten of them. That doesn't seem to make sense.

So when McConnell says "I don't have the exact number" he probably does mean just lots and lots.

We didn't need pencil and paper, let alone a calculator, to do the arithmetic—we needed one fact (the population of the United States), and then simply counted zeroes.

But before accusing McConnell of fudging the numbers, we should examine our own assumptions. Several years after this hearing *The Washington Post* reported that

> Every day the National Security Agency intercepts and stores *1.7 billion* international e-mails, phone calls, texts and other communications. [R4]

So in 2010 there were indeed billions of transactions each day if email and other electronic communications are counted along with telephone calls. There were probably fewer in 2007, quite probably still billions.

In the spring of 2013 this issue received new media attention when the extent of NSA data collection became public. The issue then wasn't intercepting and storing international communications, it was collecting and storing *metadata* about domestic telephone calls: who called whom, and when, although not what was said.

1.2 How many seconds?

Have you been alive for a thousand seconds? a million? a billion? a trillion?

Before we estimate, what's your guess? Write it down, then read on.

To check your estimate, you have to do some arithmetic. There are two ways to go about the job. You can start with seconds and work up through hours, days and years, or start with thousands, millions and billions of seconds and work backwards to hours, days and years. We'll do it both ways.

How many seconds in an hour? Easy: $60 \times 60 = 3600$. So we've all been alive much more than thousands of seconds.

Before we continue, we're going to change the rules for arithmetic so that we can do all the multiplication in our heads, without calculators or pencil and paper. We will round numbers so that they start with just one nonzero digit, so 60×60 becomes 4000. Of course we can't say $60 \times 60 = 4{,}000$; the right symbol is $\approx$, which means "is approximately." Then an hour is

$$60 \times 60 \approx 4000$$

seconds.

There are 24 hours in a day. $4 \times 24 \approx 100$, so there are

$$4{,}000 \times 24 \approx 100{,}000$$

seconds in a day.

Or we could approximate a day as 20 hours, which would mean (approximately) 80,000 seconds. We'd end up with the same (approximate) answer.

Since there are about a hundred thousand seconds in a day, there are about a million seconds in just 10 days. That's not even close to a lifetime, so we'll skip working on days, weeks or months and move on to years.

How many seconds in a year? Since there are (approximately) 100,000 in a day and (approximately) 400 days in a year there are about 40,000,000 (forty million) seconds in a year.

If we multiply that by 25 the 4 becomes 100, so a 25-year old has lived for about 1,000,000,000 (one billion) seconds.

Does this match the estimate you wrote down for your lifetime in seconds?

The second way to estimate seconds alive is to work backwards. We'll write the time units using fractions—that's looking ahead to the next chapter—and round the numbers whenever that makes the arithmetic easy. Let's start with 1000 seconds.

$$\begin{aligned} 1000\ \cancel{\text{seconds}} \times \frac{1 \text{ minute}}{60\ \cancel{\text{seconds}}} &= \frac{1000}{60} \text{ minutes} \\ &= \frac{100}{6} \text{ minutes (cancel a 0)} \\ &= \frac{50}{3} \text{ minutes (cancel a 2)} \\ &\approx \frac{60}{3} \text{ minutes (change 50 to 60—make division easy)} \\ &= 20 \text{ minutes.} \end{aligned}$$

We're all older than that.

How about a million seconds? A million has six zeroes—three more than 1,000, so a million seconds is about 20,000 minutes. Still too many zeroes to make sense of, so convert to something we can understand—try hours.

$$20{,}000\ \cancel{\text{minutes}} \times \frac{1 \text{ hour}}{60\ \cancel{\text{minutes}}} = \frac{20{,}000}{60} \text{ hours} = \frac{1{,}000}{3} \text{ hours} \approx 300 \text{ hours.}$$

There are 24 hours in a day. To do the arithmetic approximately use 25. Then $300/25 = 12$ so 300 hours is about 12 days. We've all been alive that long.

How about a billion seconds? A billion is a thousand million, so we need three more zeroes. We can make sense of that in years:

$$12{,}000\ \cancel{\text{days}} \times \frac{1 \text{ year}}{365\ \cancel{\text{days}}} = \frac{12{,}000}{365} \text{ years} \approx \frac{12{,}000}{400} \text{ years} \approx 30 \text{ years.}$$

Since a billion seconds is about 30 years, it's in the right ballpark for the age of most students.

A trillion is a thousand billion—three more zeroes. So a trillion seconds is about 30,000 years. Longer than recorded history.

1.3 Heartbeats

In *The Canadian Encyclopedia* a blogger noted that

> The human heart expands and contracts roughly 100,000 times a day, pumping about 8,000 liters of blood. Over a lifetime of 70 years, the heart beats more than 2.5 billion times, with no pit stops for lube jobs or repairs. [R5]

Should we believe "100,000 times a day" and "2.5 billion times in a lifetime"?

If you think about the arithmetic in the previous section in a new way, you may realize you've already answered this question. Since your pulse rate is about 1 heartbeat per second, counting seconds and counting heartbeats are different versions of the same problem. We discovered that there are about 100,000 seconds in a day, so the heartbeat count is about right. We discovered that 30 years was about a billion seconds, and 70 is about two and a half times 30, so 70 years is about 2.5 billion seconds. Both the numbers in the article make sense.

Even if we didn't know whether 100,000 heartbeats in a day was the right number, we could check to see if that number was consistent with 2.5 billion in a lifetime. To do that, we want to calculate

$$100{,}000\frac{\text{beats}}{\text{day}} \times 365\frac{\text{days}}{\text{year}} \times 70\frac{\text{years}}{\text{lifetime}}.$$

Since we only need an approximate answer, we can simplify the numbers and do the arithmetic in our heads. If we round the 365 up to 400 then the only real multiplication is $4 \times 7 = 28$. The rest is counting zeroes. There are eight of them, so the answer is approximately 2,800,000,000. That means the 2.5 billion in the article is about right. Our answer is larger because we rounded up.

The problems we've tackled so far don't have exact numerical answers of the sort you are used to. The estimation and rounding that goes into solving them means that when you're done you can rely on just a few *significant digits* (the digits at the beginning of a number) and the number of zeroes. Often, and in these examples in particular, that's all you need. Problems like these are called "Fermi problems" after Enrico Fermi (1901–1954), an Italian physicist famous (among other things) for his ability to estimate the answers to physical questions using very little information.

1.4 Calculators

The thrust of our work so far has been on mental arithmetic. You can always check yours with a calculator. You probably have one on the phone in your pocket. There's one on your computer. But those require pressing keys or clicking icons. If you have internet access, Google's is easier to use—simply type

```
100,000 * 365 * 70
```

into the search box. Google displays a calculator showing

```
2555000000
```

.

That 2.555 billion answer is even closer than our first estimate to the 2.5 billion approximation in the article. The Bing search engine offers the same feature.

You can click on the number and operation keys in the Google calculator to do more arithmetic. Please don't. Just type an expression in the Google search bar. Stick to the keyboard rather than the mouse. It's faster, and you can fix typing mistakes easily.

The Google calculator will do more than just the arithmetic—it can keep track of units. Although it doesn't deal with heartbeats, it does know about miles, and speeds like miles per day and miles per year. We can make it do our work for us by asking about miles instead of heartbeats. Search for

100,000 miles per day in miles per 70 years

and Google rewards you with

100000 (miles per day) = 2.55669539 × 10^9 miles per (70 years)

.

The "$\times 10^9$" means "add nine zeroes" or, in this case, "move the decimal point nine places to the right", so

$$\begin{aligned}100000 \text{ (miles per day)} &= 2.55669539 \times 10^9 \text{ miles per (70 years)}\\ &\approx 2.6 \text{ billion miles per (70 years).}\end{aligned}$$

That is again "more than 2.5 billion."

The exact answer from Google is even a little more than the 2,555,000,000 we found when we did just the arithmetic since Google knows a year is a little longer than 365 days—that's why we have leap years.

So 100,000 heartbeats per day does add up to about 2.5 billion in 70 years. We've checked that the numbers are consistent—they fit together.

But are they correct? Does your heart beat 100,000 times per day? To think sensibly about a number with lots of zeroes we can convert it to a number of something equivalent with fewer zeroes—in this case, heartbeats per minute. That calls for division rather than multiplication:

$$100{,}000 \frac{\text{beats}}{\text{day}} \times \frac{1 \text{ day}}{24 \text{ hours}} \times \frac{1 \text{ hour}}{60 \text{ minutes}}.$$

To do the arithmetic in your head, round the 24 to 25. Then $25 \times 6 = 150$—there are about 1,500 minutes in a day. Then $100{,}000/1{,}500 = 1{,}000/15$. Since $100/15$ is about 7, we can say that $1{,}000/15$ is about 70. 70 beats per minute is a reasonable estimate for your pulse rate, so 100,000 heartbeats per day is about right.

Google tells us

100 000 (miles per day) = 69.4444444 miles per minute

.

The nine and all the fours in that 69.4444444 are much too precise. The only sensible thing to do with that number is to round it to 70—which is what we discovered without using a calculator.

Sometimes even the significant digits can be wrong and the answer right, as long as the number of zeroes is correct. Informally, that's what we mean when we say the answer is "*in the right ballpark*." The fancy way to say the same thing is "the *order of magnitude* is correct." For example, it's right to say there are hundreds of days in a year—not thousands, not tens. There are billions (nine zeroes) of heartbeats in a lifetime, not hundreds of millions (eight zeroes), nor tens of billions (ten zeroes).

1.5 Millions of trees?

On May 4, 2010 Olivia Judson wrote in *The New York Times* [R6] about Baba Brinkman, who describes himself on his web page as

> ... a Canadian rap artist, playwright, and former tree-planter who worked in the Rocky Mountains every summer for over ten years, personally planting more than one million trees. He is also a scholar with a Masters in Medieval and Renaissance English Literature. [R7]

How long would it take to personally plant a million trees? Is Brinkman's claim reasonable?

To answer that question you need two estimates—the time it takes to plant one tree, and the time Brinkman may have spent planting. We have some information about the second of these—more than ten summers.

To plant a tree you have to dig a hole, put in a seedling and fill in around the root ball. It's hard to imagine you can do that in less than half an hour.

If Brinkman worked eight hours a day he would plant 16 trees per day. Round that up to 20 trees per day to make the arithmetic easier and give him the benefit of the doubt. At that rate it would take him $1{,}000{,}000/20 = 50{,}000$ days to plant a million trees. If he planted trees 100 days each year, it would take him 500 years; if he planted trees for 200 days out of the year, it would take him 250 years. So his claim looks unreasonable.

What if we change our estimates? Suppose he took just ten minutes to plant each tree and worked fifteen hour days. Then he could plant nearly 100 trees per day. At that rate it would take him 10,000 days to plant a million trees. If he worked 100 days each summer he'd still need about 100 years. That's still much more than the "more than 10 years" in the quotation. So on balance we believe he's planted lots of trees, but not "personally ... more than one million."

It's the "personally" that makes this very unlikely. We can believe the million trees if he organized tree-planting parties, perhaps with people manning power diggers of some kind. Or if planting acorns counted as planting trees.

This section, first written in 2010, ended with that unfunny joke until 2013, when Charles Wibiralske, teaching from this text, wondered if we might be overestimating the time it takes to plant a tree. To satisfy his curiosity, he found Brinkman's email address and asked. The answer was a surprising (to him, and to us) ten seconds! So our estimate of 10 minutes was 60 times too big. That means our 100-year estimate should really have been only about two years! That's certainly possible. If it took him a minute per tree rather than 10 seconds he could still have planted a million trees in ten summers.

Brinkman tells the story of Wibiralske's question and this new ending in his blog at `www.bababrinkman.com/insult-to-injury/`. When you visit you can listen to "The Tree Planter's Waltz" (`www.youtube.com/watch?v=jk-jifbpcww`).

The moral of the story: healthy skepticism about what you read is a good thing, as long as you're explicit and open minded about the assumptions you make when you try to check. That's a key part of using common sense.

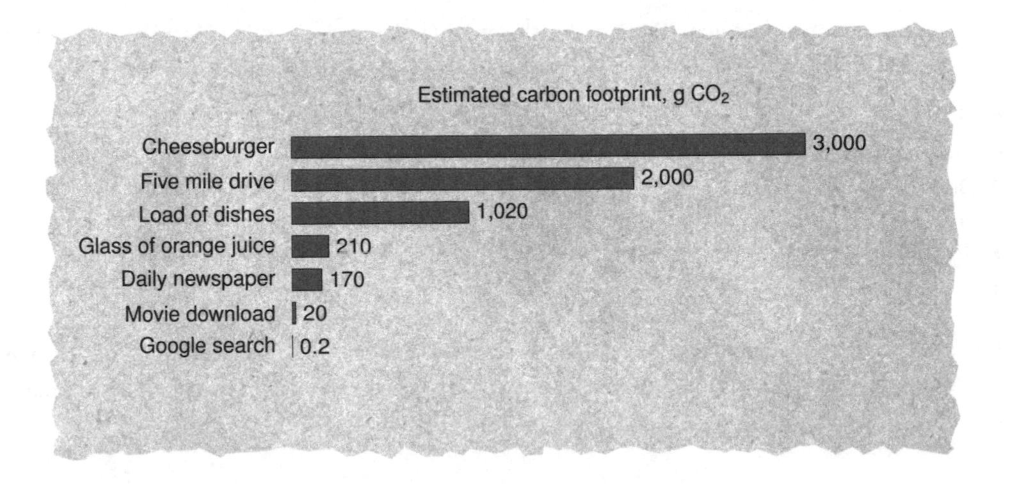

Figure 1.1. Carbon footprints [R8]

Brinkman's blog ends this way:

> Hurray! In the end it's a classic example of . . . the drunkard's walk towards knowledge. When our views are self-correcting and open to revision based on new evidence, they will continue to hone in on increasingly accurate representations of the real world. That's good honest skepticism, and when it wins over bad, knee-jerk, "it's hard to imagine" skepticism, that's a beautiful thing.

1.6 Carbon footprints

Discussions about global warming and climate change sometimes talk about the *carbon footprint* of an item or an activity. That's the total amount of carbon dioxide (CO_2) the item or activity releases to the atmosphere. An article in *The Boston Globe* on October 14, 2010 listed estimates of carbon footprints for some common activities. Among those was the 210 gram carbon footprint of a glass of orange juice. That includes the carbon dioxide cost of fertilizing the orange trees in Florida, harvesting the oranges and the carbon dioxide generated burning oil or coal to provide energy to squeeze the oranges, concentrate and freeze the juice and then ship it to its destination. It's just an estimate, like the ones we're learning to make, but much too complex to ask you to reproduce. We drew Figure 1.1 using the rest of the data from the article.

Let's look at the orange juice. How many glasses are consumed in the United States each day? If we estimate that about 5% of the 300 million people in the country have orange juice for breakfast that means 15 million glasses. So the ballpark answer is on the order of 10 to 20 million glasses of orange juice. We can use that to estimate the total orange juice carbon footprint: each glass contributes 200 grams, so 10 to 20 million glasses contribute 2 to 4 billion grams each day.

It's hard to imagine 2 billion grams. You may have heard of kilograms —a kilogram is about two pounds. In the metric system *kilo* means "multiply by 1,000" so a kilogram is 1,000 grams. Then 2 billion grams of carbon is just 2 million kilograms. That's about 4 million pounds, or 2 thousand tons.

Since there are seven activities listed in the graphic, there are six other Fermi problems like this one for you to work on:

- Google search
- Movie download
- Daily newspaper printed
- Dishwasher run
- Five miles driven
- Cheeseburger consumed

For each you can estimate the total number of daily occurrences and then the total daily carbon contribution. We won't provide answers here because we don't want to spoil a wonderful class exercise.

1.7 Kilo, mega, giga

Counting zeroes is often best done three at a time. That's why we separate groups of three digits by commas. Each step from thousands to millions to billions adds three zeroes. The metric system has prefixes for that job.

We've seen that *kilo* means "multiply by 1,000". Similarly, *mega* means "multiply by 1,000,000". A Megabucks lottery prize is millions of dollars. A megathing is 1,000,000 things, whatever kind of thing you are interested in.

New Hampshire's Seabrook nuclear power plant is rated at 1,270 megawatts. So although you may not know what a watt is, you know this power plant can generate 1,270,000,000 of them. The symbols for "mega" and for "watt" are "M" and "W" so you can write 1,270 megawatts as 1,270 MW.

How many 100 watt bulbs can Seabrook light up? Just take the hundred's two zeroes from the mega's six, leaving four following the 1,270. Putting the commas in the right places, that means 12,700,000 bulbs. Or, if you want to be cute, about 13 megabulbs.

Next after mega is *giga*: nine zeroes. The symbol is "G". When you say "giga" out loud the G is hard, even though it's soft in the word "gigantic".

You could describe Seabrook as a 1.27 gigawatt power plant.

Table 1.2 describes the metric prefixes bigger than giga. There's no need to memorize it. You will rarely need the really big ones. You can look them up when you do.

The metric system also has prefixes for shrinking things as well as these for growing them. Since division is harder than multiplication, we'll postpone discussing those prefixes until we need them, in Section 2.7.

1.8 Exercises

Notes about the exercises:

- The preface has information about the exercises and the solutions; we suggest you read it.
- One of the ways to improve your quantitative reasoning skills is to write about what you figure out. The exercises give you many opportunities to practice that. The answer to a question should be more than just a circle around a number or a simple "yes" or "no". Write complete

Table 1.2. Metric prefixes

Name	Symbol	Meaning	English name	Zeroes
Kilo	K	$\times 10^3$	thousand	3
Mega	M	$\times 10^6$	million	6
Giga	G	$\times 10^9$	billion	9
Tera	T	$\times 10^{12}$	trillion	12
Peta	P	$\times 10^{15}$	quadrillion	15
Exa	E	$\times 10^{18}$	quintillion	18
Zetta	Z	$\times 10^{21}$	sextillion	21
Yotta	Y	$\times 10^{24}$	septillion	24

paragraphs that show your reasoning. Your answer should be complete enough so that you can use a corrected homework paper to study for an exam without having to go back to the text to remember the questions.

- It often helps to write about doubt and confusion rather than guessing at what you hope will turn out to be a right answer.
- To solve Fermi problems you make assumptions and estimates. It's those skills we are helping you develop. The web can help, but you should not spend a lot of time searching for answers to the particular questions we ask. That's particularly true since many of the problems don't have a single right answer.
- Be sure to make your estimates and assumptions explicit. Then you can change them easily if necessary, as we did in Section 1.5.

Your instructor has a Solutions Manual with answers to the exercises, written the way we hope you will write them. Ask him or her to provide some for you.

Exercise 1.8.1. [S][Section 1.1] Warren Buffet is very rich.

In *The Boston Globe* on September 30, 2015 you could read that Warren Buffet is worth $62 billion, and that

> ... if [he] gave up on aggressive investing and put his money into a simple savings account, with returns at a bare 1 percent, he'd earn more in interest each hour than the average American earns in a year. [R9]

Use the data in this story to estimate the annual earnings of the average American. Do you think the estimate is reasonable?

Exercise 1.8.2. [S][Section 1.1][Goal 1.1][Goal 1.2][Goal 1.4] Dropping out.

In his May 17, 2010 op-ed column in *The New York Times* Bob Herbert noted that the dropout rate for American high school students was one every 26 seconds. [R10]

Is this number reasonable?

Exercise 1.8.3. [S][Section 1.1][Goal 1.1][Goal 1.2][Goal 1.5][Goal 1.8] Bumper sticker politics.

In the fall of 2015 at donnellycolt.com you could buy a button that said

> Every Minute 30 Children Die of Hunger and Inadequate Health Care While the World Spends $1,700,000 on War

or a small vinyl sticker with the claim

> Every Minute the World Spends $700,000 on War While 30 children Die of Hunger & Inadequate Health Care. [R11]

What are these items trying to say? Do the numbers make sense?

Your answer should be a few paragraphs combining information you find on the web (cite your sources—how do you know they are reliable?) and a little arithmetic.

Exercise 1.8.4. [S][Section 1.1][Goal 1.1][Goal 1.2][Goal 1.4][Goal 1.8] Two million matzoh balls.

In 2012 a modest restaurant in Newton Centre, MA advertised

Johnny's Luncheonette
Over 2 Million
Matzoh Balls
served!

A year or so later students thinking about whether it was reasonable found that Johnny's web site at www.johnnysluncheonette.com/ claimed "Over 1 Million Matzoh Balls Served!"

What would you believe?

[See the back of the book for a hint.]

Exercise 1.8.5. [S][Section 1.1][Goal 1.2][Goal 1.5] *Writing Your Dissertation in Fifteen Minutes a Day.*

Joan Bolker's book with that title sold about 120,000 copies in the first fifteen years since its publication in 1998.

Estimate the fraction of doctoral students who bought this book.

Exercise 1.8.6. [S][W][Section 1.1][Goal 1.1][Goal 1.2][Goal 1.4][Goal 1.5] Smartphone apps may help retail scanning catch on.

Some grocery stores are experimenting with a new technology that allows customers to scan items as they shop. Once the customer is done, he or she completes the transaction online and never has to stand in the checkout line. On March 11, 2012 *The Boston Globe* reported that

> Modiv Media's scan-it-yourself technology [is installed] in about 350 Stop & Shop and Giant stores in the United States. Many consumers have embraced the system; Stop & Shop spokeswoman Suzi Robinson said the service handles about one million transactions per month. [R12]

(a) Estimate the number of customers per day per store who use this self-scanning technology.
(b) Estimate the number of customers per day per store.
(c) Estimate the percentage of customers who use the technology.

Exercise 1.8.7. [S][Section 1.1][Goal 1.1][Goal 1.2][Goal 1.4][Goal 1.5] Health care costs for the uninsured.

On March 22, 2012 Linda Greenhouse wrote in *The New York Times* that the average cost of a family insurance policy increased by $1,000 a year because health care providers needed to recover $43 billion annually for health care costs of the uninsured. [R13]

(a) What it the annual cost per United States resident for medical care for the uninsured?
(b) Estimate (or research) how much an average person spends each year buying food at the grocery store (do not include restaurant purchases). Compare your answer to your answer from the previous question about the cost of medical care.
(c) Estimate (or research) the number of uninsured people, and then estimate the cost per uninsured person for medical care. Does your answer make sense to you?

Exercise 1.8.8. [S][Section 1.2][Goal 1.1] Is 25 the same as 30?

In Section 1.2 we showed that a 25 year old has lived for about a billion seconds. Then we estimated that a billion seconds is about 30 years.

(a) Explain why we got two different answers—25 and 30 years.
(b) Are the answers really different?
(c) Compare them to what the Google calculator says about a billion seconds.

Exercise 1.8.9. [S][Section 1.2][Goal 1.1][Goal 1.2][Goal 1.4][Goal 1.5] Spoons around the world.

According to the website `www.majorwastedisposal.com/trashtrivia.html`, Americans throw out enough plastic utensils (knives, forks and spoons) every year to circle the equator 300 times.

(a) Use the information stated at the beginning of the problem to estimate the average number of plastic utensils each person throws out each year.
(b) Is the assertion reasonable?

Exercise 1.8.10. [S][Section 1.2][Goal 1.1][Goal 1.2][Goal 1.3] 1,000,000,000,000,000,000,000,000,000.

On July 25, 2010 Christopher Shea wrote in *The Boston Globe* about "hella", a new metric prefix popular among geeks in northern California.

> Austin Sendek, a physics major at the University of California Davis, wants to take "hella" from the streets and into the lab. With the help of a Facebook-driven public relations campaign, he's petitioning the Consultative Committee on Units, a division of the very serious Bureau International des Poids et Mesures, to anoint "hella" as the official term for a previously unnamed, rather large number: 10 to the 27th power. (The diameter of the universe, by Sendek's reckoning, is 1.4 hellameters.) [R14]

(a) Does the Google calculator know about hella?
(b) Does the Bing search engine know about hella?

Exercise 1.8.11. [S][Section 1.2][Goal 1.2][Goal 1.4][Goal 1.5] Counting fish.

In *To the Top of the Continent* Frederick Cook wrote about this incident in his Alaska travels.

> The run of the hulligans was very exciting... Mr. Porter's thoughts ran to mathematics, he figured that the train of hulligans was twelve inches wide and six inches deep and that it probably extended a hundred miles. Estimating the number of fish in a cubic foot at ninety-one and one half, he went on to so many millions that he gave it up, suggesting that we try and catch some. [R15]

(a) How many millions of hulligans did Mr. Porter try to count?
(b) What's wrong with the precision in this paragraph?
(c) What's a hulligan? Cook's book was published in 1908. Are there any hulligans around today?

Exercise 1.8.12. [S][Section 1.2][Goal 1.1] Millions jam street-level crime map website.

In early 2011, the British government introduced a crime-mapping website that allows people to see crimes reported by entering a street name. The launch of the site, however, was problematic. The BBC reported that the web site was jammed by up to five million hits per hour, about 75 thousand a minute. [R16]

(a) Are the figures "five million an hour" and "75,000 a minute" consistent?
(b) Estimate the fraction of the population of London trying to look at that web site. Does your answer make sense?

[See the back of the book for a hint.]

Exercise 1.8.13. [S][Section 1.2][Goal 1.4][Goal 1.5][Goal 1.8] The popularity of social networks.

From *The New York Times*, June 28, 2011:

> In May [2011], 180 million people visited Google sites, including YouTube, versus 157.2 million on Facebook, according to comScore. But Facebook users looked at 103 billion pages and spent an average of 375 minutes on the site, while Google users viewed 46.3 billion pages and spent 231 minutes. [R17]

(a) How many webpages did the average Facebook user visit? How many webpages did the average Google user visit?
(b) On average, how many webpages per day in May did a Facebook user visit? Compare this to the average number of webpages per day for the Google user.
(c) On average, who spent more time on each page, the Facebook user or the Google user?
(d) Think about your own web behavior. Do these numbers seem reasonable to you?

Exercise 1.8.14. [S][Section 1.2][Goal 1.1][Goal 1.2][Goal 1.5][Goal 1.8] How rich is rich?

On September 19th, 2011 Aaron S. from Florida posted a comment at *The New York Times* in which he claims that if you left one of the 400 wealthy people with just a billion dollars he could not spend his fortune in 30 years at a rate of $100,000 a day. [R18]

Is Aaron's arithmetic right? Can you do this calculation without a calculator? Without pencil and paper?

Exercise 1.8.15. [S][Section 1.2][Goal 1.1][Goal 1.2][Goal 1.5][Goal 1.8] Greek debt.

In her September 26, 2011 *New York Times* review of Michael Lewis's book *Boomerang* Michiko Kakutani noted that Lewis said Greek debt of $1.2 trillion amounted to about $250,000 for each working Greek. [R19]

(a) Use Lewis's statement to estimate the population of working Greeks at the time he wrote the article.
(b) Use the web to find the national debt and population of Greece (for 2011 if you can, now if you can't).
(c) Do the answers to the previous two parts of this exercise agree? If not, what might explain any differences?
(d) Compare Greek per capita national debt to that in the United States.
(e) Here's a political question: is large national debt a bad thing? You can find both "yes" and "no" answers on the web. Here's one place to start: `www.npr.org/templates/story/story.php?storyId=99927343`.

Exercise 1.8.16. [S][Section 1.3][Goal 1.1][Goal 1.4][Goal 1.6] No Lunch Left Behind.

From *The New York Times*, Feb 20, 2009, in a column by Alice Waters and Katrina Heron with that headline:

> How much would it cost to feed 30 million American schoolchildren a wholesome meal? It could be done for about $5 per child, or roughly $27 billion a year, plus a one-time investment in real kitchens. [R20]

There are three numbers in the paragraph. Are they reasonable? Are they consistent with each other and with other numbers you know?

Exercise 1.8.17. [S][Section 1.3][Goal 1.2][Goal 1.4][Goal 1.5] Brush your teeth twice a day—but turn off the water.

The Environmental Protection Agency says on its website that

> You can save up to 8 gallons of water by turning off the faucet when you brush your teeth in the morning and before bedtime. [R21]

(a) Estimate how much water a family of four would use each week, assuming they left the water running while brushing.
(b) Estimate how much water would be saved in one day if the entire United States turned off the faucet while brushing.
(c) Put your answer to the previous question in context (compare it to the volume of water in a lake or a swimming pool, for example).
(d) Realistically, you need some water to brush your teeth because you need to get the toothbrush wet and you need to rinse the brush and your teeth. Estimate how much water that involves, per brushing, then redo the estimate in part (b).

Exercise 1.8.18. [W][S][Section 1.3][Goal 1.1][Goal 1.2][Goal 1.5] Lady Liberty.

On May 9, 2009 *The Boston Globe* reported that the Statue of Liberty's crown will reopen:

> Safety and security issues have been addressed, and 50,000 people, 10 at a time, will get to visit the 265-foot-high crown in the next two years before it is closed again for renovation, Interior Secretary Ken Salazar said yesterday. [R22]

Estimate how long each visitor will have in the crown to enjoy the view.

Exercise 1.8.19. [S][Section 1.3][Goal 1.2][Goal 1.5] Look ma! No zipper!

The bag of Lundberg Zipper Free California White Basmati Rice advertises

> We've removed the re-closable zipper from our two pound bags, which will save about 15% of the material used to make the bag, which will save 35,000 lbs. of plastic from landfills every year.

(a) How much plastic is still ending up in landfills?

(b) What fact would you need to figure out how many bags of rice Lundberg sells each year? Estimate that number, and then estimate the answer.

Exercise 1.8.20. [S][Section 1.3][Goal 1.4][Goal 1.5][Goal 1.8] Leisure in Peru.

On page 32 in *The New Yorker* on December 7, 2009, Lauren Collins wrote that late arrivals in Peru are said to amount to three billion hours each year. [R23]

We suspect that the source of Collins's assertion is an article in the July 1, 2007 edition of *Psychology Today* that commented on the campaign for punctuality and included a feature called "Tardiness by the Numbers" that provided the data:

- 107 hours: annual tardiness per Peruvian
- $5 billion: cost to the country
- 84%: Peruvians who think their compatriots are punctual only "sometimes" or "never"
- 15%: think tardiness is a local custom that doesn't need fixing [R24]

(a) According to Collins, how late are Peruvians, in hours per person per day?

(b) Is your answer to the previous question consistent with the numbers in the *Psychology Today* article?

(c) Is the $5 billion "cost to the country" a reasonable estimate?

[See the back of the book for a hint.]

Exercise 1.8.21. [R][S][Section 1.3][Goal 1.1] The white cliffs of Dover.

In his essay "Season on the Chalk" in the March 12, 2007 issue of *The New Yorker* John McPhee wrote

> The chalk accumulated at the rate of about one millimetre in a century, and the thickness got past three hundred metres in some thirty-five million years. [R25]

Check McPhee's arithmetic.

Exercise 1.8.22. [S][C][Section 1.3][Goal 1.1][Goal 1.4][Goal 1.5] Social media and internet statistics.

In January 2009 Adam Singer blogged

> I thought it might be fun to take a step back and look at some interesting/amazing social media, Web 2.0, crowdsourcing and internet statistics. I tried to find stats that are the most up-to-date as possible at the time of publishing this post. [R26]

(a) Read that blog entry, choose a few numbers you find interesting, and make sense of them. Are they reasonable? Are they consistent?

(b) Estimate (or research) what those numbers might be now (when you are answering this question.)
(c) We saw this information on the blog: in March 2008, there were 70 million videos on YouTube. It would take 412.3 years to view all of that YouTube content. Thirteen hours of video are uploaded to YouTube every minute. Can you make sense of these numbers? Are they reasonable? Are they consistent?
(d) Can you locate the source of the statistics above, or other sources that confirm them?

Exercise 1.8.23. [S][Section 1.3][Goal 1.1][Goal 1.3][Goal 1.8] Bottle deposits.

A headline in *The Boston Globe* on July 15, 2010 read "State panel OKs expansion of nickel deposit to bottled water." At the time, Massachusetts required a 5 cent bottle deposit for all bottles containing carbonated liquids. There was a debate about extending the deposit law to other liquids, including bottled water. In the article you could read that

> The Patrick administration, which supports the bottle bill, has estimated the state would raise about $58 million by allowing the redemption of an additional 1.5 billion containers a year, or about $20 million more than the state earns from the current law, and that municipalities would save as much as $7 million in disposal costs. [R27]

(a) Is it reasonable to estimate that 1.5 billion water bottles would be recycled in a year if users paid a nickel deposit on each?
(b) Is $7 million a reasonable estimate of the cost of disposing of 1.5 billion bottles (probably in a landfill) rather than recycling them?
(c) With the data you can estimate the number of water bottles potentially redeemed relative to the number of bottles and cans currently being redeemed. Does the result of the comparison seem reasonable?
(d) The article says the administration estimates that the state will collect $58 million by keeping the deposits paid by the people who don't return the bottles. Use that information to estimate the percentage of bottles that they expect will be recycled.

Note: This bill was defeated in the state legislature.

Exercise 1.8.24. [U][C][Section 1.3] [Goal 1.1][Goal 1.2][Goal 1.8] Drivers curb habits as cost of gas soars.

In *The Boston Globe* on April 21, 2011 you could read that

> [F]amilies are quickly adapting [to increasing gas prices] by carpooling, combining errands to save trips, and curtailing weekend outings, according to organizations that track gasoline consumption. Still, the US Energy Department projects that the average US household will pay $825 more for gas this year than in 2010.
>
> NPD Group Inc., a market research firm, estimates that consumers bought roughly 128 million fewer gallons of gasoline in March than a year earlier. [R28]

Combine reasonable estimates for the increase in gasoline prices, the number of miles driven annually and the average fuel economy of cars to decide whether the $825 figure in the quotation makes sense.

Exercise 1.8.25. [S][Section 1.3][Goal 1.1][Goal 1.2][Goal 1.4][Goal 1.5] The Homemade Cafe.

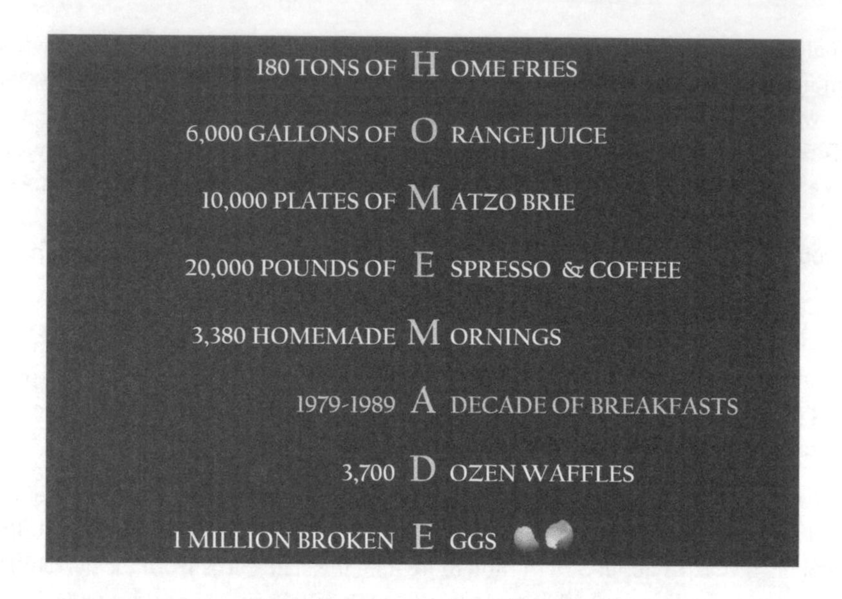

Figure 1.3. The Homemade Cafe [R29]

Figure 1.3 appeared on the back of Berkeley California's Homemade Cafe tenth anniversary tee shirt in 1989.

(a) Check that the numbers there make sense.
(b) Assume that the Homemade Cafe is still in business when you are working on this exercise. What numbers would go on this year's tee shirt?

Exercise 1.8.26. [S][Section 1.3][Goal 1.1][Goal 1.2][Goal 1.3][Goal 1.5] So Many Books, So Little Time.

On Sunday, March 4, 2012 Anthony Doerr asked in *The Boston Globe*:

> Have you ever done the math? If you're lucky enough to have 70 years of literate adulthood, and if you read one book every week, you're still only going to get to 3,640 books. Then you die.
>
> If you consider that the Harvard University Library system's collection is counted in the tens of millions, or that a new book of fiction is published every 30 minutes, 3,640 doesn't seem like so many. [R30]

(a) Confirm that 70 years of reading one book per week would amount to 3,640 books read in a lifetime.
(b) How many people reading one book per week during their lifetime would it take to read all the books in the Harvard University Library system?
(c) If you read one book of fiction each week this year, what percent of all the fiction published this year will you have read?

Exercise 1.8.27. [S][Section 1.3][Goal 1.1][Goal 1.2] Counting car crashes.

The National Safety Council estimates that "21 percent of [automobile] crashes or 1.2 million crashes in 2013 involve talking on handheld and hands-free cell phones." [R31]

(a) Use the data in the quote to estimate the total number of crashes in the U.S. in 2013.
(b) Check your answer with a web search.
(c) If crashes were evenly distributed across the population, how many would you expect in your community? Does your answer seem reasonable?

Exercise 1.8.28. [R][S][Goal 1.3][Section 1.4] Do parentheses matter?

What does the Google calculator tell you if you accidentally leave out the parentheses in the computation $12/(2 * 3)$?

Exercise 1.8.29. [U][R][Section 1.4][Goal 1.4] Should the U.S. Really Try to Host Another World Cup?

> The proposed budget for the 2010 [Soccer World Cup] games was about $225 million for stadiums and $421 million overall. Expenses have far exceeded those numbers. Reported stadium expenses jumped from the planned level of $225 million to $2.13 billion, and overall expenses jumped similarly from $421 million to over $5 billion. [R32]

How many orders of magnitude off were these estimates? That is, how many places were the decimal points away from where they should have been?

Exercise 1.8.30. [U][Section 1.5][Goal 1.5][Goal 1.1] Check our arithmetic, please.

When we found out in Section 1.5 that Baba Brinkman needed just ten seconds to plant a tree, we decided that his claim was possible.

(a) Verify the statement in the section that "... our 100 year estimate should really have been only about two years!"
(b) Verify the next statement: "If it took him a minute per tree rather than 10 seconds he could still have planted a million trees in ten summers."

Exercise 1.8.31. [S][Goal 1.2][Goal 1.6][Goal 1.8][Section 1.6] The tooth fairy.

How many visits per day does the tooth fairy make in the United States? What's the daily transaction volume (in dollars) in the tooth fairy sector of the economy?

Exercise 1.8.32. [S][Section 1.6][Goal 1.2][Goal 1.4][Goal 1.8] Paying for college.

In 2006, the ABC program 20/20 told the story of a couple on Los Angeles who put their children through college by collecting and redeeming soda cans and bottles. Their oldest son went to MIT and their two other children attended California state schools. According to the article, the Garcias collected cans and bottles for 21 years with the goal of saving for their children's college tuition. [R33]

Is this possible?

[See the back of the book for a hint.]

Exercise 1.8.33. [S][Section 1.6][Goal 1.2][Goal 1.4][Goal 1.8] Low flow toilets.

In 1994, a U.S. federal law went in to effect that required all new residential toilets to be "low-flow", using just 1.6 gallons of water per flush instead of the five gallons per flush of older toilets.

Estimate how much water a household could save in one year by switching to low flow toilets.

Exercise 1.8.34. [S][W][Section 1.6][Goal 1.1][Goal 1.5] Vet bills add up.

The inside back cover of the September/October 2008 issue of *BARk* magazine carried an ad for pet insurance asserting that every ten seconds a pet owner faced a $1000 vet bill.

Is this claim reasonable?

Exercise 1.8.35. [U][Section 1.6][Goal 1.2][Goal 1.6][Goal 1.8] Americans love animals.

In the November 9, 2009 issue of *The New Yorker* Elizabeth Kolbert wrote in a review of Jonathan Safran Foer's *Eating Animals* that there were 46 million dog-owning households, 38 million with cats and 13 million aquariums with more than 170 million fish.

> Collectively, these creatures cost Americans some forty billion dollars annually. (Seventeen billion goes to food and another twelve billion to veterinary bills.) [R34]

Is the twelve billion dollar figure she quotes for veterinary bills consistent with the numbers in the previous problem?

Exercise 1.8.36. [U][Section 1.6][Goal 1.2][Goal 1.4][Goal 1.5][Goal 1.8] Total carbon footprint.

Use the estimates for the seven tasks discussed in Section 1.6 to rank those tasks in order of *total* daily carbon footprint.

Exercise 1.8.37. [S][Section 1.7][Goal 1.1][Goal 1.5][Goal 1.6] How may internet ads?

An employee from Akamai claimed that

> There are 4.5×10^{12} internet advertisements annually. That's two thousand ads per person per year.

(a) Are the figures for the total number of ads and the number per person consistent?
(b) Do you think two thousand ads per person per year is a good estimate?

Exercise 1.8.38. [R][S][Section 1.7][Goal 1.7][Goal 1.8] Metric ton.

A *metric ton*, also known as a *tonne*, is 1000 kilograms.

(a) Is a metric ton a megagram or a gigagram?
(b) How many grams are there in a kilotonne?
(c) How many grams are there in a megatonne?

Exercise 1.8.39. [U][Section 1.7] [Goal 1.4][Goal 1.7][Goal 1.8] e-reading.

In December, 2014 Amazon offered a Kindle e-reader with

> **Storage:** 16GB (10.9GB available to user) or 32GB (25.1GB available to user), or 64 GB (53.7GB available to user)

(a) Compare the percentage of storage available to the user for each of these options.
(b) Estimate the number of e-books you could store on the 64 GB Kindle.
(c) Estimate the size of *Common Sense Mathematics* in MB.

Exercise 1.8.40. [U][Section 1.7][Goal 1.4][Goal 1.7][Goal 1.8] Personal storage.

(a) How many bytes of storage are there on the hard drive of your computer (or tablet or smartphone, or some device you use regularly)? Is that best measured in megabytes or gigabytes?
(b) If you have a thumb drive or flash memory stick, what's its capacity?

Exercise 1.8.41. [U][Section 1.7][Goal 1.4][Goal 1.7][Goal 1.8] Backing up the Library of Congress.

How many 200 gigabyte computer memories would you need to store the books in the Library of Congress?

Exercise 1.8.42. [U][Section 1.7][Goal 1.6][Goal 1.7] giga-usa.

The website `www.giga-usa.com/` advertises itself as an

> Extensive collection of 100,000+ ancient and modern quotations, aphorisms, maxims, proverbs, sayings, truisms, mottoes, book excerpts, poems and the like browsable by 6,000+ authors or 3,500+ cross-referenced topics. Extensive collection of 100,000+ ancient and modern quotations, aphorisms, maxims, proverbs, sayings, truisms, mottoes, book excerpts, poems and the like browsable by 6,000+ authors or 3,500+ cross-referenced topics. [R35]

Is the web site properly named?

Exercise 1.8.43. [R][S][Section 1.7][Goal 1.4][Goal 1.5][Goal 1.7][Goal 1.8] Data glut.

In the article from *The Boston Globe* on February 24, 2003 with the long headline

> Data glut as gene research yields information counted in terabytes. Researchers struggle to visualize and process it while technology businesses scramble to profit from it.

you could read that

> [Peter Sorger's] bioengineering lab produces a terabyte of data in a typical month. [R36]

(a) At what rate in bytes per minute is the lab producing data? Write your answer with the appropriate metric prefix and the appropriate level of precision.
(b) If the lab has been producing data from the time the article appeared to the present, how much has accumulated now?
(c) When will a petabyte of data have accumulated? Do you believe your prediction?
(d) When will an exabyte of data have accumulated?

Exercise 1.8.44. [S][C][Section 1.7][Goal 1.7][Goal 1.6] Zettabytes.

On September 7, 2010 an editorial in *The Boston Globe* said that

> The total amount of digital storage worldwide is approaching 1 zettabyte, or 1 million times the contents of the Earth's largest library. [R37]

A zettabyte is 1,000,000,000,000,000,000,000 bytes. The editor provides the comparison to help her reader make sense of that nearly incomprehensible number.

Check her arithmetic. Are the numbers consistent—that is, do they agree with each other when you compare them? Do they make sense?

Exercise 1.8.45. [U][C][Goal 1.6][Section 1.7] [Goal 1.7][Goal 1.8] Zettabytes redux.

On August 6, 2011 Kari Kraus wrote in *The New York Times* that

> We generate over 1.8 zettabytes of digital information a year. By some estimates, that's nearly 30 million times the amount of information contained in all the books ever published. [R38]

Are the two estimates in this quotation (1.8 million zettabytes, 30 million times ...) consistent with each other?

Exercise 1.8.46. [W][S][Section 1.1][Goal 1.1] Waiting for the light to change.

In the Pooch Cafe comic strip on August 27, 2012 Poncho the dog is sitting in the car with his master Chazz. He says "Did you know the average person spends six months of their life waiting at red lights?" (You can see the strip at `www.gocomics.com/poochcafe/2012/08/27`)

What do you think of Poncho's estimation skills?

Exercise 1.8.47. [U][Section 1.1][Goal 1.1][Goal 1.2] Killer cats.

In the article "The impact of free-ranging domestic cats on wildlife of the United States" in January 2013 in *Nature Communications* Scott R. Loss, Tom Will and Peter P. Marra offered an

> ... estimate that free-ranging domestic cats kill 1.4–3.7 billion birds and 6.9–20.7 billion mammals annually. Un-owned cats, as opposed to owned pets, cause the majority of this mortality. [R39]

Make sense of those numbers. Consider kills per cat or kills per day, or kills per day in your community.

Exercise 1.8.48. [S] Viagra, anyone?

A 2014 Viagra ad on TV stated that more than 20 million men already use Viagra. Use the U.S. population pyramid in Figure 1.4 to argue whether or not this claim seems reasonable. Be explicit about any assumptions you make about the age groups of men who might typically use this drug.

Exercise 1.8.49. [U] Lots of olives?

On August 23, 2014 the Associated Press reported that

> In the 1980s, American Airlines chief executive Robert Crandall famously decided to remove a single olive from every salad. The thought was: passengers would not notice and American would save $40,000 a year. [R41]

(a) About how many olives could Crandall buy for $40,000?
(b) Is your answer to the previous question in the same ballpark as the number of American Airlines passengers in 1980?

Exercise 1.8.50. [U][C][Section 1.3][Goal 1.1][Goal 1.2][Goal 1.5][Goal 1.8] Nuclear bombs.

"Kiloton" and "megaton" are terms you commonly hear when nuclear bombs are being discussed. In that context the "ton" refers not to 2,000 pounds, but to the explosive yield of a ton of TNT.

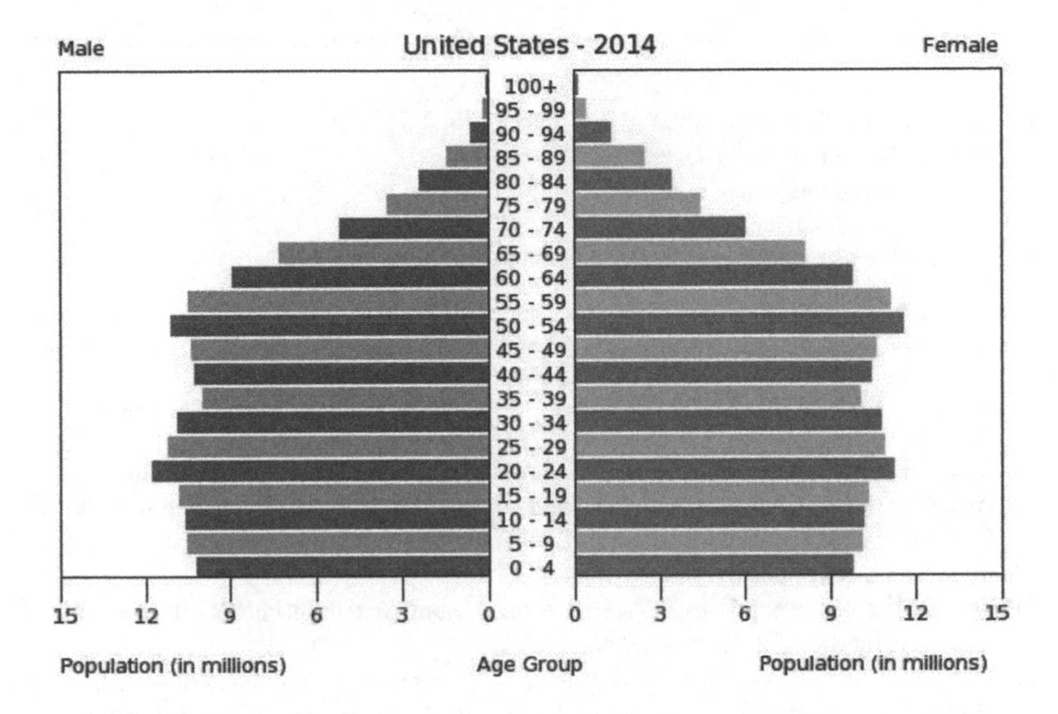

Figure 1.4. U.S. Population pyramid [R40]

(a) What was the explosive yield of the only (two) atomic bombs ever used in war?
(b) How does the explosive yield of the hydrogen bombs in the current arsenals of the United States and Russia (and other countries) compare to that of those first atomic bombs?
(c) Estimate the destructive power of the world's current stockpile of nuclear weapons, in terms easier to grasp than kilotons or megatons or gigatons or ....

Review exercises

Exercise 1.8.51. [A] Do each of these calculations by counting zeroes. Use the Google or Bing calculator (or another calculator) to check your answers.

(a) One million times one billion.
(b) Four hundred time three thousand.
(c) Two billion divided by two hundred.
(d) One-tenth times five thousand.
(e) Four thousand divided by two hundred.
(f) Twelve thousand times two hundred.
(g) $10{,}000 \times \frac{2}{1000}$.
(h) $450{,}000 \times 100$.
(i) $\frac{50{,}000}{200}$.
(j) $\frac{4{,}000{,}000{,}000{,}000}{2{,}000{,}000}$.

Exercise 1.8.52. [A] Use rounding to estimate each of these quantities. Then check your answers.

(a) The number of feet in ten miles.
(b) The number of minutes in a week.
(c) The number of inches in a mile.
(d) The number of yards in a mile.
(e) The number of seconds in a month.

Exercise 1.8.53. [A] Answer each of these questions without using pencil and paper (or a calculator).

(a) There are twelve cans of soda in a case. If I buy five cases of soda, how many cans will I have in total?
(b) A case of bottled water contains 24 bottles. About how many bottles are there in four cases?
(c) At the market my cart contains: three bags of cereal at $2.19 each; a gallon of milk for $2.99; a pound of potatoes that costs $3.50 and a $3.99 bag of apples. Will my total be more than twenty dollars?
(d) At our local deli, a bagel costs $1.19. How much would five bagels cost? If the deli offers six bagels for $6.99, is that a better deal?
(e) It costs $2.50 to ride the New York City subway. A seven-day unlimited pass costs $29. How many rides in a seven day period should you take before it's a better deal to buy the pass?
(f) I have a $100 gift card for a department store. Can I buy a pair of shoes for $19.99, two shirts that are $12.59 each and three pairs of pants that each cost $20.50?

Exercise 1.8.54. [A] Convert

(a) One kilometer into meters.
(b) One megameter into kilometers.
(c) One terabyte into megabytes.
(d) Three megawatts into kilowatts.
(e) Five gigabytes into kilobytes.
(f) One thousand kilograms into megagrams.

2

Units and Unit Conversions

In real life there are few naked numbers. Numbers usually measure something like cost, population, time, speed, distance, weight, energy or power. Often what's measured is a rate, like miles per hour, gallons per mile, miles per gallon, dollars per gallon, dollars per euro or centimeters per inch.

Chapter goals

Goal 2.1. Explicitly manipulate units in expressions.

Goal 2.2. Explore metric prefixes.

Goal 2.3. Read and write scientific notation.

Goal 2.4. Understand how units of length determine units of area and volume.

Goal 2.5. Learn to use the Google calculator to make unit conversions.

Goal 2.6. Compare information by converting to similar units (dollars per ounce, for example, or births per one million people).

Getting the units wrong can have significant consequences:

> *NASA's metric confusion caused Mars orbiter loss*
> September 30, 1999
> (CNN)—NASA lost a $125 million Mars orbiter because one engineering team used metric units while another used English units for a key spacecraft operation, according to a review finding released Thursday. [R42]

On a less serious note, if in Canada you see a speed limit sign with the number 100 and don't read the units you may think you may drive 100 miles per hour. That puts you at risk for an expensive speeding ticket, since in Canada distance is measured in kilometers. We'll see soon how to find out what 100 kilometers/hour corresponds to in miles/hour.

2.1 Rate times time equals distance

If you drive at 60 miles per hour for an hour and a half how far do you travel?

Since $60 \times 1.5 = 90$ the answer is 90 miles. You don't need the machinery we're inventing to get it right, but this easy problem is a good place to learn that machinery, which shows how to multiply the units along with the numbers:

$$60 \frac{\text{miles}}{\text{hour}} \times 1.5 \text{ hours} = 90 \text{ miles}.$$

When the same factor appears on the top and the bottom of a fraction you can cancel it, as in these calculations we made up to help you remember that fact.

With numbers:

$$3 \times 11 \times \frac{7}{5 \times 11} = 3 \times \cancel{11} \times \frac{7}{5 \times \cancel{11}} = 3 \times \frac{7}{5} = \frac{3 \times 7}{5},$$

particularly for zeroes:

$$\frac{1{,}000{,}000}{100} = \frac{1{,}000{,}0\cancel{00}}{1\cancel{00}} = 10{,}000$$

where cancelling a zero corresponds to cancelling a factor of 10, and in algebraic expressions:

$$3x \frac{y}{wx} = 3\cancel{x}\frac{y}{w\cancel{x}} = 3\frac{y}{w} = \frac{3y}{w}.$$

When you write numbers with their units you can treat the words expressing the units just as you treat the numbers and letters—cancel them when they match. Revisiting our original example, we see that the hours cancel, leaving just miles as the units of the answer:

$$60 \frac{\text{miles}}{\cancel{\text{hour}}} \times 1.5 \cancel{\text{hours}} = 90 \text{ miles}, \quad (2.1)$$

which illustrates an important habit to develop:

Always write the units along with the numbers.

2.2 The MPG illusion

What's a better gas saving choice—adding a second electric power source to a gas guzzling SUV to increase its fuel efficiency from 12 to 14 miles per gallon (MPG), or replacing an ordinary sedan that gets 28 miles per gallon by a hybrid that gets 40 miles per gallon?

Before you read on, jot down your first guess.

In June, 2008 Richard Larrick and Jack Soll wrote in *Science* that

> Many people consider fuel efficiency when purchasing a car, hoping to reduce gas consumption and carbon emissions. However, an accurate understanding of fuel efficiency is critical to making an informed decision. We will show that there is a systematic misperception in judging fuel efficiency when it is expressed as miles per gallon (MPG), which is the measure used in the U.S.A. [R43]

In their article they calculated that to drive 10,000 miles you'd need 833 gallons at 12 miles per gallon but only 714 at 14 miles per gallon. Then they showed that saves more gas than replacing a car that gets 28 miles per gallon with an efficient hybrid getting 40 miles per gallon.

Let's check the calculations. To figure out how much gas is used driving 10,000 miles in a car that gets 12 miles per gallon we must do something with the numbers 10,000 and 12. Don't just guess whether to multiply or divide, write the equation with just the units first. We're starting with miles and want to end up with gallons, so we need to multiply by a fraction with units gallons/mile:

$$\text{miles} \times \frac{\text{gallons}}{\text{mile}} = \cancel{\text{miles}} \times \frac{\text{gallons}}{\cancel{\text{mile}}} = \text{gallons}.$$

That equation tells us that to fill in the numbers we need the fuel efficiency in gallons/mile rather than in miles/gallon. 12 miles/gallon means driving 12 miles uses one gallon of gas. That says one gallon takes us 12 miles, which is a rate of (1 gallon) per (12 miles). Now we know where to put the numbers:

$$10{,}000\,\cancel{\text{miles}} \times \frac{1\text{ gallon}}{12\,\cancel{\text{miles}}} = \frac{10{,}000}{12}\text{gallons} = 833\text{ gallons},$$

just as Larrick and Soll say.

It's important to know how to do the calculation, but also important to know how to ask someone to do it for you—in this case, Google (or Bing). Enter

12 miles per gallon in gallons per 10000 miles

in the search box. Google tells you

12 miles per gallon = 833.333333 US gallons per (10 000 miles)

.

We'll use Google to finish checking Larrick and Soll's arithmetic.

14 miles per gallon = 714.285714 US gallons per (10 000 miles)

.

So the gas saved retrofitting the SUV is $833 - 714 = 119 \approx 120$ gallons.

Then

28 miles per gallon = 357.142857 US gallons per (10 000 miles)

and

40 miles per gallon = 250 US gallons per (10 000 miles)

.

So the gas saved swapping the sedan for the hybrid is just $357 - 250 = 107$ gallons—significantly less than the 120 gallons saved by the SUV fix.

The paper in *Science* reported on the authors' survey showing that most people didn't know this. The reason is psychological: the change in MPG from 28 to 40 looks much more significant than the change from 12 to 14. But looks can be deceiving. We found the truth by using units carefully to think through the arithmetic.

Larrick and Soll go on to argue that the psychological problem can't be solved by asking people to do the work we just did. They conclude that instead that

> These studies have demonstrated a systematic misunderstanding of MPG as a measure of fuel efficiency. Relying on linear reasoning about MPG leads people to undervalue small improvements on inefficient vehicles. We believe this general misunderstanding of MPG has implications for both public policy and research on environmental decision-making. From a policy perspective, these results imply that the United States should express fuel efficiency as a ratio of volume of consumption

Figure 2.1. Gasoline vehicle label (2015) [R44]

to a unit of distance. Although MPG is useful for estimating the range of a car's gas tank, gallons per mile (GPM) allows consumers to understand exactly how much gas they are using on a given car trip or in a given year and, with additional information, how much carbon they are releasing. GPM also makes cost savings from reduced gas consumption easier to calculate.

To convert MPG to GPM, we simply turn the fraction upside down:

$$\frac{1}{\frac{12\text{ miles}}{\text{gallon}}} = \frac{1}{12}\,\frac{\text{gallons}}{\text{mile}} = 0.083\,\frac{\text{gallons}}{\text{mile}}.$$

The old fashioned fraction rule $\frac{1}{a/b} = b/a$ handles the units just fine.

If (like most people) you're not comfortable with small numbers like 0.083, write this as

$$8.3\,\frac{\text{gallons}}{100\text{ miles}}.$$

Larrick and Soll argue persuasively that this figure should appear on the fact sheets for new cars, as it does in Europe (in liters per 100 km). The EPA was convinced. Figure 2.1 shows the new car window sticker design as of 2013. Item 5 gives the fuel efficiency in gallons per hundred miles. Too bad it's in smaller type than item 2, the more familiar miles per gallon.

Visit `www.epa.gov/carlabel/gaslabel.htm` to find out what the other numbers mean.

2.3 Converting currency

When you visit a foreign country you will have to exchange your dollars for that country's currency. There are web sites that will do the arithmetic for you, but you may not always have access to the internet. And you'll find it easier to travel if you can make quick estimates without consulting your smart cell phone (which may not work abroad). Here's how to use units to understand the conversions.

Suppose your trip is to France, where they use the euro. The value of the euro (in dollars) varies from day to day—in fact, from minute to minute. When we wrote this section the going rate was

$$€1 = \$1.38. \tag{2.2}$$

That means you could buy 1 euro for 1.38 dollars. In fraction form, with units, the conversion rate is

$$\frac{1.38\ \$}{1\ €} = 1.38\ \frac{\$}{€}. \tag{2.3}$$

If you always write the conversion factor with its units you will never have to worry about whether to multiply or divide by 1.38.

Suppose you are tempted by a pair of shoes that costs €49. You want to know the dollar cost. To estimate the answer, think "49 is close to 50" and visualize the conversion factor with its units. That tells you that if you know the number of euros you must multiply by (dollars per euro) to make the euros cancel, leaving you with dollars. Since the conversion factor is $1.38\ \$/€ \approx 1.4\ \$/€$, the cost is about $50 \times 1.4 = 70$ dollars.

If you're comfortable with percentage calculations you can read the conversion factor of $1.38\ \$/€$ as telling you that the cost in dollars will be about 40% larger than the cost in euros. Since 40% more than 50 is 70, the €49 pair of shoes will cost about $70.

To make the conversion precisely, with units carefully displayed, first write

$$€ \times \frac{\$}{€} = \$. \tag{2.4}$$

Then fill in the numbers:

$$49\ € \times 1.38\frac{\$}{€} = 49 \times 1.38\frac{\$}{€} = 68.05\ \$. \tag{2.5}$$

The equals sign in Equation 2.2 says that €1 and $1.38 are *the same*. Since dividing something by itself always gives the number 1, the expressions on both sides of Equation 2.3 are just two ways to write the number 1. Multiplying by 1 doesn't change anything. That's another way to see the truth of Equation 2.5

Our $70 estimate was a little high. But that's probably a good thing for two reasons. First, we haven't considered the fee the bank (or credit card company) charges each time you exchange currency. Second, when considering a purchase it's always better to overestimate than underestimate what it will cost.

In France you may be driving as well as shopping. You need to plan for the cost of gas. It's sold there in liters, not gallons. The unit price at the pump when we wrote this section was $1.30\frac{€}{\text{liter}}$. Looks like a bargain—let's find out, by converting the units to dollars per gallon.

We know how to convert between euros and dollars. We can look up the conversion between gallons and liters: there are 0.264 gallons in 1 liter.

As usual, set up the computation first just with units:

$$\frac{€}{\cancel{\text{liter}}} \times \frac{\$}{€} \times \frac{\cancel{\text{liters}}}{\text{gallon}} = \frac{\$}{\text{gallon}}.$$

Then fill in the numbers and do the arithmetic:

$$\begin{aligned} 1.30 \frac{€}{\cancel{\text{liter}}} \times \frac{1.38\ \$}{€} \times \frac{1\ \cancel{\text{liter}}}{0.264\ \text{gallon}} &= \frac{1.30 \times 1.38}{0.264} \frac{\$}{\text{gallon}} \\ &= 6.79545455 \frac{\$}{\text{gallon}} \\ &\approx 6.80 \frac{\$}{\text{gallon}}. \end{aligned}$$

Not such a bargain after all. When we wrote this section gas in France cost more than twice what it cost here.

When a question has a surprising answer first check your work. We did. The answer is right. Gas was really twice as expensive in France. Why?

The first two answers that many people try are

- It's because dollars are worth less than euros.
- It's because liters are smaller than gallons.

Both of these statements are correct, but neither explains the high cost of gas in Europe. Those facts were *precisely what we took into account* when we converted 1.30 $\frac{€}{\text{liter}}$ to 6.80 $\frac{\$}{\text{gallon}}$. You need to search further for the real reason: gasoline taxes in Europe are much higher than they are in the United States. In Exercise 2.9.50 we ask you to research that question further.

Once you understand how the units work you can ask the Google calculator to do the work for you. For this calculation it will tell you something like

1.3 (Euros per liter) = 6.60550571 U.S. dollars per US gallon

.

The $6.60 per gallon here does not match the $6.80 per gallon in the text above. They differ because the Google calculator looks up the current exchange rate for the euro. We wrote this paragraph a month or so after we found the $6.80. The dollar value of the euro changed in the interim.

Google will tell you the current exchange rate (rounded to the nearest penny). We searched for

1 euro in dollars

and were told

1 Euro equals 1.33 US Dollar

which is a little less than the $1.38/€ we used above.

2.4 Unit pricing and crime rates

Grocery shopping is full of decisions. Suppose you're staring at two boxes of crackers—the brand you like. The 16-ounce box costs \$1.99; the 24-ounce box costs \$2.69. Which is the better deal?

To decide, you need to know the *unit price*. The units of the unit price will be $\frac{\$}{\text{ounce}}$.

A quick estimate helps: 24 ounces is 50% more than 16 ounces. The 16-ounce box costs \$2, so if the larger box costs less than \$3 then it's a better buy. It is in this case.

Writing out the calculation with the units, we see that for the small box the unit price is

$$\frac{1.99\ \$}{16\ \text{ounces}} = 0.124\ \frac{\$}{\text{ounce}}$$

while for the big box it's

$$\frac{2.69\ \$}{24\ \text{ounces}} = 0.112\ \frac{\$}{\text{ounce}}.$$

Our estimate gave us the right answer: the larger box is the better buy, ounce for ounce.

Some states require stores to list unit prices on the shelves. But they don't require consistency in the units used, so you may see unit prices in dollars per ounce next to those in dollars per pound or dollars per serving. You still need to think.

The supermarket isn't the only place where converting absolute quantities to rates (*something per something*) helps make sense of numbers. The FBI compiles and reports crime statistics for cities across the United States. The 2011 data recorded 3,354 violent crimes for Sacramento, CA and 3,206 for San Jose, CA. [R45]

Which city had more crime?

They seem to be about equally dangerous. But for a true comparison we should take into account the sizes of the two cities. Sacramento's population was 471,972 while San Jose's was 957,062: San Jose is about twice the size of Sacramento. Since the number of violent crimes in each city was about the same, the crime *rate* in San Jose was about half the rate in Sacramento.

Formally, with units and numbers, for Sacramento

$$\frac{3{,}354\ \text{crimes}}{471{,}972\ \text{people}} = 0.00711\ \text{crimes per person}.$$

If we make the same calculation for San Jose we find a rate of 0.00335 crimes per person.

It's hard to think about small numbers like those with zeroes after the decimal point. For that reason it's better to report the crime rate with units crimes per 1,000 people rather than crimes per person. We do that by extending our calculation using units this way:

$$\frac{3{,}354\ \text{crimes}}{471{,}972\ \text{people}} \times \frac{1{,}000\ \text{people}}{\text{thousand people}} = 7.11\ \text{crimes per thousand people}.$$

The second fraction in the computation, $\frac{1{,}000\ \text{people}}{\text{thousand people}}$, is just another way to write the number 1. Multiplying by 1 doesn't change the value of the number, just its appearance. Multiplying by 1,000 in the numerator moves the decimal point three places, turning the clumsy 0.00711 into the easy to read 7.11.

The crime rate for San Jose is just 3.35 crimes per thousand people. Now we have easy to read numbers to compare. San Jose's crime rate just under half of Sacramento's.

2.5 The metric system

Most of the world uses the *metric system*. Although it's not common in the United States, sometimes it pops up in unexpected places. When we used the Google calculator to check our solution to the problem in Section 2.1 the answer was weird. Asking for

1.5 hours * 60 miles per hour

we were surprised to see

1.5 hours times (60 miles per hour) = 144.84096 kilometers

.

We've no idea why Google decided to use kilometers instead of miles. We did make it tell us what we really wanted to know this way:

1.5 hours * 60 miles per hour in miles

leads to

(1.5 hours) * 60 miles per hour = 90 miles

.

We'll spend the rest of this section learning just a little about some of the metric units and how to convert among them and between them and English units.

To carry on everyday quantitative life we use small, medium and large units so that we don't have to deal with numbers that are very small or very large. For example, we measure short lengths (the size of your waist) in inches, medium lengths (the size of a room) in feet or yards, and large lengths (the distance you commute) in miles. In the metric system the small, medium and large units for length are the centimeter, the meter and the kilometer.

In the metric system the units of different sizes are always related by a conversion factor that's a power of 10. Since it's easy to multiply or divide by a power of 10 by moving the decimal point, it's easy to move among the units. In contrast, converting units in the English system means knowing (or looking up) awkward conversion factors like 12 inches/foot or 5,280 feet/mile, and then actually multiplying or dividing by them.

A *centimeter* is about the width of your fingernail. There are about two and a half centimeters in an inch. Your index finger is about 10 centimeters (four inches) long. The metric system has nothing like the foot for intermediate lengths. It relies on the *meter*: like a yard, but 10% longer. For longer distances the metric equivalent of the mile is the *kilometer*—about 60% of a mile.

If you lived in a metric world you would think in metric terms rather than in our English system (even England has mostly abandoned the English system).

You would understand "meter" directly, not as "about a yard." You'd never need to convert meters to yards or kilometers to miles any more than you'd need to translate French to English if you lived in France and spoke the language like a native.

But in our world you may occasionally need to convert, with the following factors. Remember the approximations; look up the exact values up when you need them.

$$2.54 \frac{\text{centimeters}}{\text{inch}} \approx 2.5 \frac{\text{centimeters}}{\text{inch}},$$
$$0.393700787 \frac{\text{inches}}{\text{centimeter}} \approx 0.4 \frac{\text{inches}}{\text{centimeter}},$$

so 10 centimeters is about four inches.

$$0.9144 \frac{\text{meters}}{\text{yard}} \approx 0.9 \frac{\text{meters}}{\text{yard}}.$$
$$1.0936133 \frac{\text{yards}}{\text{meter}} \approx 1.1 \frac{\text{yards}}{\text{meter}}.$$
$$1.609344 \frac{\text{kilometers}}{\text{mile}} \approx 1.6 \frac{\text{kilometers}}{\text{mile}}.$$
$$0.621371192 \frac{\text{miles}}{\text{kilometer}} \approx 0.6 \frac{\text{miles}}{\text{kilometer}}.$$

The last of these approximations leads to an easy answer to the question asked implicitly at the start of this chapter. 100 kilometers is about 60 miles, so 100 kilometers per hour is about 60 miles per hour.

The small, medium and large units for weight in the metric system are the gram, kilogram and metric ton.

A paperclip weighs about a *gram*—about a thirtieth of an ounce. So there are about 30 grams in an ounce.

A *kilogram* is just over two pounds—about 2.2, or 10% more. So a pound is just under half a kilogram (about 10%).

A *metric ton* (1000 kilograms) is about 10% larger than a ton (2000 pounds); a ton is about 10% smaller than a metric ton.

The small and medium metric units for volume are the cubic centimeter and the liter.

It takes about 30 cubic centimeters to fill an ounce, so the small volume unit in the metric system is a lot smaller than the corresponding unit in our English system. (One of many annoying feature of the English system is that there are two different kinds of ounces—one for weight and one for volume. You can't convert between them.)

The name "cubic centimeter" is a clue to a nice feature of the metric system. The different kinds of units are related in as sensible a way as possible. The unit used for small volumes was created from the unit for small lengths. The gram (the unit for small weights) is just the weight of a cubic centimeter of water.

The metric unit for volumes of medium size is the liter. It's just about 5% larger than a quart. A quart is about 95% of a liter. We encountered liters earlier in the chapter when thinking about buying gasoline in France.

2.6 Working on the railroad

On April 28, 2010 *The Boston Globe* reported that "Urgent fixes will disrupt rail lines—T to spend $91.5 m to repair crumbling Old Colony ties":

> Starting in August, T officials say, they plan to tear up 150,000 concrete ties along 57 miles of track and to replace them with wooden ties. [R46]

There are three numbers in this quotation and the headline: 150,000 *ties*, 57 *miles* and 91.5 million *dollars*.

Using them two at a time we can figure out both how far apart the ties are, and how much each tie costs.

Railroad ties are fairly close together, so we should measure the distance between them using feet, not miles: we want an answer with units feet/tie. Since we have miles of track, the units equation will be

$$\frac{\cancel{\text{miles}}\text{ (of track)}}{\text{(number of) ties}} \times \frac{\text{feet}}{\cancel{\text{mile}}} = \frac{\text{feet}}{\text{tie}}.$$

Putting in the numbers and doing the arithmetic:

$$\begin{aligned}\frac{57\ \cancel{\text{miles}}}{150{,}000\text{ ties}} \times \frac{5{,}280\text{ feet}}{\cancel{\text{mile}}} &= \frac{57 \times 5{,}280}{150{,}000}\,\frac{\text{feet}}{\text{tie}}\\ &= 2.0064\frac{\text{feet}}{\text{tie}}\\ &\approx 2\frac{\text{feet}}{\text{tie}}\end{aligned}$$

so the ties are about two feet apart.

Here's a way to trick the Google calculator into doing the unit calculations as well as the arithmetic. It doesn't know about railroad ties, but we can pretend ties are gallons. Ask it for

57 miles per 150000 gallons in feet per gallon

and it tells us

(57 miles) per (150 000 US gallons) = 2.0064 feet per US gallon

.

That answer is suspiciously close to the round number of 2 feet. That suggests that the T estimated the number of ties they'd need by assuming the two foot separation. They knew that they were laying 57 miles of new track. They converted 57 miles to $57 \times 5{,}280 = 300{,}960$ feet. With ties two feet apart they need half that many ties, which rounds nicely to 150,000.

To compute the cost in dollars per tie, the units tell you to find

$$\frac{91.5\text{ million \$}}{150{,}000\text{ ties}}.$$

If you ask Google for

91.5 million / 150000

you find out that

91.5 million / 150 000 = six hundred ten

so the ties cost about $600 each. That includes material and labor.

We were surprised and interested to see Google write 610 as "six hundred ten". (That's the right way to say "610" aloud. "six hundred and ten" is common, but wrong.) We realized that was because our search used the word "million". Had we asked instead for

91.5 * 1,000,000 / 150000

or

91,500,000 / 150000

we'd have been told

91 500 000 / 150 000 = 610 .

2.7 Scientific notation, milli and micro

In the exercises in Chapter 1 we found out that 70 years was about 2.5 billion seconds. Since 70 is about two and a half times 30, 30 years is about a billion seconds. If we search Google for

30 years in seconds

we're told

30 years = 946 707 779 seconds

so our estimate of one billion is quite good.

If we ask for the number of seconds in 60 years we expect an answer just twice as big:

60 years = 1 893 415 558 seconds

but see instead

$$60 \text{ years} = 1.89341556 \times 10^9 \text{ seconds} .$$

What is going on?

Google chose to express the answer in scientific notation. It's telling you to multiply the number 1.89...56 by 10, nine times. Multiplying by 10 moves the decimal point to the right. If we follow those instructions,

$$1.89341556 \times 10^9 = 1{,}893{,}415{,}600$$

which is quite close to the 1,893,415,558 we expected to see.

When you read "$1.893\ldots \times 10^9$" aloud, you say

"one point eight nine three ... times ten to the ninth".

A common error (which we hope you will never make) is to say instead "one point eight nine three ... to the ninth". To help you remember that the exponent 9 is a power of 10 you have to *say* the 10.

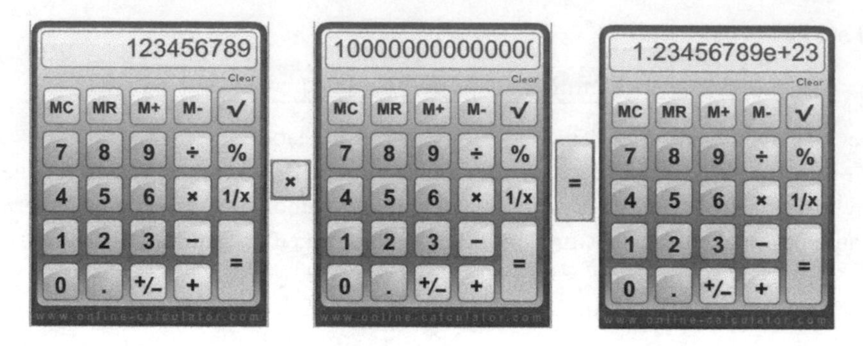

Figure 2.2. Stressing the calculator [R47]

The Google calculator can read powers of 10 as well as write them. To enter an exponent, use the caret (^) on your keyboard. It's the upper case 6. The 91.5 million we worked with in the previous section is 91.5×10^6. Ask Google for

91.5 * 10 ^6

and you will see

91.5 * (10 ^6) = 91 500 000

.

Google's isn't the only calculator that switches to scientific notation for really large numbers. Figure 2.2 shows what happens at `www.online-calculator.com/` when you ask for

$$123{,}456{,}789 * 1{,}000{,}000{,}000{,}000{,}000.$$

Why? 24 digits won't fit on its screen, so it can't show

123,456,789,000,000,000,000,000

(even without the commas). Moreover, it hasn't the height to display exponents, so it can't show

$$1.23456789 \times 10^{23}.$$

Instead it uses the symbols "`e+23`" to tell you the exponent that it can't write is (positive) 23. If you do need to say this aloud, you put the "ten" with its power back: "1 point 23 . . . 89 times 10 to the 23rd", *not* "1.12 . . . 89 to the 23rd".

You don't often encounter scientific notation in daily life. But we want to show you what it looks like so that you can read it if you stumble on it.

A number written in scientific notation always has just one nonzero digit before the decimal point, as many after the decimal point as appropriate for the precision desired, and then instructions about how many places to move the decimal point right or left. The power of 10 tells you the number of places.

You can make numbers bigger by adding zeroes but you can't always make them smaller by removing zeroes. Scientific notation provides a way. For example,

$$2.5 \times 10^{-4} = 0.00025;$$

Table 2.3. Metric prefixes for shrinking

Name	Symbol	English name	Power of 10
deci	d	one tenth	−1
centi	c	one hundredth	−2
milli	m	one thousandth	−3
micro	μ	one millionth	−6
nano	n	one billionth	−9
pico	p	one trillionth	−12
femto	f	one quadrillionth	−15

the decimal point has been moved four places to the *left* because

$$10^{-4} = \frac{1}{10^4}.$$

It's no surprise that 2.5×10^{-4} is a very small number.

We've seen the metric prefixes kilo, mega, giga, ... that make units larger. There are also prefixes for fractions of a unit. The most common is *milli*, which means "divide by 1000" or "multiply by 10^{-3}" or "move the decimal point three spaces to the left". The symbol is "m", in lower case. We've already seen it at work: a milliliter is one one-thousandth of a liter. The thickness of lead in a pencil is measured in millimeters.

In Greek mythology, Helen's beauty led to the Trojan war — hers was "the face that launch'd a thousand ships" [R48] so a milliHelen is the unit of beauty sufficient to launch one ship.

For microscopic things you need the next prefix, micro. It tells you to multiply by 10^{-6}— one one millionth. The symbol is the Greek letter *mu*: "μ". There are few calls for micro in everyday use.

Although most metric prefixes move the decimal point three places, one common prefix moves it just two: *centi* means "one one-hundredth of" or "divide by 100." So there are 100 centimeters in a meter, and 10 millimeters in a centimeter.

Table 2.3 shows the metric prefixes for shrinking things. These days "nano" sometimes means just "very small" rather than "one billionth", as in "nanotechnology" or even "ipod nano".

2.8 Carpeting and paint

How much carpeting does it take to cover the floor of a 12 foot by 15 foot room?

To find the floor area of the room you multiply the length by the width. Multiply the units along with the numbers:

$$12 \text{ feet} \times 15 \text{ feet} = 180 \text{ feet} \times \text{feet} = 180 \text{ feet}^2$$

which we read as "180 square feet".

Carpeting is sold by the square yard. How many square yards will you need? How do you convert from square feet to square yards? We know

$$1 \text{ yard} = 3 \text{ feet}$$

so

$$(1 \text{ yard})^2 = (3 \text{ feet})^2 = 9 \text{ feet}^2.$$

In other words, the fraction

$$\frac{1 \text{ yard}^2}{9 \text{ feet}^2}$$

is just another way to write the number 1, so multiplying by it doesn't change the value, just the way the answer looks:

$$180 \text{ feet}^2 \times \frac{1 \text{ yard}^2}{9 \text{ feet}^2} = 20 \text{ yards}^2.$$

Once you see that simple answer you may realize that the problem would have been even easier if you'd worked in yards from the start: the room is 4 yards wide by 5 yards long, so has an area of 20 square yards.

Before you lay the new carpeting you want to paint the room. How much paint will you need?

To continue computing, you must know the height of the ceiling—suppose it's nine feet. To find the area to cover, think four walls: two are 12 feet long and 9 feet high, two are 15 feet long and 9 feet high, so the total area to be covered is

$$(2 \times 12 \text{ feet} + 2 \times 15 \text{ feet}) \times 9 \text{ feet} = 486 \text{ feet}^2.$$

A gallon of paint covers 400 square feet, so you will need about a gallon and a quarter. That's convenient, because you can buy a gallon and a quart.

If you want to paint the ceiling too you'll need more paint. The area of the ceiling is the same as the area of the floor to be carpeted: 180 square feet. So buy two quarts of ceiling paint. (You can't simply get two gallons of the wall paint and use two quarts of that for the ceiling since the ceiling requires a different kind of paint even if it's the same color.)

Finally, if you plan to paint a real room the real computation is more complicated. Some colors (particularly yellow) require two coats, or a primer. The doors and windows represent wall area you don't have to cover, but you do need to paint the woodwork that surrounds them—that's yet another kind of paint. And you can buy a gallon of paint for the same price as three quarts—be sure to get enough. When you're done, carefully label and close the partially full paint cans. Be sure to paint the ceiling before you paint the walls, and do all the painting before you install the carpet.

We've used this discussion of paint and carpet to show how units for length determine square units for area. Next we look at cubic units for volume.

You happen to have a nice wooden box one foot on each side. You want to fill it for your grandchildren with those nice kids' alphabet blocks each one inch on a side. How many blocks will you need? A lot more than you might think.

A row of 12 blocks will be a foot long. 12 of those rows, or 144 blocks, will fill one layer on the bottom of the box. Then you need 12 layers of 144, or 1,728 blocks in all, to fill the box.

The size of that answer—nearly 2,000 blocks—surprises many people. Now that you know it you will remember that volumes grow quickly. The volume of the box is one cubic foot: 1 foot^3. The volume of each block is one cubic inch. Then

$$1 \text{ foot}^3 = (12 \text{ inches})^3 = 12^3 \text{ inches}^3 = 1{,}728 \text{ inches}^3.$$

2.9 Exercises

Exercise 2.9.1. [Section 2.1][Goal 2.1] Units in the news.

Find a current news or magazine article that uses unit conversions either explicitly or implicitly. Verify the calculation. Include a copy of the article (or a link to it on the web) when you turn in your work.

[See the back of the book for a hint.]

Exercise 2.9.2. [S][Section 2.1][Goal 2.1] Drive carefully!

Suppose you drive from Here to There at a speed of 30 miles/hour and then drive back from There to Here at 60 miles/hour.

What is your average speed?

[See the back of the book for a hint.]

Exercise 2.9.3. [R][S][Section 2.1][Goal 2.1] Airline ticket taxes.

On July 22, 2011 the Associated Press quoted Transportation Secretary Ray LaHood saying that a partial shutdown of the Federal Aviation Administration (FAA) would cost the government about $200 million a week in airline ticket taxes. [R49]

(a) Calculate the total amount of lost revenue from the partial shutdown of the FAA that began on July 23, 2011 and ended on August 4, 2011.
(b) One of the issues at stake in the debate over the shutdown was the $16.5 million the FAA provided in federal subsidies to rural airports. House Republicans wanted to eliminate these subsidies. Estimate how many hours of collecting federal aviation taxes would be needed to support these subsidies.

Exercise 2.9.4. [S][Section 2.1][Goal 2.1] A race through the tree of life.

Jessica Bolker teaches biology at the University of New Hampshire. She writes in an email:

```
If I were going to cover all the animal groups in proportion
to their extant (living) species diversity in a single
50-minute lecture, how long would I spend on each?

Start animals at 10:10  (50 minutes to cover ~1,255,000 species;
~0.0024 seconds each)

Talk about arthropods (1,000,000) for 40 minutes, until 10:50 am

mollusks (110,000) from 10:50 till about 10:54

all the other non-chordates, combined (94,000, but perhaps
really a lot more) for about 3.75 minutes, until almost 10:58

chordates beginning at 10:58 (remember class ends at 11)

fish (31,000) for a minute and a quarter (75 seconds)

reptiles (7000) for 17 seconds
```

```
amphibians (5500) for 13 seconds

mammals (4000) for almost 10 seconds

birds (3000) for a bit over 7 seconds

- and we're done.
```

For the first few questions, assume that her data about species counts are correct.

(a) Is her estimate of about 0.0024 seconds per species correct?
(b) Do her species counts by group add up to approximately 1,255,000?
(c) Does the way she apportions lecture time match the apportionment of species?

Now think about the data.

(d) The website `www.biologicaldiversity.org/species/birds/` says there are in fact about 10,000 bird species, not the 3000 in Professor Bolker's letter. How would you rearrange the last two minutes of her class to take that new information into account?
(e) You probably know about reptiles, amphibians, mammals and birds. What are arthropods, mollusks, chordates? Don't just copy and paste a definition from the internet—provide one that shows that you understand what you've written. Did you learn anything doing this part of the problem?
(f) While you were answering the previous question you probably found web sites for each of those groups. Do the number of species of each kind match Professor Bolker's assumptions? (Full sentences, and citations, please, not just "yes" or "no".)
(g) How is Professor Bolker related to one of the authors of this book?

Exercise 2.9.5. [S] [Section 2.1][Goal 2.1][Goal 2.3] The penny stops here.

On June 4, 2008 *The Seattle Times* reported on MIT physicist Jeff Gore's research on the cost of a penny. He estimated that dealing with pennies makes transactions take two to two and a half seconds longer, costing each of us four hours every year. At $15 per hour "that's $15 billion a year lost nationwide annually." [R50]

(a) How many cash transactions per person per year involving pennies did Gore assume when he made his estimate?
(b) On the average, how many cash transactions per day does each of these people participate in?
(c) How many people did Gore assume were making those transactions?
(d) Use the answers to these questions to decide whether Gore's estimate of $15 billion worth of wasted time is reasonable.

Exercise 2.9.6. [S][Section 2.1][Goal 2.1] Scrap the penny?

The U.S. General Accounting Office reported in February 2012 that

> replacing $1 notes with $1 coins could potentially provide $4.4 billion in net benefits to the federal government over 30 years. The overall net benefit was due solely to increased seigniorage and not to reduced production costs. Seigniorage is the difference between the cost of producing coins or notes and their face value. [R51]

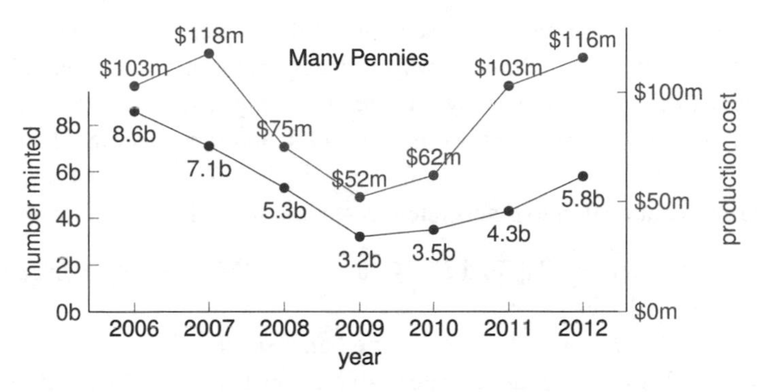

Figure 2.4. What's a penny worth these days? [R52]

The graphs in Figure 2.4 display data from the U.S. Mint on the number of pennies produced each year and the cost to make them.

(a) How much did it cost to manufacture one penny in 2012?
(b) How much money did the United States lose for each penny minted in 2012?
(c) How can you tell from the graphs without doing any arithmetic that it cost more to make a penny in 2007 than in 2006?
(d) In which of the years 2006-2012 was the total cost of producing pennies the largest?
(e) In which of the years was the cost per penny the largest?
(f) What fraction of the estimated $4.5 billion in savings would come from the money saved in seignorage by scrapping the penny?

Exercise 2.9.7. [S][Section 2.1][Goal 2.1] A literal ballpark estimate.

The Boston Globe reported on June 21, 2012 that

> It was so hot at Fenway that Dave Mellor, the Red Sox groundskeeper, said he could probably fry enough eggs on the Green Monster to feed the entire city. Not possible, mathematically, but he could probably fry enough for everyone in the stadium. [R53]

Verify the writer's "Not possible, mathematically, but he could probably . . .". Write up your argument in the form of a letter to the editor.

There's enough information on the web to estimate the area of the Green Monster.

Exercise 2.9.8. [S][Section 2.1][Goal 2.1][Goal 2.5] Counting red blood cells.

On page 240 of the April 2011 issue of the *Bulletin of the American Mathematical Society* Misha Gromov wrote that

> Erythrocytes are continuously produced in the red bone marrow of large bones . . . at the rate $\approx$ 2.5 million/second or $\approx$ 200 billion/day. Adult humans have 20–30 trillion erythrocytes, $\approx$ 5 million/ cubic millimeter of blood. [R54]

(a) What is an erythrocyte?
(b) Check that 2.5 million/second is about 200 billion/day.
(c) Use the numbers in the second sentence to estimate the volume of blood in an adult human, in liters.
(d) Check your answer to the previous question with a web search.

(e) (Warmup for the next question.) Suppose your hen lays an egg a day, and you always have seven eggs in your refrigerator. Explain why each egg is on average a week old when you eat it. (This is easy to explain if you always eat the oldest egg first. It's a little harder to explain if you eat the eggs in random order and have to think about the average. Don't bother with that.)

(f) Use the data in the quotation to determine the average lifetime of an erythrocyte.

Exercise 2.9.9. [S][C][Section 2.1][Goal 2.1][Goal 2.2][Goal 2.3] Global warming opens Arctic for Tokyo-London undersea cable.

On January 21, 2010 *The Seattle Times* carried an Associated Press report about plans to connect London and Tokyo by an underwater cable through the Northwest Passage, in order to reduce the time a message takes to 88 milliseconds from 140 milliseconds. "The proposed system would nearly cut in half the time it takes to send messages from the United Kingdom to Asia." [R55]

(a) Estimate the distance from London to Tokyo via the Northwest Passage.
(b) Use that estimate to estimate the speed of the transmission signal, in appropriate units.
(c) Compare the speed of transmission to the speed of light.

Exercise 2.9.10. [S][Section 2.1][Goal 2.1][Goal 2.6] Maple syrup.

The March page of the 2008 Massachusetts "Celebrating the Seasons of Agriculture" calendar contains the quote

> Farm Fact #3
> In 2007, Massachusetts maple producers set 230,000 taps to collect sap, which, after boiling produced 30,000 gallons of maple syrup. It takes about 40 gallons of sap to produce one gallon of syrup.

And take a look at `www.gocomics.com/frazz/2015/09/25`.

(a) How many gallons of syrup did each tap yield?
(b) How many gallons of sap did each tap yield?
(c) What extra information would you need to estimate the number of sugar maple trees that were tapped?
(d) Suppose the sap is collected in gallon buckets that hang under each tap, and that the maple syrup season lasts one month. How many times a day will the buckets need to be emptied?
(e) Visit a local supermarket and compare the price of 100% maple syrup to the price of some "pancake syrup." Read the label on the pancake syrup to find out how much maple syrup it contains. (If you can find this information on the web without visiting a supermarket, more power to you.)

What fraction of the cost of the generic syrup can you attribute to the actual maple syrup it contains?

Exercise 2.9.11. [Section 2.2][Goal 2.1] The MPG illusion.

(a) Take the quiz at `www.fuqua.duke.edu/news/mpg/mpg.html`. Then ask two friends to take the quiz, and watch over their shoulder as they do. If they are willing to think out loud for you, so much the better. If they get the wrong answer, explain the correct one.

Write a brief summary of how you did on the quiz and how your friends did.

(b) When you ask Google for

12 miles per gallon in gallons per 10000 miles

you get more than the answer we used in the text—you get lots of links. Follow some of them that look particularly interesting—please not just the first two—and write about what you find.

Exercise 2.9.12. [R][S][Section 2.2][Goal 2.1][Goal 2.5] Fuel economy.

The government's 2011 *Fuel Economy Guide* (`www.fueleconomy.gov/feg/pdfs/guides/FEG2011.pdf`) lists the fuel economy for a Chevrolet Malibu as 22 MPG for city driving and 33 MPG for highway driving, with an annual fuel cost of $2,114. The *Guide* says that figure is computed this way:

> Combined city and highway MPG estimates ... assume you will drive 55% in the city and 45% on the highway. Annual fuel costs assume you travel 15,000 miles each year and fuel costs $3.66/gallon for regular unleaded gasoline and $3.90/gallon for premium.

(a) Convert these ratings to gallons per mile and gallons per hundred miles, using a calculator just for the arithmetic, not Google's for the units.
(b) Check your answers with the Google calculator.
(c) Verify the annual fuel cost figure. (The Malibu does not require premium gasoline.)
(d) Explain why the computation in the previous part of the exercise would have been easier if the fuel economy was reported in GPM instead of (or in addition to) MPG.
(e) What would the annual fuel cost be at current gas prices?
(f) (optional) Look at the *Guide*, find something interesting there, and write briefly about it.

Exercise 2.9.13. [S][Section 2.2][Goal 2.1] Sticker shock.

An editorial in *The New York Times* on Sunday, June 4, 2011 discussed the redesigned fuel economy stickers to be required on new cars starting in 2013.

> Consider a new midsize Ford Fusion S model. At $4-a-gallon gas, annual fuel costs would be about $2,200 for 15,000 miles driven. ... And that's for a car with a rated mileage of 34.9 m.p.g. (actual highway mileage, as with all vehicles, is lower, in this case about 27 m.p.g.). Industry clearly needs to do better. [R56]

Did the *Times* use the rated or the actual mileage in its computation?

Exercise 2.9.14. [W][S][Section 2.2][Goal 2.1] Hybrids vs. nonhybrids: The 5-year equation.

On February 23, 2011 Matthew Wald blogged at *The New York Times* about a study in *Consumer Reports* saying that

> A car buyer who lays out an extra $6,200 extra [sic] to buy the hybrid version of the Lexus RX will get the money back in gas savings within five years, according to Consumer Reports magazine, but only if gasoline averages $8.77 a gallon. Otherwise, the nonhybrid RX 350 is a better buy than the Hybrid 450h. [R57]

Wald notes that the study assumes

- the car will be driven 12,000 miles a year.
- gas will cost $2.80 a gallon.
- the hybrid gets 26 miles per gallon, the nonhybrid, 21.

(a) Show that the computation is wrong—that at $8.77 per gallon of gas you can't save $6,200 in 60,000 miles of driving.
(b) Show that you can save that much with that much driving if gas costs $8.77 per gallon more than $2.80 per gallon.
(c) The *Times* blogger was reporting on a study from *Consumer Reports* magazine. Do you think the error was the blogger's, or the magazine's? What would you have to do to find out which?
(d) Write a response to post as a comment on the blog.

Exercise 2.9.15. [S][Section 2.2][Goal 2.1][Goal 2.6] Lots of gasoline.

> In [the U.S. in] 2011, the weighted average combined fuel economy of cars and light trucks combined was 21.4 miles per gallon (FHWA 2013). The average vehicle miles traveled in 2011 was 11,318 miles per year. [R58]

(a) Convert the weighted average combined fuel economy to gallons per hundred miles.
(b) Compute the amount of gasoline the average vehicle used in 2007.
(c) Estimate the number of cars and light trucks on the road in the U.S. in 2007.
(d) Estimate the total amount of gasoline used by cars and light trucks in the U.S. in 2011.
(e) At `www.americanfuels.net/2014/03/2013-gasoline-consumption.html` you could read that the U.S. consumed 134,179,668,000 gallons of gasoline in 2011. Is that number consistent with your answer to the previous part of the problem?

Exercise 2.9.16. [S][Section 2.2][Goal 2.1] Yet another problem on gas mileage.

On February 18, 2012 ScottW of Chapel Hill, NC responded to an article in *The New York Times* saying that a car that averaged 35 mpg instead of 20 would cut your gas bill almost in half. [R59]

Is ScottW's arithmetic right?

Exercise 2.9.17. [S][Section 2.3][Goal 2.1] Refining oil.

When a barrel of oil is refined it produces 19.5 gallons of gasoline, 9 gallons of fuel oil, 4 gallons of jet fuel and 11 gallons of other products (kerosene, etc.)

(a) How many gallons of oil are there in a barrel before any of it is refined?
(b) Determine the fraction (or percentage) of the barrel that gasoline represents.
(c) Find the current price for a barrel of oil and the current price for a gallon of gasoline at the pump.
(d) What percentage of the price of a gallon of gas at the pump is accounted for by the cost of the oil from which it is made? What do you think accounts for the rest?
(e) In 2005, the United States consumed oil at an average rate of about 22 million barrels per day. What is the corresponding consumption of gasoline in gallons per day?
(f) Suppose each gallon can is about one foot tall. If we stacked them up would those cans of gasoline reach the moon?

Exercise 2.9.18. [S][Section 2.3][Goal 2.1] Harry Potter.

The last Harry Potter book went on sale at 12:01 a.m. on July 21, 2007. The publisher Scholastic reported the next day that it broke all records,

> ... selling an unprecedented 8.3 million copies of *Harry Potter and the Deathly Hallows* in its first 24 hours on sale. [R60]

(a) On the average, how many books per minute were sold during that first day?
(b) Estimate how many bookstores were selling the book.
(c) In Harry Potter's wizard world currency is measured in galleons, sickles and knuts, where

$$1 \text{ galleon} = 17 \text{ sickles}, \quad 1 \text{ sickle} = 29 \text{ knuts}.$$

Scholastic Books, the publisher of the Harry Potter series in the United States, has issued paperback copies of books that (they claim) Harry Potter used at Hogwarts. The price is listed as

```
$5.99 US (14 Sickles 3 Knuts)
```

Use this information to figure out the number of dollars per galleon.
(d) The list price of the hardback version of *Harry Potter and the Deathly Hallows* was \$34.99. Convert this price to galleons. (You may write your answer with decimal fractions of a galleon. No need to convert the fractional part to sickles and knuts—but you can if you want to.)

Exercise 2.9.19. [R][Section 2.4][Goal 2.1][Goal 2.6] The national debt.

According to the United States Department of the Treasury, the national debt on January 1, 2013 was $\$16{,}432{,}730{,}050{,}569.12 \approx \$16.5$ trillion.

(a) Estimate the average share of the debt for each person in the United States on January 1, 2013.
(b) By some estimates, the debt increases an average of just over \$2.5 billion per day. Use the web to find the current national debt of the United States and comment on the accuracy of those estimates.
(c) Find the current population of the United States and update the average share of the debt for each person.
(d) Choose a different country and use the web to find its current national debt. Estimate the average share of the debt for each person in that country and compare this to what you calculated for the United States.

Exercise 2.9.20. [S][Section 2.4][Goal 2.6][Goal 2.3] Medicare fraud.

An Associated Press story in Easton, Maryland's *The Star Democrat* on June 3, 2010 reported that

> All told, scam artists are believed to have stolen about \$47 billion from Medicare in the 2009 fiscal year, nearly triple the toll a year earlier. Medicare spokesman Peter Ashkanaz said that ... charges have been filed against 103 defendants in cases involving more than \$100 million in Medicare fraud. [R61]

(a) What percentage of the Medicare fraud has been targeted by filed charges?
(b) What is the average claim in each fraud charge?
(c) How many of these average size claims would need to be filed to recover the entire $47 billion?
(d) The article suggests that the administration is vigorously pursuing Medicare fraud. Do the numbers support that suggestion?

Exercise 2.9.21. [S][Section 2.4][Goal 2.6][Goal 2.5] Mom blogs and unit prices.

`carrotsncake.com/2011/01/grocery-shopping-101-unit-price.html` is just one of many "mom blogs" that discuss shopping wisely using unit prices. There you will find the quote

> In the photo [on the website], the unit price is listed in orange on the left side of the label. In this case, it's how much you pay per gallon of olive oil. The cost per gallon is $50.67, but you're only buying 16.9 ounces of olive oil, so you pay $6.69. It might seem like a deal because you're only paying $6.69, but the cost per gallon is high. [R62]

(a) Verify the unit price calculation in the quotation.
(b) Visit the blog. There you will find some comments in each of the categories
"I've always looked at unit prices."
"This is new to me, and cool. I'll do this from now on."
"My husband/boyfriend told me about this."
Estimate the percentage or fraction of comments in each category—perhaps "About a third of the comments are from people who learned about unit pricing on this blog."
(c) If you were to post a comment, what would it be? (Optional: post it.)

Exercise 2.9.22. [S][Section 2.4][Goal 2.6] Greenies.

A 12 ounce box of Greenies costs $12.99. A 30 ounce box costs $26.99.

(a) What is the cost in dollars per ounce in each box? Which size should you buy?
(b) Suppose you have a coupon worth $5 toward the price of any box of Greenies. If you use the coupon, which size should you buy? Is it always better to buy the large economy size?
(c) Answer the previous question if your coupon gives you a 10% discount rather than $5 off.
(d) What's a Greenie? Are these prices reasonable?

Exercise 2.9.23. [S][R][Section 2.4][Goal 2.6][Goal 2.5] Race to the moon.

On July 22, 2011 *The Seattle Times* reported that Astrobiotics Technology planned to charge $820,000 a pound to send scientific experiments to the moon in a spacecraft that could carry 240 pounds of cargo. They would collect two hundred million dollars for the delivery. [R63]

In August 2014 the company advertised on its website that the cost to deliver a payload to the lunar surface for $1.2M per kilo. [R64]

(a) Verify the arithmetic in the 2011 article from *The Seattle Times*.
(b) How did the quoted price change between then and 2015?

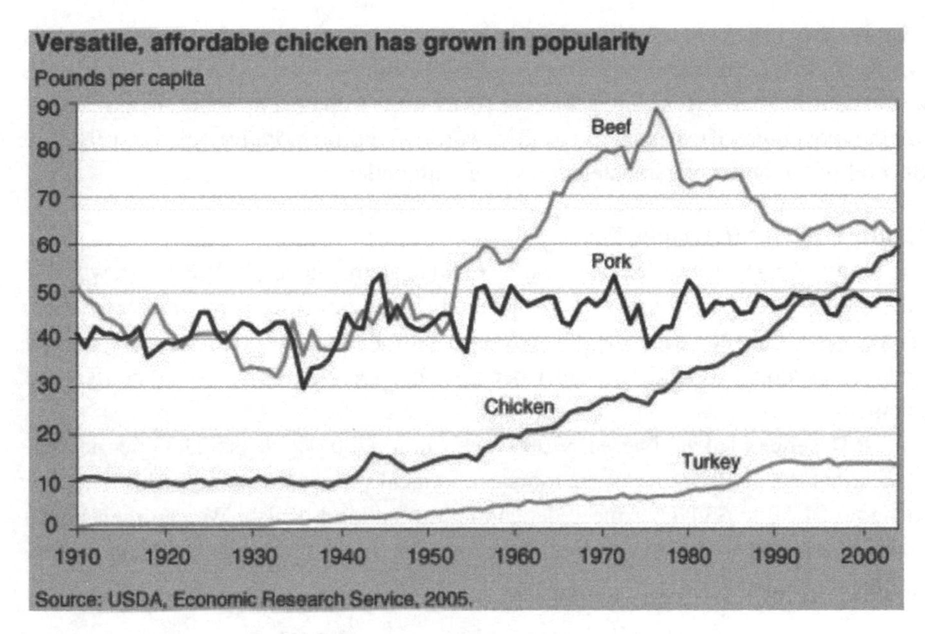

Figure 2.5. Beef, chicken, pork, turkey [R65]

Exercise 2.9.24. [S][Section 2.4][Goal 2.6] When's the beef?

Use the graph in Figure 2.5 to answer the following questions. You may need to supplement the data with information from the web, but don't ask the web for answers to the questions, except (if you wish) to check your computations.

[See the back of the book for a hint.]

(a) What does "per capita" mean?
(b) Estimate the number of pounds per capita of beef, chicken, pork and turkey consumed in the year 2000.
(c) The graph stops at about 2004. Use the trends it shows to estimate the number of pounds per capita of beef, chicken, pork and turkey that are being consumed this year.
(d) The peak year for per capita beef consumption was about 1975. Was more total beef consumed that year than at any other time in the past century?

Exercise 2.9.25. [S][C][Section 2.4][Goal 2.6][Goal 2.5] Apple's app store.

A March 5 press release from the Apple Corporation heralded the 25 billionth download from their App Store.

> . . . by the users of the more than 315 million iPhone®, iPad® and iPod touch®devices worldwide. [R66]

(a) Find out when the App Store opened, and calculate the rate at which apps have been downloaded since then, assuming the rate was constant. Choose units that best express your answer: apps/year? apps/day? apps/second?
(b) Find estimates for the number of devices (phones, tablets, pads, pods) sold. Use those estimates to estimate the number of apps per device.
(c) Do you think these rates have been constant since the store opened?

Exercise 2.9.26. [S][Section 2.4][Goal 2.6][Goal 2.5] Seven thousand dollars for sushi?

On April 5, 2012 in *The New York Times* Gail Collins wondered "Why sushi rolls for 300 people cost $7,000" at a "four-day gathering [that] wound up costing more than $820,000. . . . Among the most notable excesses was $6,325 worth of commemorative coins in velvet boxes."

Several of the online commenters did some arithmetic:

> **Truth Will Out (Olympia, WA)**
> No biggie. This is the misleading tyranny of large aggregate totals. Break it down, and it is only $683 per person per day. Not exactly off the wall for conference expenses, especially if this included room (Gail doesn't say). Even those commemorative coins only break down to a $20 chachka per person. Pretty cheap as far as conference mementos go.
>
> **P Desenex (Tokyo)** I agree. $7000 for sushi for 300 people is $23.33 per person. At a kaiten sushi place (one of the more economical venues for eating sushi) the price is about $1.50 to $3.00 per piece depending on the kind of fish. We can then figure each person had from 7 to 15 pieces. A reasonable price and a reasonable amount to eat for lunch. [R67]

(a) Check the unit pricing arithmetic in these comments.
(b) What does Truth Will Out mean by "the misleading tyranny of large aggregate totals"?
(c) What's a "chachka"?

Exercise 2.9.27. [U][Section 2.4][Goal 2.6] Splash!

The Tournament Players Club at Sawgrass is a golf course in Florida and is considered one of the most difficult courses in the world. The most famous hole on the course is the 17th, known as the "Island Green", which is completely surrounded by water. The Wikipedia page for the course notes that

> It is estimated that more than 100,000 balls are retrieved from the surrounding water every year . . . . [R68]

(a) Estimate the number of hours per year golf is played at that course in Florida.
(b) If the estimate in the article is correct, how many golf balls enter the water per minute?
(c) Do you think the estimate in the article makes sense?

Exercise 2.9.28. [S][Section 2.4][Goal 2.6] Automobile accident statistics.

Here are two quotations from the web:

> According to the National Highway Traffic Administration, car accidents happen every minute of the day. Motor vehicle accidents occur in any part of the world every 60 seconds. And if it's all summed up in a yearly basis, there are 5.25 million driving accidents that take place per year. [R69]
>
> Car Crash Stats: There were nearly 6,420,000 auto accidents in the United States in 2005. The financial cost of these crashes is more than 230 billion dollars. 2.9 million people were injured and 42,636 people killed. About 115 people die every day in vehicle crashes in the United States—one death every 13 minutes. [R70]

(a) The first quotation gives accident rates in two ways, with two different time units. Are the rates the same when you convert one to the other? If not, what mistake might the writer have made?
(b) Have the approximate death rates in the second quotation been computed correctly from the total number of deaths?
(c) Are the accident rates in the two quotations consistent? Can you find other evidence on the internet to confirm them?

[See the back of the book for a hint.]

Exercise 2.9.29. [S][Section 2.4][Goal 2.6][Goal 2.5] Gas is cheap.

B.P. Phillips of Broomfield, Colorado conducted "An Impartial Price Survey of Various Household Liquids as Compared to a Gallon of Gasoline". You can find it in the *Annals of Improbable Research*.

> With the price of oil surging past $100 per barrel recently, consumers and the media in the United States have been in an uproar over the outrageous price of gasoline. In my opinion, however, they have failed to place this cost in context to other common, household liquids. Thus, I decided to perform an informal price survey.
>
> I visited my local grocery store, where I recorded the prices and respective volumes of 22 off-the-shelf fluids. Then, after returning home, I computed the price per gallon of each item, and arranged them in ascending order by cost, as shown in Table 2.6.
>
> Clearly, all of the fluids I recorded are at least as expensive, if not more so, than gasoline, which is currently hovering around $3.25 per gallon in Broomfield, Colorado. In fact, the only two products with prices that are even comparable are Coca-Cola and milk, each costing less than four dollars per gallon. All of the other products range anywhere from $6.38 per gallon of orange juice all the way up to a staggering $400.46 per gallon of Wite-Out Correction Fluid.
>
> After being made aware of my results, it should be painfully obvious to consumers and the media that gasoline is still relatively inexpensive compared to other liquids. In closing, I ask readers to provide us with additional examples of outstandingly expensive fluids from the U.S. or other countries. [R71]

(a) Verify at least four of the numbers in Table 2.6.
(b) Do some research (either on the web at an online grocery store or at a local store) and see if you can find items less expensive than Coca-Cola and more expensive than Wite-out Correction Fluid.
(c) Investigate the cost of bottled water in dollars per gallon. The answer will depend on where you buy it, and in what quantity. Discuss where your answers fit in this list.

Exercise 2.9.30. [S][R][A][Section 2.5][Goal 2.2][Goal 2.5] Mph and kph.

(a) Convert 100 km/hour to miles per hour, so you'll know how fast you may drive in Canada.
(b) Convert 65 miles/hour to km/hour.

Write your answers using just two significant digits, not lots of decimal places.

Table 2.6. Unit prices for common fluids

Product	Price	Price per Gallon
Coca-Cola	$1.89 / 2 liters	$3.58
Lucerne 2% Milk	$3.79 / 1 gal.	$3.79
Tropicana Orange Juice	$3.19 / 64 fl. oz.	$6.38
Tree Top Apple Juice	$3.55 / 64 fl. oz.	$7.10
Evian Water	$1.89 / 1 liter	$7.16
Bud Light	$1.99 / 24 fl. oz.	$10.61
Heinz Ketchup	$4.19 / 46 fl. oz.	$11.66
Crisco Vegetable Oil	$5.69 / 48 fl. oz.	$15.17
Tide Laundry Detergent	$14.49 / 100 fl. oz.	$18.55
Windex Glass Cleaner	$3.99 / 26 fl. oz.	$19.64
Aunt Jemima Maple Syrup	$4.39 / 24 fl. oz.	$23.41
Starbucks Frappuccino	$7.29 / 38 fl. oz.	$24.56
Hidden Valley Ranch Dressing	$3.75 / 16 fl. oz.	$30.00
Red Bull Energy Drink	$1.99 / 8.3 fl. oz.	$30.69
Pert Shampoo	$3.49 / 13.5 fl. oz.	$33.09
Lucky Clover Honey	$4.45 / 12 fl. oz.	$47.47
Pepto-Bismol	$5.99 / 16 fl. oz.	$47.92
A1 Steak Sauce	$7.34 / 15 fl. oz.	$62.63
Listerine Mouthwash	$3.59 / 5.8 fl. oz.	$79.23
Bertolli Extra Virgin Olive Oil	$11.49 / 17 fl. oz.	$86.51
Vick's Nyquil	$7.49 / 10 fl. oz.	$95.87
Wite-Out Correction Fluid	$2.19 / 0.7 fl. oz.	$400.46

Exercise 2.9.31. [U][Section 2.5][Goal 2.1][Goal 2.2] Roll your own.

Invent a problem based on something that catches your fancy on Wikipedia's list of humorous units of measurement: `en.wikipedia.org/wiki/List_of_humorous_units_of_measurement`.

Exercise 2.9.32. [S][Section 2.5][Goal 2.1][Goal 2.2] The chron

The World Time Organization is proposing that we convert to a metric time measurement system. The basic unit would be the chron; a day would be 10 chrons long. Use this information to answer the following questions.

(a) How many kilochrons are there in a year?
(b) How many seconds in a millichron?
(c) If we switch to this new system, the Department of Transportation will need to replace its speed limit signs. What would the following sign look like if miles per hour were converted to kilometers per chron?
How would the Federal Highway Administration change this regulation?

This sign is used to display the limit established by law, ordinance, regulation, or as adopted by the authorized agency. The speed limits shown shall be in multiples of 10 km/h or 5 mph. [R72]

(d) French revolutionary metric time (optional)
The chron isn't entirely our own invention. The metric system of measurement for lengths and weights was a byproduct of the French Revolution. There were proposals then for a metric time system too:
- 10 metric hours in a day
- 100 metric minutes in a metric hour
- 100 metric seconds in a metric minute
- 10 days in a metric week (called a dekade)

There are several web sites devoted to debating whether this or something like it would be a good idea. Find some and write about the question.

Exercise 2.9.33. [R][S][Section 2.5][Goal 2.1][Goal 2.2][Goal 2.5] How old is the United States?

The United States was created in 1776. How old is it in seconds? Write your answer in metric terms with the proper prefix (Kiloseconds? Petaseconds? Decide what works best here). Remember not to use too many significant digits.

Exercise 2.9.34. [Section 2.5][Goal 2.2] Metric prefixes as metaphors.

The metric prefixes are often used to suggest that quantities are large or small without any particular quantitative meaning. What is microfinance? micromanagement? nanotechnology? megalomania? What about the Mega Society at `megasociety.org/about.html`? Find other similar usages.

Exercise 2.9.35. [S][Section 2.4][Goal 2.1] World vital events per time unit.

The data in Table 2.7 are from `www.census.gov/population/international/data/idb/worldvitalevents.php`

(a) Check that the rates in each column are consistent.
(b) Check that the rates in each row are consistent.
(c) Which of the rates is the easiest to understand the impact of? Why do you think so?

Table 2.7. World vital events per time unit: 2014
(Figures may not add to totals due to rounding)

Time unit	Births	Deaths	Natural increase
Year	134,176,254	56,605,700	77,570,553
Month	11,181,355	4,717,142	6,464,213
Day	367,606	155,084	212,522
Minute	255	108	148
Second	4.3	1.8	2.5

Exercise 2.9.36. [S][C][S][Section 2.5][Goal 2.2] Pumping blood.

We haven't done anything with the "8,000 liters of blood" in Section [Section 1.3]. What does it mean? Clearly it's not the amount of blood in the body—the number is way too large for that. It's the amount pumped by the heart, in liters per day.

(a) Verify the assertion that the human body contains about 5 liters of blood.
(b) Use that fact to estimate how many times a day (on average) each red blood cell passes through the heart on its way to the rest of your body, and how long the round trip takes.

Exercise 2.9.37. [S][W][Section 2.6][Goal 2.1][Goal 2.4] Arizona using donations for border fence.

On November 24, 2011 *The Washington Times* published an article from the Associated Press with that headline. There you could read that the state was seeking donations to run a fence along the border with Mexico.

> ... [State Senator] Smith acknowledges he has a long way to go to make the fence a reality. The $255,000 collected will barely cover a half mile of fencing. Mr. Smith estimates that the total supplies alone will cost $34 million, or about $426,000 a mile. Much of the work is expected to be done by prisoners at 50 cents an hour. [R73]

(a) Use the information in the quotation above to estimate the length of Arizona's border with Mexico, in miles.
(b) Find another estimate of the length of Arizona's border with Mexico. Is the answer you find consistent with your answer to the previous question?
(c) If $255,000 pays for labor and supplies for half a mile of fence, how much does the labor cost?
(d) How many hours will it take prisoners to construct that half mile? Does your answer make sense?

Exercise 2.9.38. [S][Section 2.6][Goal 2.5] U.S. gallons.

Explain why Google says

12 miles per gallon = 833.333333 US gallons per (10 000 miles)

instead of simply

12 miles per gallon = 833.333333 gallons per (10 000 miles)

.

Exercise 2.9.39. [S][Section 2.8][Goal 2.4] Comet 67P/Churyumov-Gerasimenko.
On August 5, 2014 *The New York Times* reported that

> In June, the [Rosetta] spacecraft measured the flow of water vapor streaming off the comet [67P/Churyumov-Gerasimenko] at a rate of about two cups a second, which would fill an Olympic-size swimming pool in about 100 days. [R74]

Check this volume calculation.

Exercise 2.9.40. [U][Section 2.8][Goal 2.4] A table for your puzzle.

You are about to start work on your new 1000 piece jigsaw puzzle and you need to know how much space to clear for it on the table. Each piece is approximately square, about 3/4 of an inch on each side. The finished puzzle is a rectangle about one and a half times as wide as it is high. How big is the puzzle?

(Optional) Answer the same question if the finished puzzle is circular.

Exercise 2.9.41. [U][Section 2.8][Goal 2.2][Goal 2.4] A pint's a pound the whole world round.
Discuss the accuracy of that well known saying.
[See the back of the book for a hint.]

Exercise 2.9.42. [S][Section 2.8][Goal 2.1][Goal 2.4] An ark full of books.

In March of 2012 *The New York Times* reported on a warehouse in California storing forty-foot shipping containers full of books. Each week twenty thousand new books arrive. There were 500,000 stored at the time the article was written. The site plans to expand to store 10 million. [R75]

(a) How long had the collection been accumulating when the article appeared?
(b) Estimate how many volumes are in each shipping container (you can assume that these are standard forty-foot shipping containers, which means you will have to look up the dimensions of this type of container).
(c) How many shipping containers would be needed to store 10 million volumes?

Exercise 2.9.43. [S][Section 2.8][Goal 2.1][Goal 2.4] Floating trash threatens Three Gorges Dam.

A story with that headline in the August 2, 2010 *China Daily* reported about a flood in China piling garbage 60 cm thick over an area of 50,000 square meters every day. To protect the Three Gorges Dam 3,000 tons of garbage were being collected each day. [R76]

(a) Check that 60 cm is approximately two feet and that 50,000 square meters is approximately half a million square feet.
(b) The article suggests that collecting 3,000 tons a day will keep the pile under control. Is that reasonable?
(c) Estimate how long it would take to clear the garbage at the rate of 3,000 tons a day.
(d) How long would it take the city you live in to produce a layer of garbage covering half a million square feet two feet deep?

Exercise 2.9.44. [R][U][Section 2.6][Goal 2.5][Goal 2.3] Stressing your calculator.

(a) How many digits are there in the largest number your calculator can display normally?
(b) When you ask it for a number bigger than that does it switch to scientific notation or complain?

[See the back of the book for a hint.]

Exercise 2.9.45. [U] Gold.

In a comment in *The New York Times* online on December 23, 2013, Adam wrote that

> "Gold has no real practical use except making jewelry and filling cavities and the like." Not true. Gold has many industrial uses, and it coats the terminals of the best high-fidelity electrical connections. A ton of cell phones contains about 300 grams of gold—just 300 parts per million by weight, but there are a lot of cell phones. [R77]

(a) Is Adam's "ton" a metric ton or an English ton?
(b) How much gold is there in one cell phone? Write your answer in grams using an appropriate metric prefix.
(c) Estimate the number of cell phones in the world.
(d) Estimate the amount of gold in the world's cell phones.

Exercise 2.9.46. [U] Expensive solar energy.

In an article headlined "Solar use will push energy costs up in Mass. 20-year rise put at $1 billion" in *The Boston Globe* on February 12, 2014 you could read that the added cost for residential customers would be about $1 to $1.50 more a month. [R78]

(a) Use the data to estimate the number of residential customers in Massachusetts.
(b) Is your answer to the previous question a reasonable approximation of the actual number? If not, what might explain the discrepancy?
(c) The headline and the quote convey the same information, but with very different psychological content. In one the added cost seems enormous, in the other insignificant. Write a letter to the editor politely criticizing the way the *Globe* presented the numbers.

Exercise 2.9.47. [S] Boca negra.

Charlotte Seeley's boca negra cake recipe calls for 12 ounces of chocolate. She has a 17.6 ounce bar of Trader Joe's Pound Plus Belgian Chocolate that's divided into 40 squares.

(a) How many of the squares should she use?
(b) Why do you think the package comes in such a peculiar size?
(c) What's a boca negra cake?

Exercise 2.9.48. [S] Smoots

If you follow Mass Ave from Boston to Cambridge you will cross the Charles River via the Harvard Bridge. Looking down you may notice that the bridge is marked off in *smoots*. From one end to the other the length of the bridge is 364.4 smoots. Wikipedia, reliable in this case, says a smoot is "a nonstandard unit of length created as part of an MIT fraternity prank. It is named after Oliver R. Smoot, a fraternity pledge to Lambda Chi Alpha, who in October 1958 lay on the Harvard Bridge . . . and was used by his fraternity brothers to measure the length of the bridge." [R79]

(a) The length of the Harvard Bridge in standard English measure is 2035 feet. How tall was Oliver R. Smoot, in feet and inches?
(b) How tall was Oliver R. Smoot, in meters? (Use only the right number of decimal places in your answer.)
(c) In 2014 Kelly Olynyk was the 7 foot tall starting center for the Boston Celtics. How long would the Harvard Bridge be in olynyks?
(d) What is the conversion rate between smoots and olynyks?

Exercise 2.9.49. [S] Looking into the nanoworld.

In October, 2014 Eric Betzig, Stefan W. Hell and William E. Moerner won the 2014 Nobel Prize in Chemistry for inventing a microscope using fluorescence to see things as tiny as the creation of synapses between brain cells.

In the Reuter's report of the award you could read that

> In 1873, Ernst Abbe [a German scientist] stipulated that resolution could never be better than 0.2 micrometers, or around 500 times smaller than the width of a human hair. [R80]

(a) Use the data in this quote to estimate the width of a human hair. Choose the appropriate metric unit of length: meter, millimeter, micrometer or nanometer.
(b) Use the web to confirm the accuracy of your estimate.
(c) The Nobel Prize award statement said that "Due to their achievements the optical microscope can now peer into the nanoworld."
   - Express the width of a human hair in nanometers.
   - How much smaller than a human hair are the things scientists can now see?

Exercise 2.9.50. [Section 2.3][Goal 2.1][Goal 2.2] Gasoline elsewhere.

Compare the cost of gasoline in a foreign country of your choice with the cost where you live.

You will need to find the price at the pump and the tax rates in each place. Then use appropriate unit conversions and subtract the tax from the price at the pump to find the cost of the gasoline itself.

Exercise 2.9.51. [S][Section 2.8][Goal 2.1] How thick is paint?

In Section 2.8 you can read that a gallon of paint covers about 400 square feet.

How thick is the layer of paint on the wall? What are good units for the answer?

Exercise 2.9.52. [S][Section 2.8][Goal 2.4] Molasses!

The story headlined in the *The Boston Post* on the cover of this book tells of a January 15th, 1919 disaster in Boston's North End. Chuck Lyons retold the story in *History Today* in 2009, describing how a tank ruptured, "releasing two million gallons of molasses." [R81]

Wikipedia adds that "Several blocks were flooded to a depth of 2 to 3 feet." [R82]

Lyons goes on to say that several years later the auditor appointed to decide on compensation recommended

> . . . around $300,000 in damages, equivalent to around $30 million today, with about $6,000 going to the families of those killed, $25,000 to the City of Boston, and $42,000 to the Boston Elevated Railway Company.

(a) Estimate the number of blocks that could be covered two to three feet deep by that much molasses.
(b) Is Lyons's inflation calculation correct?
(c) Estimate the number of families that lost loved ones. How much would each receive, in today's dollars? How does that amount compare to damage awards in similar cases today?

[See the back of the book for a hint.]

Review exercises

Exercise 2.9.53. [A]

(a) A car travels for 3 hours at an average speed of 50 miles per hour. How far has it gone?
(b) A car travels for 2 hours at an average speed of 55 miles per hour, then travels for 30 minutes at an average speed of 40 miles per hour, then travels for 30 minutes at an average speed of 55 miles per hour. How far has it gone?
(c) If you plan to drive between two cities that are 250 miles apart, and you expect to go about 55 miles per hour, how long will it take you?
(d) If you earn $9.10 per hour and you work a 37-hour week, what are your gross (before tax) earnings?
(e) Apples sell for $1.39 per pound. If you buy 4.5 pounds, how much will you have to pay?

Exercise 2.9.54. [A] The exchange rate between euros and U.S. dollars is 1 € = $1.23.

(a) How much will it cost to buy one hundred euros?
(b) How many euros can you buy with one hundred dollars?
(c) The price of gas in a town I visited in France was 1.75 euros per liter. Convert this to dollars per gallon.
(d) A nice restaurant meal in Germany cost 32.50 euros. Convert this to dollars.

Exercise 2.9.55. [A] Suppose the exchange rate between US dollars and GBP is £1=$2. If a meal in a London pub costs £13 and a pint costs £5, how much would a meal and 3 pints cost in US dollars?

Exercise 2.9.56. [A] Write each of the following expressions in standard notation (that is, using all of the zeros).

(a) 3.2×10^5
(b) 4.666×10^8
(c) -3.3×10^6
(d) 5.3×10^{-3}
(e) 6.2×10^{-5}
(f) -3.222×10^{-6}

Exercise 2.9.57. [A]

(a) If my favorite cereal costs $2.99 for a 14-ounce box, what is the price per ounce?
(b) A gallon of milk costs $2.89. What is the price per ounce?
(c) A three-pound bag of apples is on sale for $2.49. What is the price per pound?

(d) In one store, a 36-pack of diapers costs $12.99. Another store is selling bonus packs, containing 42 diapers, for $14.59. Which is the better deal?

Exercise 2.9.58. [A] Do each conversion.

(a) One kilogram into centigrams.
(b) Two millimeters into nanometers.
(c) Three inches into centimeters.
(d) Ten feet into decimeters.
(e) Forty gallons into liters.
(f) Forty gallons into milliliters.
(g) Forty gallons into cubic centimeters.
(h) Forty gallons into picoliters.
(i) 1000 nanometers into micrometers.
(j) 100 milliliters into liters.
(k) 40 centiliters into liters.

Exercise 2.9.59. [A] Find each of these areas:

(a) A 12 foot by 18 foot carpet.
(b) A small rug measuring 4 feet by 18 inches.
(c) A deck measuring 5 meters by 7 meters.
(d) A football field in the United States measuring 120 yards by 110 yards.
(e) An NBA basketball court measuring 94 feet by 50 feet.

Exercise 2.9.60. [A] Find each of these volumes:

(a) A room measuring 12 feet by 18 feet by 9 feet.
(b) A bathtub measuring 6 feet by 2 feet by 3 feet.
(c) A wooden block measuring 4 centimeters by 2 centimeters by 1.5 centimeters.

Exercise 2.9.61. [A] Provide sensible answers with just one or two digits.

(a) One yard per second in miles per hour.
(b) One foot per second in miles per hour.
(c) One meter per second in kilometers per hour.

3

Percentages, Sales Tax and Discounts

The focus of this chapter is the study of relative change, often expressed as a percent. We augment an often much needed review two ways—stressing quick paperless estimation for approximate answers and, for precision, a new technique: multiplying by 1+(percent change).

Chapter goals

Goal 3.1. Understand absolute and relative change.

Goal 3.2. Work with relative change expressed as a percentage.

Goal 3.3. Master strategies for deciding how to arrange percentage calculations.

Goal 3.4. Learn (and appreciate) the "1+" trick for computing with percentages.

Goal 3.5. Calculate successive percentage changes.

Goal 3.6. Calculate percentage discounts.

Goal 3.7. Understand the difference between percentages and percentage points.

3.1 Don't drive and text

On January 13, 2010 an article in *The Washington Post* said in part

> Twenty-eight percent of all traffic accidents are caused when people talk on cellphones or send text messages while driving, according to a study released yesterday by the National Safety Council. The vast majority of those crashes—1.4 million of them—are caused by cellphone conversations, while 200,000 are blamed on text messaging, the council report said. [R83]

This quote suggests several questions. First, is the statement "the vast majority . . ." reasonable? Second, and more interesting, given this information, can we estimate how many accidents there were altogether? Finally, can we check that estimate?

There are three numbers to work with: 28% of all accidents, 1.4 million cellphone accidents and 200,000 texting accidents.

To answer the first question we have to compare the number of crashes caused by cellphone conversations to the total number of "those crashes," not to the number of "all traffic accidents".

Since the first paragraph reports on crashes caused by cellphone talk *or* text messaging, we can find the total we need by adding up the number of each kind: 1.4 million and 200,000 respectively. So there 1.6 million of "those accidents".

The fraction of the total due to cell phone conversations is

$$\frac{1.4 \text{ million}}{1.6 \text{ million}} = \frac{1.4}{1.6} = \frac{14}{16} = \frac{7}{8},$$

which certainly qualifies as "the vast majority." The convenient round numbers meant we didn't need a calculator to find the answer.

To report the answer as a percentage, remember that $3/4$ is 75%. Since $7/8$ is halfway from $3/4$ to 1, it's halfway from 75% to 100%, so it's $87.5\% \approx 88\%$.

By the summer of 2013, 41 states in the U.S. banned texting while driving. These statistics suggest that a complete cell phone ban would prevent many more crashes.

The second question asks for the number of "all traffic accidents". We're told that 28% of that unknown number is 1.6 million. To find the answer we need to do something with 28% and 1.6 million. Multiply? Divide? Try to guess which? Guessing is not a good idea—you'll be right only half the time. Thinking is better. Start by writing the equation in words. Then you can figure out whether to multiply or divide.

$$\frac{\text{cellphone or texting accidents}}{\text{all accidents}} = \frac{1.6 \text{ million}}{\text{all accidents}} = 28\% = 0.28.$$

That means

$$1.6 \text{ million} = 0.28 \times \text{all accidents}$$

so there were about

$$\frac{1.6 \text{ million}}{0.28} \approx 5.7 \text{ million}$$

accidents of all kinds.

We can check that answer with a ballpark estimate. Since 28% is just about 25%, which is one fourth, the quotation says that just about one accident in four was caused by texting. Then

$$4 \times 1.6 \text{ million} = 6.4 \text{ million}.$$

That's a little larger than our precise answer, but in the right ballpark.

Does this answer make sense? Treat it as a Fermi problem. There are about 300 million people in the United States so $5.7 \approx 6$ million accidents means that about 2% of the population, or one person out of every 50 was involved in one. If you have 50 acquaintances, on average one of them will have had an accident. Does that number seem to you to be in the right ballpark? You can confirm it with a web search—or work Exercise 2.9.28.

This story has two morals. First, percentages are a good way to think usefully about numbers. Second, don't use your cell phone (at all) while driving, and don't let the driver do it either when you're a passenger.

3.2 Red Sox ticket prices

The front page of *The Boston Globe* on September 9, 2008 provided the information on Red Sox ticket prices shown in Figure 3.1.

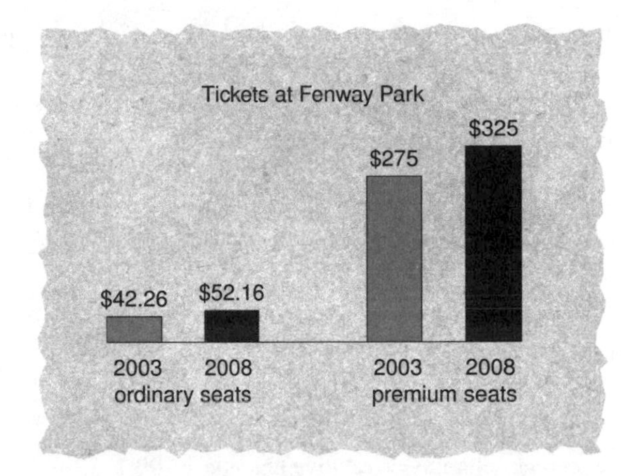

Figure 3.1. Red Sox ticket prices [R84]

Which ticket price increased the most?

The natural way to compare two numbers is to subtract. The absolute increases in ticket prices were \$52.16 − \$42.26 = \$9.90 for ordinary seats and \$325 − \$275 = \$50 for the most expensive seats. So the price of the most expensive seats increased much more in absolute terms.

But you often learn more when you compare a change to its starting value. A calculator will tell you that the relative change in the cost of an ordinary seat was

$$\frac{\$9.90}{\$42.26} = 0.23426108 \approx 0.23 = 23\%.$$

Most of the digits in the exact answer make no sense when talking about the ticket price, so we rounded that answer to two decimal places: 0.23. That's 23/100, which is 23 percent in everyday language.

Percentages are just fractions with 100 in the denominator. We use them instead of the fractions they stand for because people are uncomfortable with numbers less than one. It's easier to think about 23 than about 0.23.

This answer makes sense. 23% is nearly 25%, which is 1/4. The \$9.90 increase in the ticket price is indeed just a little less than one fourth of the original \$42.26.

For the most expensive seats the relative change was $\$50/\$275 \approx 0.18$, or about 18%—so the percent change for premium tickets was less than that for ordinary seats even though the dollar change was greater.

3.3 Patterns in percentage calculations

At the risk of encouraging you to mark parts of this text with a highlighter instead of thinking about and absorbing what they have to say, we'll summarize the idea behind our study of Red Sox ticket price increases. Then we will show how to work with the same numbers from different points of view.

- Subtraction is the obvious way to compare two numbers. The difference tells you the *absolute change*.
- Often division is more informative, even though it's less natural. It tells you the *relative change*. Relative change is frequently reported as a percentage.

$$\begin{aligned} \textit{absolute}\text{ change} &= \text{new value} - \text{reference value} \\ \textit{relative}\text{ change} &= \frac{\text{absolute change}}{\text{reference value}} \\ &= \frac{\text{new value} - \text{reference value}}{\text{reference value}} \end{aligned}$$

and

$$\begin{aligned} \text{relative } \textit{percent}\text{ change} &= \text{relative change} \times 100 \\ &= \frac{\text{new value} - \text{reference value}}{\text{reference value}} \times 100. \end{aligned}$$

When thinking about the price change for nonpremium Red Sox tickets there are four numbers that matter:

- The original value: $42.26. (Sometimes this is called the old value, or the reference value.)
- The new value: $52.16.
- The absolute change: $9.90.
- The relative change: 0.23, or, written as a percent, 23%.

If you know any two of those four numbers you can find the other two. The job when reading a paragraph is to understand which two are given, and which others you might want to know.

So far we've seen how to find the absolute and relative changes if you know the original and new values. Those are the easiest problems.

It's also easy to find the new value when you know the original value and the absolute change. Just add.

Suppose you're told that tickets now sell for $52.16 and that they've gone up by $9.90 and you want to know what they used to cost. Then you have the new value and the absolute change, so subtract to find the original value $42.26. (You can do the subtraction in your head if you notice that $9.90 is just a dime less than $10.00.)

Now suppose instead that you know the original value and the relative change ($42.26 and 23% in this example). It's a little harder (but still straightforward) to find the new value. First find the absolute change, which is 23% of the original value:

$$0.23 \times \$42.26 = \$9.7198 \approx \$9.72.$$

Then add:

$$\$42.26 + \$9.72 = \$51.98.$$

That's not quite the known correct answer of $52.16 since the 23% we're working with was rounded from the true percentage.

The hardest question is the one that asks you to find the original value when you know the new value and the relative change (as a percent). Here's a made up version of the problem we're working on that calls for that.

> The price of a plain old ordinary Red Sox ticket has increased a whopping 23% to \$52.16. That's outrageous! We should go back to the good old days.

What did tickets cost in the good old days, before the increase? Reading the first sentence tells you the meaning of the two numbers in it. The 23% is the relative change; the \$52.16 is the new value. We'd like to know the original value before the change.

Many people start by finding 23% of \$52.16, hoping that will tell them the absolute increase, so they can subtract. The numbers tell us that's not right: $0.23 \times \$52.16 = \$11.9968$, which is \$12, but we know the absolute increase is about \$10!

You can't undo a 23% increase with a 23% decrease!

Remember that every time you're tempted. Now read on to find out a correct way to solve the problem.

3.4 The one-plus trick

We found the relative change in Red Sox ticket prices by subtracting to find the absolute change (\$9.90), then dividing by the reference value (\$9.90/\$42.16). Here's another way—just divide:

$$\frac{\text{new value}}{\text{reference value}} = \frac{\$52.16}{\$42.26} \approx 1.23. \tag{3.1}$$

The ".23" in "1.23" is the 23% relative increase. The "1" is 100%; it represents the original ticket price.

To understand (3.1) better, we can rewrite it without fractions.

$$\begin{aligned}\$52.16 &\approx 1.23 \times \$42.26\\ &= (1.00 + 0.23) \times \$42.26\\ &= 1.00 \times \$42.26 + 0.23 \times \$42.26\\ &= \text{ original value} + \text{absolute change.}\end{aligned}$$

Note that the absolute change in ticket prices is a number of dollars while the relative change has no units: the dollars in (3.1) cancel, leaving the 1.23 as a naked number.

We've now seen two ways to find the relative change. The straightforward way finds the absolute change first (by subtracting), then dividing by the original value. The new way divides the new value by the original value and then "throws away the 1".

Which way is better? That depends. The first way is familiar, and easy to understand. Those are good reasons to rely on it. The second way is new, and confusing. Those are good reasons to reject it. But in this case there is are several reasons to take the time to learn the new way, which we call the "one plus trick".

The first time you see something like this it's a trick. After a while it becomes a technique. In mathematician and teacher George Pólya's words, "What is the difference between method and device? A method is a device which you used twice." [R85]

The one-plus trick simplifies many percentage calculations.

When you use it to find the relative change when you start with the old and new values you need just one operation—division—rather than two when you subtract to find the absolute change and then divide. That saves time, and allows fewer places to make a mistake.

Second, the trick points to an easy way to answer the question we left hanging at the end of the last section: if ticket prices increase by 23% to a (new) value of $52.16, what were they before? The question tells us that

$$\text{new value} = \$52.16 = 1.23 \times \text{ original value}$$

so

$$\text{original value} = \frac{\$52.16}{1.23} = \$42.41.$$

This is pennies off the correct original value of $42.26 because the 23% is a rounded percentage increase.

Here's another problem where we're told the result of a percentage increase and want to find the starting value. Suppose the enrollment in your school has increased 25 percent over the past few years to 15,000 students.

What was the enrollment before the increase?

The "1+" trick tells us that

$$1.25 \times \text{ enrollment before increase} = 15{,}000 \text{ students.}$$

Dividing both sides of this equation by 1.25:

$$\text{enrollment before increase} = \frac{15{,}000 \text{ students}}{1.25} = 12{,}000 \text{ students.}$$

We can check this answer. The absolute increase from 12,000 to 15,000 is 3,000, which is in fact 25% of 12,000.

If you try to do the problem without the trick your intuition may lead you astray. You might be tempted to compute the enrollment before the 25% increase by finding 75% of the new 15,000. But 3,000 is only 20% of the new 15,000 enrollment. That means the old enrollment was 80% of the new, not 75%.

The *25 percent increase* from the old to the new value corresponds to a *20 percent decrease* from the new to the old—and you can know that without knowing either the new value (15,000 students) or the old (12,000 students). That's because

$$\frac{1}{1.25} = 0.80 = 1 - 0.20,$$

or, with fractions instead of decimals,

$$\frac{1}{5/4} = \frac{4}{5} = 1 - \frac{1}{5}.$$

To fix this kind of calculation firmly in mind, imagine an extreme case. If some quantity increases by 100% it has doubled. To get it back to where it started, you don't decrease it by 100% (which would make it vanish). You cut it in half. That's a 50% decrease.

3.5 Two raises in a row

Lucky you. You've just received a 20% salary increase. Last year your raise was 10%. How far ahead of the game are you now?

Almost everyone's first guess is 30%—they add the two percentage increases. We'll use the 1+ trick to find the correct answer and see why the 30% guess is wrong.

If working with naked percentages makes you uncomfortable, here's a general strategy that may help you through this kind of problem. Imagine a particular starting salary and work with that actual number to find out what happens. Suppose that two years ago you were making $100,000. (Since it's your salary, why not imagine a nice large one that's easy to work with?) Then after the first raise the 1+ trick tells you your salary was

$$1.10 \times \$100{,}000 = \$110{,}000.$$

Then the second raise kicks in, and you're earning

$$1.20 \times \$110{,}000 = \$132{,}000.$$

Wow! That's a 32% increase overall—even better than the first (wrong) guess of 30%. That's because the second raise was applied to your salary after the first raise took effect. Here's how to combine the calculations:

$$\begin{aligned} \text{new salary} &= 1.20 \times (1.10 \times \$100{,}000) \\ &= 1.20 \times \$110{,}000 \\ &= \$132{,}000. \end{aligned}$$

Even better, we can do the multiplication in the other order:

$$\begin{aligned} \text{new salary} &= 1.20 \times (1.10 \times \text{old salary}) \\ &= (1.20 \times 1.10) \times \text{old salary} \\ &= 1.32 \times \text{ old salary}. \end{aligned}$$

Then the 1+ trick tells us the combined increase is 32%, and we didn't even have to imagine actual dollar values to work with.

The 1+ trick tells you how to combine percentage increases. To work this kind of problem correctly you have to remember first what *not* to do:

Don't add percentage increases! Use the 1+ trick and multiply.

3.6 Sales tax

You've decided on a new $149.99 cell phone. In Massachusetts (in 2014) the sales tax was 6.25%. What will you pay for the phone?

First a quick approximation. That $149.99 is really $150. The sales tax used to be 5%, not 6.25%. Then the approximation was easy. Five percent is half of ten percent—half of $15 is $7.50. Estimating the new tax is a little harder 6.25% is more than half of 10%, so the tax will be more than half of $15. Call that about $9 and the phone should cost about $159.

You can find the cost exactly with the 1+ trick and your calculator. Recall that "percent" means "divide by 100"—because "per" means "for each" (or often, "divide by") and "cent" is the Latin root for "hundred". So 6.25% is $6.25/100 = 0.0625$.

The leading zero (before the decimal point) in 0.0625 doesn't change the value of the number. Writing .0625 would be just as good. But that extra zero helps you read the number. Without it you might not see the decimal point. We will always use it. You should learn to.

The total cost is 6.25% more than the list price of $149.99 so

$$\text{total cost} = 1.0625 \times \$149.99 = \$159.364375 \approx \$159.36.$$

3.7 20 percent off

Suppose you have a coupon that offers you a 20% discount on that $149.99 phone. What is the discounted cost?

The mental arithmetic is easy: you save twice 10%, which is $30, so you will pay only $119.99.

With a calculator you can check the answer and see if the pennies change:

$$149.99 \times 0.20 = 29.998;$$

then

$$149.99 - 29.998 = 119.99200 \approx 119.99.$$

They don't.

With a version of the 1+ trick you can do this computation in one step without computing the discount first and subtracting. After a 20% discount, you will pay 80% of the sticker price. Estimating mentally, 10% is $15 so the phone will cost $8 \times 15 = 4 \times 30 = 120$ dollars.

Since $100\% - 20\% = 80\%$ is the same as $1 - 0.2 = 0.8$, this is really the "1+" trick in disguise. The "1" is invisible because the percent change is a decrease, so subtracted rather than added.

$$\begin{aligned} \text{discounted price} &= (1.00 - 0.20) \times \text{list price} \\ &= 0.80 \times \$149.99 \\ &= \$119.99. \end{aligned}$$

With sales tax, your phone will cost

$$1.0625 \times 0.80 \times \$149.99 = \$127.49.$$

3.8 Successive discounts

Because you also get your cable TV and internet access from the telephone company, you have a second coupon that gives you an additional 30% off on your new cell phone. How should you calculate the result of combining your 20% and 30% discounts?

Don't succumb to temptation. It's easy to add the percentages and think you'll get the phone at half price (50% off). You won't, because the second discount applies only to the reduced cost from the first discount, so the final cost will be 70% of (80% of \$149.99):

$$0.70 \times (0.80 \times \$149.99) = \$83.994 \approx \$84.99 \tag{3.2}$$

so you will pay about \$85. That's more than half the sticker price. Your total discount is \$149.99 – \$83.99 = \$65, so the savings as a percent (in other words, the effective percentage discount) is

$$\frac{65}{150} = 0.44 \text{ or } 44\%.$$

We could have discovered that right away by rearranging the multiplication in (3.2):

$$0.70 \times (0.80 \times \$149.99) = (0.70 \times 0.80) \times \$149.99 = 0.56 \times \$149.99.$$

You pay 56% of the list price so the discount is $1.00 - 0.56 = 0.44 = 44\%$. You don't have to compute the final cost to know that.

Would the answer be different if we took the 30% discount first, then the 20% discount? No, because $0.70 \times 0.80 = 0.80 \times 0.70$. The order of the discounts doesn't matter.

3.9 Percentage points

There are a few occasions when adding or subtracting percentages is correct. Here is one.

The Federal Reserve Board sets the prime interest rate (in part to try to control inflation, but that's a complicated story), which in turn determines interest rates for savings accounts and for loans such as mortgages. To describe a rise in the prime rate from 2.25% to 2.75% you say "The interest rate has risen half a *percentage point*."

Subtract to find the half: $2.75 - 2.25 = 0.5$. Remember to express the answer using percentage *points*, not as a percent. This is an absolute difference.

In this example the *percent* increase in the prime rate is $0.5/2.25 = 0.22 = 22\%$.

3.10 Exercises

Exercise 3.10.1. [S][Goal 3.2][Section 3.2] Ordinary *vs.* premium.

Figure 3.1 gives some data about prices of different levels of seats at Fenway Park.

(a) Compare the cost of an ordinary seat to the cost of a premium seat in 2003. You need to decide which comparison is the most informative. Consider absolute and relative change, perhaps with percentages.
(b) Compare the cost of an ordinary seat to the cost of a premium seat in 2008.
(c) Use your calculations to make a statement about the cost of ordinary seats compared to premium seats between 2003 and 2008.

Exercise 3.10.2. [S] The shrinking rainy day fund.

On October 5, 2015 *The Boston Globe* published a story headlined "State's rainy day fund has dwindled over past decade" that said (in part):

> In the summer of 2007, before the massive recession began, the rainy day fund had \$2.3 billion—a cushion of about 7.8 percent of total state spending, according to the Taxpayers Foundation. This summer, the rainy day balance stood at \$1.1

> billion—about 2.7 percent of total state spending, a fraction small enough to raise a red flag for analysts. [R86]

(a) Calculate total state spending as of the summer of 2007 and this past summer (2015).
(b) What is the percentage change in total state spending between the summer of 2007 and this past summer?
(c) Write a short argument supporting the statement "Today's rainy day fund is only about half what it was in 2007". (One good sentence will do the job.)
(d) Write a short argument supporting the statement "Today's rainy day fund is only about a third of what it was in 2007".
(e) Which of the previous two statements is a better description of the situation? (Why do you think so? Don't just say it's one or the other.)

Exercise 3.10.3. [S][R][Goal 3.2][Section 3.2] The New York marathon.

46,795 runners finished the 2011 New York marathon. The 1981 marathon had 13,203 finishers.

What was the percentage increase in runners who finished?

Exercise 3.10.4. [R][A][S][Section 3.1][Goal 3.4][Goal 3.2] A raise at last!

In a recent negotiation, the union negotiated a 1.5% raise for all staff. If a staff member's annual salary is $35,000, what is her salary after the raise takes effect?

Exercise 3.10.5. [S][R][Section 3.1][Goal 3.3] The cell phone market.

In *The New Yorker* on March 29, 2010 James Surowicki wrote that the 25 million iPhones Apple sold in 2009 represented 2.2 percent of world cell phone purchases. [R87]

(a) Use Surowicki's numbers to estimate how many people bought a cell phone in 2009.
(b) Estimate the percentage of the world population that bought a cell phone in 2009.
(c) Estimate the percentage of the United States population that bought a cell phone in 2009.

Exercise 3.10.6. [S][Section 3.1][Goal 3.3] The 2010 oil spill.

Reuters, reporting in June on the April 2010 Gulf of Mexico oil spill, noted that

> BP continues to siphon more oil from the blown-out deep-sea well. It said it collected or burned off 23,290 barrels (978,180 gallons/3.7 million liters) of crude on Sunday, still well below the 35,000 to 60,000 barrels a day that government scientists estimate are gushing from the well. [R88]

(a) Check that the reporters are describing the same amount of oil independent of the units (barrels, gallons or liters) used.
(b) Criticize the article for its inconsistent use of significant digits (precision) in reporting these numbers.
(c) What percentage of the oil is being collected or burned off? Your answer should be a range, not a number, expressed with just the appropriate amount of precision.

Exercise 3.10.7. [S][C][W][Goal 3.3] New taxes?

On January 14, 2013 *The Boston Globe* reported that Massachusetts could raise $1 billion a year by increasing the income tax rate from 5.25 to 5.66 percent. [R89]

(a) Find the 2013 taxable income in Massachusetts.

(b) Find the total revenue from this income tax at the 5.25 percent rate.
(c) Compare your answer to the state budget. Are the numbers consistent?

[See the back of the book for a hint.]

Exercise 3.10.8. [W][S][Section 3.1][Goal 3.3] Youth sports head injuries.

The Massachusetts Department of Public Health collects annual data on brain injuries in school athletics. In 2011, the department reported that in their survey of middle and high school students, about 18 percent of students who played on a team in the previous 12 months reported symptoms of a traumatic brain injury while playing sports. These symptoms include losing consciousness, having memory problems, double or blurry vision, headaches or nausea. During that year, about 200,000 Massachusetts high school students participated in extracurricular sports. [R90]

(a) How many reported injuries were there?
(b) The survey reports the number injured as a percentage of the number of students participating in extracurricular sports. What is the number injured as a percentage of the population of Massachusetts?
(c) How many injured high school students would you expect in the town in which you live?
(d) Discuss the reliability of the reported 18% figure. What factors might make the true value greater? What factors might make it less?

Exercise 3.10.9. [S][Section 3.1][Goal 3.3] Listen up on public broadcasting.

In 2011, the United States government was facing a \$1.5 trillion federal budget deficit. Numerous cuts were proposed, including cutting federal funding of the Corporation for Public Broadcasting (CPB). The Western Reserve Public Media website argued against the cuts, saying:

> Public television is America's largest classroom, . . . all at the cost of about \$1.35 per person per year. . . .
>
> Eliminating the . . . investment in CPB would only reduce the \$1.5 trillion federal budget deficit by less than 3 ten-thousandths of one percent, but it would have a devastating impact on local communities nationwide. [R91]

(a) Use the information in the first paragraph to estimate annual federal spending on public broadcasting.
(b) Find evidence online showing that your answer to the previous question is in the right ballpark.
(c) Show that the information in the last paragraph of the quotation leads to an estimate of annual federal spending on public broadcasting that is wrong by two orders of magnitude.
(d) How would you rewrite the last paragraph so that it was correct?
(e) The article discusses the spending on public broadcasting as a percentage of the federal *deficit*. That's unusual—most expenditures are reported as a percentage of the federal *budget*.
What percentage of the federal budget is that spending?
(f) (Optional, no credit.) Do you listen to public radio or public television? If so, do you contribute when they ask for money?

Exercise 3.10.10. [R][S][Section 3.1][Goal 3.3] Coming as tourists, leaving with American babies.

From *The Arizona Republic* on August 27, 2011:

> In 2008, slightly more than 7,400 children were born in the U.S. to non-citizens who said they lived outside the country, according to the National Center for Health Statistics. . . .
>
> Although up nearly 50 percent since 2000, the 7,462 children are still just a tiny fraction of the 4,255,156 babies born in the U.S. that year. [R92]

(a) What percentage of the total births were to foreign residents? Report your answer with the proper number of decimal places.
(b) Criticize the precision of the numbers in the article.
(c) About how many children were born in 2000 in the U.S. to non-citizens living elsewhere?

Exercise 3.10.11. [S][Section 3.1][Goal 3.3] Mercury in fog.

A researcher at the University of California at Santa Cruz studied the level of methyl mercury in coastal fog. The UCSC student newspaper reported that

> methyl mercury concentrations ranged from about 1.5 parts per trillion to 10 parts per trillion, averaging at 3.4 parts per trillion—five times higher than concentrations formerly observed in rainwater. [R93]

The Green Blog at *The New York Times* notes that fish is safe to eat if it has less than 0.3 parts per million of methyl mercury. [R94]

(a) What are the highest levels of methyl mercury recorded in rainwater?
(b) Express the methyl mercury content of the fog in parts per million.
(c) Express the methyl mercury content of the fog as a percent.
(d) Compare the methyl mercury content of the fog to the safe threshold for methyl mercury in fish.

[See the back of the book for a hint.]

Exercise 3.10.12. [S][Section 3.1][Goal 3.3] For-profit colleges could be banned from using taxpayer money for ads.

In April 2012, Senators Tom Harkin and Kay Hagan introduced a Senate bill to prohibit colleges from using federal education dollars for advertising or marketing. In March 2013 they reintroduced the bill. The Senate committee on Health, Education, Labor and Pensions reported that

> - Fifteen of the largest for-profit education companies received 86 percent of their revenues from federal student aid programs—such as the G.I. Bill and Pell grants.
> - In Fiscal Year 2009, these for-profit education companies spent $3.7 billion dollars, or 23 percent of their budgets, on advertising, marketing and recruitment, which was often very aggressive and deceptive. [R95]

(a) What were the total revenues of those fifteen for-profit colleges in fiscal year 2009?
(b) How much did those colleges receive in Federal student aid in fiscal year 2009?

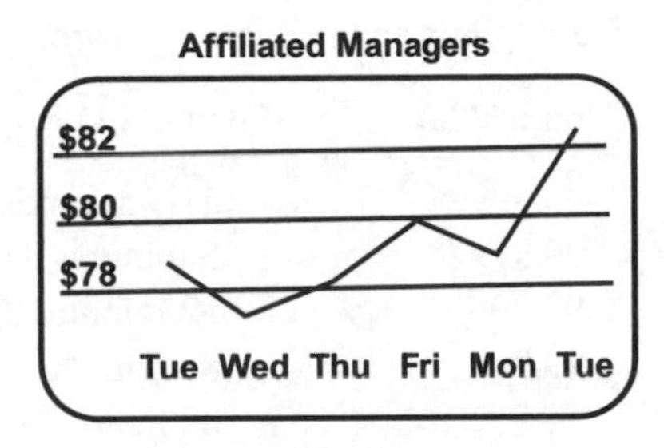

Figure 3.2. AMG stock

(c) How much of the money they spent on advertising, marketing and recruitment could be considered as coming from the Federal government?
(d) Could the colleges maintain the same level of advertising, marketing and recruiting expenses if they did not use any Federal dollars for those purposes?
(e) What was the fate of Harkin and Hagan's bill?

Do check out *Doonesbury* for February 19, 2014, at `www.gocomics.com/doonesbury/2014/02/19`.

Exercise 3.10.13. [S][R][Section 3.1][Goal 3.3] Paper jams.

A box of 20 reams of computer printer paper advertises "99.99% jam free." How many paper jams would you expect from that box?

Exercise 3.10.14. [R][S][Section 3.2][Goal 3.1] 300 million?

We regularly use 300 million as an estimate for the population of the United States.

(a) Find the absolute and relative errors when you use that figure instead of the correct one.
(b) Answer the previous question for the years 2000, 2005 and 2010.

Exercise 3.10.15. [R][S][Section 3.2][Goal 3.1] The minimum wage.

The federal minimum wage increased from $5.85 per hour to $7.25 per hour in 2009. Describe this increase in absolute and relative terms.

Exercise 3.10.16. [S][R][Section 3.2][Goal 3.1] Health care spending.

The January 2010 National Health Expenditures report stated that overall health care spending in the United States rose from $7,421 per person in 2007 to $7,681 per person in 2008.

(a) Calculate the absolute change in health care spending per person from 2007 to 2008.
(b) Calculate the percentage change in health care spending per person from 2007 to 2008.
(c) Estimate the total amount spent on health care in the United States in 2007 and in 2008.

Exercise 3.10.17. [S][Section 3.2][Goal 3.1][Goal 3.2] AMG.

On October 6, 2010 you could find a graph like the one in Figure 3.2 in *The Boston Globe*, along with the assertion that stock for Affiliated Managers jumped 4.4 percent to close at its highest point since May 4. [R96]

(a) What were the absolute and relative changes in the AMG share price between Monday and Tuesday of the week?

Table 3.3. Uptime and maximum downtime

Uptime	Uptime (%)	Maximum Downtime per Year
Six nines	99.9999	31.5 seconds
Five nines	99.999	5 minutes 35 seconds
Four nines	99.99	52 minutes 33 seconds
Three nines	99.9	8 hours 46 minutes
Two nines	99	87 hours 36 minutes
One nine	90	36 days 12 hours

(b) Is the "4.4 percent jump" in the article consistent with the data in the graph?

(c) Why does the graph seem to show the AMG share price more than doubling between Monday and Tuesday even though the change was just 4.4 percent?

(d) What can you say about the AMG share price on May 4 (without trying to look up the exact value)?

Exercise 3.10.18. [S][Section 3.2][Goal 3.2] Five nines.

The web posting titled "Five nines: chasing the dream?" at `www.continuitycentral.com/feature0267.htm` discusses the cost to industries when their computer systems are down. The "five nines" in the title refers to 99.999% availability.

Tables 3.3 and 3.4 appear there along with this quotation:

> Let us examine the math of it, first. . . . [T]he maximum downtime permitted per year may be calculated as reflected in Table 3.3. Please do not debate with me leap years, lost seconds or even changes to Gregorian Calendar. Equally let us not debate time travel! To quote a Cypriot saying: "I am from a village: I know nothing." The figures [in the table] are sufficiently accurate to make the points this article is trying to get across. [R97]

(a) Check the arithmetic in Table 3.3.

(b) Why do the numbers in the second column increase by a factor of 10 from each line to the next?

(c) How much would an energy sector company save if it increased its computer system availability from four nines to five? From five nines to six?

Table 3.4. Cost of downtime

Industry Sector	$000's Revenue / hour
Energy	2,818
Telecom	2,066
Manufacturing	1,611
Finance	1,495
IT	1,344
Insurance	1,202
Retail	1,107

(d) Compare the savings in these two scenarios (move from four nines to five, or from five nines to six). Which upgrade is likely to cost more? Which is more likely to be worth the cost? (The web posting discusses the estimated costs of these upgrades, and compares them to the savings, but that's more complex quantitatively than what we can do in this book.)

Exercise 3.10.19. [R][W][S][Section 3.3][Goal 3.2][Goal 3.3] How many master's degrees?

The National Center for Educational Statistics reported that in 2008–2009, 178,564 master's degrees in education were awarded in the United States, and that this represented 27% of all master's degrees awarded. [R98]

How many master's degrees were awarded in 2008–2009?

(a) Make a quick back-of-an-envelope estimate for the answer.
(b) Calculate to find a more precise result, appropriately rounded.
(c) Compare your estimate and your calculation.

Exercise 3.10.20. [S][Section 3.3][Goal 3.3] Rent in Boston.

According to a survey conducted in spring 2006 by Northeast Apartment Advisors, an Acton, MA research firm, the average monthly rent in Greater Boston was $1,355, an increase of 3.6 percent from 2005. What was the average monthly rent in 2005?

Exercise 3.10.21. [R][S][Section 3.3][Goal 3.3] Counting birds.

In *The Washington Post* on April 25, 2010 Juliet Eilperin wrote that birder Timothy Boucher said he had 4,257 species on his "life list" of birds he'd seen; 43 percent of all the bird species there are. [R99]

(a) If this report is correct, about how many bird species are there?
(b) Can you find independent evidence that there are that many species?

Exercise 3.10.22. [S][Section 3.3][Goal 3.1][Goal 3.2] 60 years in seconds.

In Chapter 2 we calculated the number of seconds in 60 years two ways, and found 1.89341556×10^9 and $1,893,415,558$.

(a) What is the absolute difference between these two figures?
(b) What is the relative change, expressed as a decimal and as a percentage?
(c) Which way of describing the difference is likely to be more useful?

Exercise 3.10.23. [S][Section 3.3][Goal 3.2] How long is a year?

In Fermi problems we often use 400 days/year to make the arithmetic easier.

(a) What is the relative error when you use 400 days/year instead of 365 days/year?
(b) What does the Google calculator tell you when you ask it for

| one year in days | ?

(c) Why isn't the answer above "365 days"?
(d) What are the relative and absolute errors when you use Google's answer instead of 365 days?
(e) Why isn't Google's answer just "365.25 days" to take leap years into account?

[See the back of the book for a hint.]

Exercise 3.10.24. [C][W][S][Section 3.3][Goal 3.3] Improving fuel economy.

If your car's fuel economy increases by 20% will you use 20% less gas?

(a) Show that the answer to this question is "no" if you measure fuel economy in miles per gallon.
(b) Show that the answer to this question is "yes" if you measure fuel economy in gallons per mile.

[See the back of the book for a hint.]

Exercise 3.10.25. [S][Section 3.3][Goal 3.2] Bytes.

In Chapter 1 we saw that a kilobyte is 1024 bytes, not the 1000 bytes you expect.

(a) What is the percentage error if you use 1000 bytes instead of 1 kilobyte?
(b) How many bytes in a megabyte (exact answer, please)? In a gigabyte?
(c) What is the percentage error if you work with 1,000,000 bytes per megabyte?

Exercise 3.10.26. [S][W][Goal 3.5] [Section 3.3][Goal 3.4][Goal 3.3] Hard times.

The boss says "Hard times. In order to avoid layoffs, everyone takes a 10% pay cut." The next day he says "Things aren't as bad as I thought. Everyone gets a 10% pay raise, so we're all even."

(a) Is the boss right?
(b) What if the 10% pay raise comes first, followed by a 10% cut?

Exercise 3.10.27. [S][Section 3.3][Goal 3.3][Goal 2.3] Oil spill consequences.

In a June 19, 2010 article headlined "Gulf oil spill could lead to drop in global output" *The Denver Post* quoted the Dow Jones Newswire:

> Global oil output could decline up to 900,000 barrels a day from projected levels for 2015 if oil-producing countries follow the U.S. lead and impose moratoriums on development of new offshore oil reserves. . . . [that] would represent a mere 1 percent or so of global oil output. [R100]

(We consider other numbers related to the Gulf oil spill in Exercise 3.10.6.)

(a) Use the figures in the quotation to estimate the projected global output for 2015.
(b) Do enough research to verify the order of magnitude of your answer.

Exercise 3.10.28. [S][Section 3.3][Goal 3.3][Goal 3.2] Taking the fifth.

Several years ago the liquor industry in the United States started selling wine bottles containing 0.75 liters instead of bottles containing 1/5 of a gallon ("fifths"). They charged the same amount for the new bottle as the old. What was the percentage change in the cost of wine?

Exercise 3.10.29. [S][Section 3.3][Goal 3.3] Tax holiday.

Occasionally a state designates one weekend as a tax-free holiday. That is, consumers can purchase some items without paying the sales tax. Many stores advertise the savings in the days before the weekend, to encourage customers to shop with them. The sales tax in South Carolina (in 2013) was 6%. Would it be right for an ad to read, "No sales tax this weekend—save 6% on your purchases!"

Exercise 3.10.30. [S] [W][Section 3.3][Goal 3.3][Goal 3.6] Measuring markups.

The standard markup in the book industry is 50%: the retail price of a book is one and one half times the wholesale price.

(a) What percentage of the retail price of a book is the bookstore's markup?
(b) The New England Mobile Book Fair advertises its bestsellers as "30% off retail." What is their markup on bestsellers, as a percentage of the wholesale price?
(c) Answer the previous question if the Book Fair discounts bestsellers by 40%. Would they ever do that?

[See the back of the book for a hint.]

Exercise 3.10.31. [S][C][Section 3.3][Goal 3.2][Goal 3.3] Goldman Sachs bonuses.

On January 21, 2010 the *Tampa Bay Times* carried an Associated Press report headlined "Goldman Sachs limits pay, earns $4.79 billion in fourth quarter" reporting that in 2009 the bank gave out salaries and bonuses worth $16.2 billion. That figure was 47 percent more than the year before.

> In all, compensation accounted for 36 percent of Goldman's $45.17 billion in 2009 revenue. . . . In 2008, Goldman set aside 48 percent of its revenue to pay employees. [R101]

(a) Are the numbers $16.2 billion, 36% and $45.17 billion in the report consistent?
(b) How much did Goldman Sachs pay out in bonuses in 2008?
(c) What were Goldman Sachs's revenues in 2008?
(d) What do you make of the discrepancy between the headline and the text in the article? Read the article to find out.

Exercise 3.10.32. [S][C][Section 3.3][Goal 3.2][Goal 3.3][Goal 3.7] More than 40 m now use food stamps.

> The number of Americans receiving food stamps rose to a record 41.8 million in July as the jobless rate hovered near a 27-year high, the government said.
>
> Recipients . . . jumped 18 percent from a year earlier.
>
> . . .
>
> An average of 43.3 million people, more than an eighth of the population, will get food stamps each month in the year that began Oct. 1, according to White House estimates. [R102]

(a) About how many Americans were receiving food stamps in 2009 ("a year ago" when this article appeared)?
(b) How did the percentage of the population receiving food stamps in July change between 2009 and 2010?
(c) Was 43.3 million people more than an eighth of the population?

Exercise 3.10.33. [S][C][Section 3.3][Goal 3.3][Goal 3.2][Goal 3.6] Gamblers spending less time, money in AC casinos.

In December 2010 NBC News reported that compared to 2006 gamblers were spending 22 percent less time in Atlantic City casinos and spending almost 30 percent less. As a result, revenue in 2009 was just $3.9 billion—down from $5.2 billion in 2006.

> Gross gaming revenue fell from $9.13 per hour in 2006 to $6.42 [in 2010].
> Gross operating profit per visitor hour went from $2.74 in 2006 to $1.05 in the third quarter of this year. [R103]

(a) The units "per hour" in the first quoted line are wrong. What should they be?

(b) Calculate the percentage change in gross operating profit per visitor hour from 2006 to 2009. Does this match the assertion in the article?

(c) Calculate the percentage change in gaming revenue per visitor hour from 2006 to 2009. Does this match the assertion in the article?

(d) The article says the total number of hours gamblers spent in the third quarter is down 22% and the gross revenue per hour is down 30%. What percentage decrease in revenue would this combination lead to?

(e) Calculate the percentage change in revenue. Is your answer consistent with the numbers in the article?

Exercise 3.10.34. [S][Section 3.3][Goal 3.1][Goal 3.2] Tracking Harvard's endowment.

On September 23, 2011 *The Boston Globe* published data from Harvard University on the performance of its endowment from 2007 to 2011. We've displayed that information in Figure 3.5. [R104]

The graph on the left shows the percentage change from year to year; the one on the right the actual value.

(a) Between 2007 and 2008 the graph on the left goes down while the one on the right goes up. How is that possible?

(b) For the years 2008, 2009, 2010 and 2011 use the second graph to compute the absolute and relative changes in endowment assets from the previous year

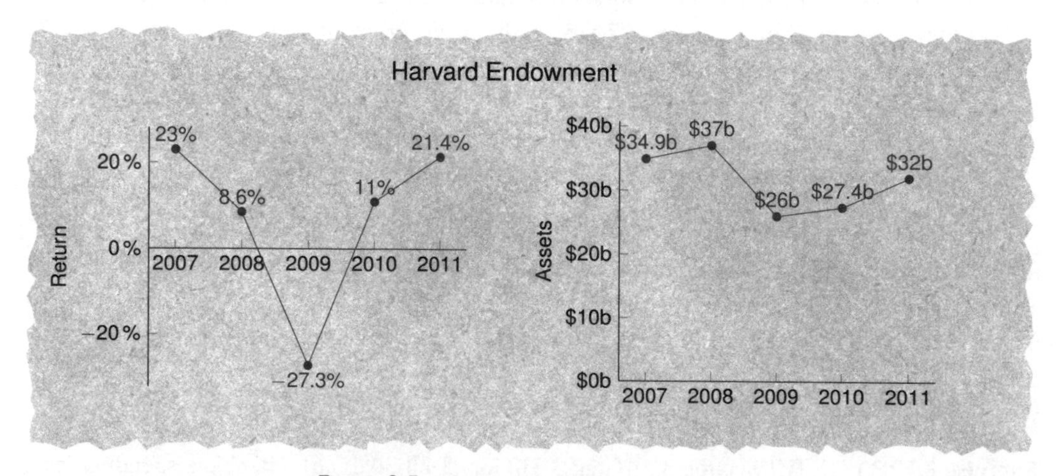

Figure 3.5. Tracking Harvard's endowment

Exercise 3.10.35. [S][Section 3.3][Goal 3.2][Goal 3.3][Goal 3.7] Banks to make customers pay fee for using debit cards.

From *The New York Times*, Friday, September 30, 2011:

> Until now, [debit card] fees have been 44 cents a transaction, on average. The Federal Reserve in June agreed to cut the fees to a maximum of about 24 cents. While the fee amounts to pennies per swipe, it rapidly adds up across millions of transactions. The new limit is expected to cost the banks about $6.6 billion in revenue a year. [R105]

(a) Use the data in the article to estimate the number of debit card transactions in a year.

[See the back of the book for a hint.]

(b) What assumptions did you make in arriving at your estimate?
(c) Is the phrase "millions of transactions" a good way to describe the order of magnitude of your answer?
(d) What percentage of a $10 debit card transaction does a merchant have to pay to the bank that issued the card? How will that percentage change when the new rule takes effect?
(e) Rewrite your estimate in (a) in transactions per day.
(f) Estimate the number of people in the United States who use a debit card. Then rewrite your answer to the previous problem in transactions per person per day.
(g) (Optional) Do you use a debit card? Did you know how much your use costs the merchant? Will you change your behavior if your bank charges $5/month for debit card use?

Exercise 3.10.36. [S][Section 3.3][Goal 3.1][Goal 3.3] Taxing gasoline.

The federal tax on gasoline at the end of 2011 was 18.4 cents per gallon. That hadn't changed since 1993, when gasoline cost just $1.16 per gallon (tax included).

(a) What percentage of the 1993 cost of gasoline was the federal tax in 1993?
(b) If the average cost of gas in 2011 was $3.40 per gallon, what percentage of the cost of gasoline then was the federal tax?
(c) What would the cost of a gallon of gasoline have been at the end of 2011 if the percentage rather than the amount of federal tax was the same then as in 1993?
(d) Gasoline consumption in the United States was estimated to be about 175 million gallons per day in 2011. How much revenue was generated by the federal gas tax at its then current rate? How much would have been generated if tax were computed using your answer to part (c)?

Exercise 3.10.37. [S][Section 3.3][Goal 3.1][Goal 3.3] How long is a microcentury?

The mathematician John von Neumann is often identified as the source of the fact that a 50 minute lecture lasts about as long as a microcentury.

(a) Check this fact by doing your own arithmetic.
(b) What are the absolute and relative errors in the "about"?
(c) If you could listen to lectures one after another day in and day out for a century about how many would you hear?
(d) You can find several answers on the web to the question "how long is a microcentury?" Critique them.

(e) Estimate how many years it would take for professors at your school to have delivered a century's worth of classes. (Don't assume the classes were back to back year round.)

Exercise 3.10.38. [S][W] [Section 3.4][Goal 3.2][Goal 3.3][Goal 3.4] Currency conversion. When you read this question you will see that you need at least two days to answer it.

(a) Suppose your company is sending you to France for business, and asks you to convert $1250 (U.S. dollars) to euros in preparation for your trip. Find a currency conversion calculator on the web and use it to find out how many euros your $1250 will buy. Your answer should clearly identify the web site, the date and time and the conversion factor used (in euros/dollar) as well as the amount of euros you would get. Imagine that you have done the exchange, so that you now have euros instead of dollars.

(If you'd rather your company arranged your business trip to some other country, feel free to change the problem accordingly.)

(b) Two days after you changed your money into euros, your company calls off the trip. You need to change your money back into dollars. Go back to the web to figure out what you would get back in dollars for the euros that you have. Again identify the web site, date and time, and conversion factor, this time in dollars/euro. Once you've done the conversion, compare the amount of money you now have with the original amount. Calculate the percentage gain (or loss) due to the change in the exchange rate.

(c) If you actually converted dollars to euros and back the bank would charge a fee each time. Suppose that fee is 2% of the amount converted. Go through the problem again, but this time account for the 2% fee. That is, figure out how many euros you would have gotten in the first conversion, with the 2% fee included, then how many dollars back in the second, with another 2% fee. Finally, calculate the total percentage gain (or loss).

(d) If the conversion rate was exactly the same each time (unlikely, but imagine it) would your loss in the conversion to euros and back be 4%?

Travel note: If you're planning a trip to a foreign country, explore the cheapest way to convert your money. Getting cash at a local bank is probably not the answer. Your debit and credit cards will probably work world-wide—ask your credit card company about the fees. And tell them that you will be using the card elsewhere so they don't think it's been stolen.

Exercise 3.10.39. [S][Section 3.4][Goal 3.4] Town weighs future with fossil fuel.

The Bend, Oregon *Bulletin* reported on April 20, 2012 that Boardman, Oregon is one of several port cities thinking about shipping coal mined in Wyoming and Montana to Asia.

If all the cities decided to do this

> ... as much as 150 million tons of coal per year could be exported from the Northwest, nearly 50 percent more than the nation's entire coal export output last year. [R106]

(a) Use the data in the quotation (not a web search) to estimate how many tons of coal the United States exported in 2011.

(b) Use the web to confirm (or not) your estimate.

Exercise 3.10.40. [W][S][Section 3.4][Goal 3.1][Goal 3.4][Goal 3.6] Brand loyalty.

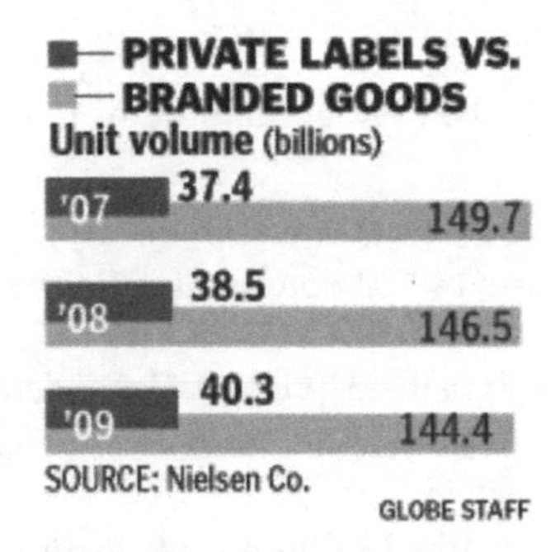

Figure 3.6. Private labels, branded goods

An article headlined "Seeking savings, some ditch brand loyalty" appeared in *The Boston Globe* on January 29, 2010, saying (among other things) that

> Unit sales of private label goods have jumped 8 percent since 2007, while brand names have declined roughly 4 percent, according to Nielsen Co. [R107]

You can see the data in Figure 3.6.

(a) Find the absolute and relative changes in unit sales of private label and brand name goods in the years since 2007.
(b) Are the percentages in the article consistent with the numbers in the graphic?
(c) How have total sales changed between 2007 and 2009?
(d) How has the percentage of private label sales changed in this period?

Exercise 3.10.41. [S][Section 3.4][Goal 3.2][Goal 3.3][Goal 3.4] Research funding.
On February 10, 2011 *The Boston Globe* carried an article that said

> The University of Massachusetts received a record amount of funding for research in fiscal year 2010, taking in $536 million, according to preliminary figures. That was an increase of $47 million, or 9.5 percent, over the previous year. [R108]

Use the information in the article to calculate the total research funding for the UMass system in fiscal year 2009 two ways, by subtracting, and using the "1 + " idea. Do you get the same answer each way?

Exercise 3.10.42. [S][Section 3.4][Goal 3.4][Goal 3.3] Long gone?
On March 4, 2011 columnist Alex Beam wrote in *The Boston Globe*:

> PBooks? No one has read a "p-book"—that's industry jargon for "print books"—in years, according to nice Mr. Bezos of Amazon, who would like to sell you e-books on his Kindle gizmo. How counterintuitive, then, that total U.S. books sales increased 3.9 percent to almost $12 billion in 2010, according to the Association of American Publishers.
>
> I just read two p-books: "The Last Crossing," by Guy Vanderhaeghe, and Mordecai Richler's hilarious "Barney's Version," which I am told is 20 times better than the movie. And I read them for free, thanks to my local library.
>
> Wait—aren't libraries obsolete? The subject of my next column, perhaps. [R109]

(a) What were book sales in 2009?
(b) Use the data given to estimate the average number of dollars spent and books bought by each adult in the U.S. in 2010.
(c) Discuss how reasonable you find your answers to the previous question.
(d) Can you make quantitative sense of "20 times better" in the second paragraph?

Exercise 3.10.43. [S][Section 3.4][Goal 3.3][Goal 3.4] Student loan growth.
On March 21, 2011 John Sununu wrote in *The Boston Globe*

> For-profit schools ... served nearly 2 million students in 2010, training nurses, technicians, chefs, and just about any other profession imaginable. They account for 10 percent of American higher education.
>
> ...
>
> Since 2003, the total debt burden from education loans to private and public institutions has grown 18 percent per year and now stands at over $500 billion. [R110]

(a) Use the information in the first paragraph to calculate the total number of students in American higher education in 2010.
(b) What was the average student debt in 2010? (Assume the $500 billion figure is for 2010.)
(c) Use the information in the second paragraph to calculate total student loan debt burden for each of the years between 2003 and 2009.
(d) Estimate the average student debt for each of the years between 2003 and 2010.

[See the back of the book for a hint.]

Exercise 3.10.44. [U][Section 3.4][Goal 3.3][Goal 3.4] One thousand three hundred percent.
On September 29, 2011 Hiawatha Bray wrote in *The Boston Globe* that

> The social networking team at Google Inc. has a great sense of timing. They picked last week to tear down the gates at their new Google+ service, which was previously accessible only by invitation. It worked. Google+ visits surged by nearly 1,300 percent on the week of September 24, with 15 million users coming to the site. [R111]

How many users were there before the gates fell?

Exercise 3.10.45. [S][W][Section 3.5][Goal 3.4][Goal 3.5] What's a good basketball player worth these days?
In *The Boston Globe* on June 30, 2010 you could read that

> [A] maximum deal for [LeBron] James would start at $16.56 million (at the minimum) and increase by 8 percent each season over the five-year period. [R112]

(a) Use the 1+ trick four times to calculate James's salary in the fifth season.
(b) In the five seasons, James will get four eight percent salary increases. Compare the result of your calculation to a single $4 \times 8\% = 32\%$ increase.

Exercise 3.10.46. [S][R][Section 3.7][Goal 3.3][Goal 3.4][Goal 3.6] An unexpected bill.

On December 17, 2010 the Wilmington, North Carolina *StarNews* carried an Associated Press report headlined "13 million get unexpected tax bill from Obama tax credit". The report said that

> The Internal Revenue Service reported that the average tax refund was $2,892 in the 2010 filing season, up from $2,663 in 2009. However, the number of refunds dropped by 3.5 percent, to 93.3 million. [R113]

(a) What was the percentage increase in the average tax refund from 2009 to 2010?
(b) How many refunds were there in 2009 and in 2010?
(c) The number of refunds decreased while the average refund increased, so you need some arithmetic to decide how the total amount refunded changed from 2009 to 2010. Find that change, in both absolute and percentage terms.
(d) Does the headline match the quote in accuracy? In tone?

Exercise 3.10.47. [S][R][Section 3.7][Goal 3.2][Goal 3.6] Bargain bread.

A local supermarket offers

Special
Pain au Levain
2 for $5
Regularly $3.99
Over 30% off.

(a) Is the claim true?
(b) How much more than 30% is the actual savings?

Exercise 3.10.48. [S][A][Section 3.8][Goal 3.3][Goal 3.4][Goal 3.6] Tax and discount.

Suppose sales tax is 5% and a $117 item is discounted 40%. Compute the final price and the sales tax two ways:

(a) First find the discounted price, then the sales tax, then the final price.
(b) First find the sales tax, then the discounted (final) price.
(c) Does it matter to you which way the computation is made? Does it matter to the state? (Remember: the state collects the sales tax!)

Exercise 3.10.49. [S][Section 3.8][Goal 3.6][Goal 3.4][Goal 3.5] Free!?

If a shirt on the clearance rack is marked "50% off original price" and you have a preferred customer coupon giving you 50% off the sale price of any item, do you get the shirt free?

Exercise 3.10.50. [S] [Section 3.9][Goal 3.7][Goal 3.2] Paying off your student loan.

We found this information about student loans on the web at `www.myfedloan.org/make-a-payment/ways-to-pay/direct-debit.shtml` [R114]

> Benefits of Direct Debit
>
> - Direct Debit is the most convenient way to make your student loan payments—on time, every month.

- Direct Debit is a free service. And you qualify for a 0.25% interest rate reduction when actively making payments on Direct Debit.

(a) Explain how direct debit works. (You may need to look this up. You may want to look it up even if you think you know.)
(b) If you have a student loan on which you pay interest at a rate of 6.8% what will your interest rate be if you sign up for direct debit?
(c) Why should the second bullet in the list say "0.25 percentage point interest rate reduction" instead of "0.25% interest rate reduction"?
(d) If that bullet really means what it says, what will the interest rate be on your 6.8% loan if you sign up for direct debit?

Exercise 3.10.51. [S][W][Section 3.9][Goal 3.7][Goal 3.2] Benefits take hit in Patrick budget.

In *The Boston Globe* on January 13, 2008, reporter Matt Viser wrote

> Looking for ways to trim a looming $1.3 billion state budget gap, Governor Deval Patrick will propose shifting more of the cost of health insurance premiums onto tens of thousands of state employees.
>
> Under his plan, about 37,000 employees would see their monthly premiums increase by 10 percent.
>
> ...
>
> Right now, most employees pay 15 percent, and the state covers 85 percent. [In the plan] those making more than $50,000 would pay 25 percent.
>
> ...
>
> For the 37,000 employees who would face the 10 percent increase, that would mean additional monthly costs of $51 for an individual plan and $120 for a family plan. [R115]

(The Commonwealth of Massachusetts went ahead and made this change in 2009.)

(a) Explain why this increase of 10 percentage points is really a 67 percent increase.
(b) Why might Governor Patrick prefer to publicize the 10 instead of the 67?
(c) Write a letter to the editor pointing out the error in the article. Your letter should be short, accurate and pointed or funny in some way if possible.

Exercise 3.10.52. [S][C] Slow down; save gas.
A U.S. government webpage on fuel economy said in June 2015 that

> While each vehicle reaches its optimal fuel economy at a different speed (or range of speeds), gas mileage usually decreases rapidly at speeds above 50 mph.
>
> You can assume that each 5 mph you drive over 50 mph is like paying an additional $0.19 per gallon for gas.
>
> Observing the speed limit is also safer. [R116]

and provided Figure 3.7

(a) Use the graph to estimate the percentage decrease in fuel economy that results when you increase your average speed from 50 mph to 55 mph. (Think about whether you should

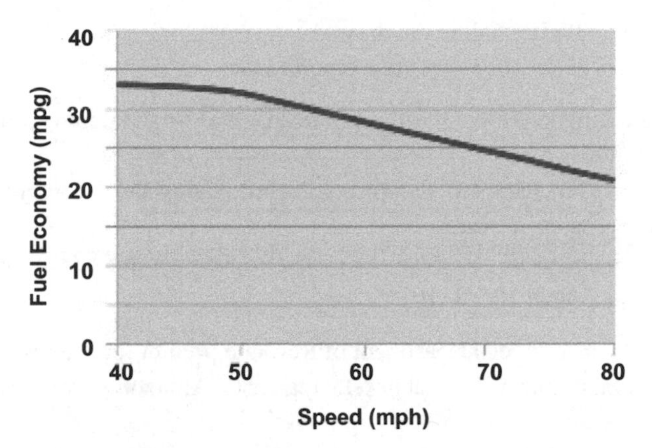

Figure 3.7. Slow down, save gas [R117]

measure fuel economy in miles per gallon or in gallons per mile. You might want to look at Section 2.2.)

(b) Use the graph to estimate the percentage decrease in fuel economy that results when you increase your average speed from 55 mph to 60 mph.

(c) Use the graph to estimate the percentage decrease in fuel economy that results when you increase your average speed from 60 mph to 65 mph.

(d) Are these percentages the same? Do your calculations support the common assumption that (once you are at a high enough speed) every time you increase your speed by 5 mph you decrease your fuel efficiency by about 5%?

(e) What assumption is being made about the price of gas in these calculations?

Exercise 3.10.53. [S][Section 3.3][Goal 3.2][Goal 3.3] Holiday shows bringing box office joy. *The Boston Globe* published the data in Figure 3.8 on December 27, 2012.

(a) Verify the four percentage calculations.

(b) For both venues, revenue increased faster than attendance between 2009 and 2012. That means the average ticket price must have increased. Compute the average ticket price in each year for each venue, and the percentage change in each case.

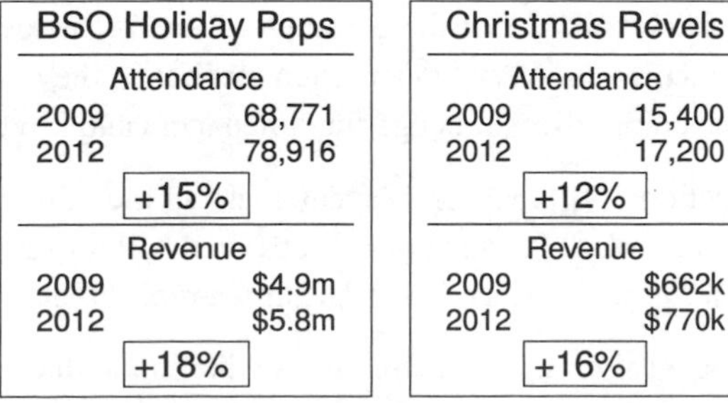

BSO Holiday Pops		Christmas Revels	
Attendance		Attendance	
2009	68,771	2009	15,400
2012	78,916	2012	17,200
+15%		+12%	
Revenue		Revenue	
2009	$4.9m	2009	$662k
2012	$5.8m	2012	$770k
+18%		+16%	

Figure 3.8. Good news at the box office [R118]

Exercise 3.10.54. [U] Taxing online purchases.

On October 28, 2013 *The Boston Globe* reported that

> If you're considering making a big-ticket purchase on Amazon.com, you might want to do it before Friday.
>
> That's when Massachusetts shoppers will start paying the 6.25 percent sales tax on items they buy from Amazon.
>
> The state Department of Revenue expects the tax to raise $36.7 million before the fiscal year ends June 30. [R119]

(a) What total taxable sales does Department of Revenue predict for the rest of the fiscal year?
(b) Estimate the average (mean) Massachusetts resident's Amazon purchases during that time period.
(c) Estimate the number of Massachusetts residents who shop online. Then estimate their Amazon purchases during that time period.

Exercise 3.10.55. [U][S] How hot was it?

On Friday, March 23, 2012 *The Boston Globe* claimed that the 82 degree record temperature on March 22 was about 35% higher than the historical average of 61 degrees.

(a) Convert these Boston record and average temperatures to their equivalents in degrees Celsius.
(b) Ben Bolker lives in Hamilton, Ontario, Canada, where they measure temperature in degrees Celsius. What percentage change would his newspaper, the *Hamilton Spectator*, have reported for those temperatures?
(c) The different answers to the previous two questions show that talking about a percentage change in a temperature makes no sense. Write a letter to the editor explaining what's wrong with the claim.

Exercise 3.10.56. [S] What should you do after you graduate?

According to Vox.com, "the top 25 hedge fund managers earned a collective $21.1 billion this year." [R120]

Vox.com put this figure into context, saying "it's about 0.13 percent of total national income for 2013 being earned by something like 0.00000008 percent of the American population."

(a) Check the calculation that the 25 hedge fund managers are the percent of the population claimed.
(b) Use the data from Vox.com to estimate the total national income in 2013.
(c) The same Vox.com article stated that the earnings of the 25 hedge fund managers were "about 2.5 times the income of every kindergarten teacher in the country combined." Can you verify this income comparison of hedge fund managers and kindergarten teachers?

Exercise 3.10.57. [S][Section 3.9][Goal 3.7] Payday loans in the military.

In their study "In Harm's Way? Payday Loan Access and Military Personnel Performance" in the *Review of Financial Studies* S. Carrell and J. Zinman report that

> Access [to payday loans] significantly increases the likelihood that an airman is ineligible to reenlist by 1.1 percentage points (i.e., by 3.9%).We find a comparable decline in reenlistment. Payday loan access also significantly increases the likelihood that an

airman is sanctioned for critically poor readiness by 0.2 percentage points (5.3%). [R121]

(a) What is the likelihood (as a percent) that an airman is ineligible to reenlist when no payday loans are available? What is the percentage when they are?
(b) What is the likelihood (as a percent) that an airman is sanctioned for critically poor readiness when no payday loans are available? What is the percentage when they are?
(c) What is a payday loan? Should servicemen and women be paid so poorly that they need them?

Exercise 3.10.58. [S][R] Who's ahead?

On September 10, 2014 the race for the American League Central Division was close. Both the Kansas City Royals and the Detroit Tigers had a winning percentage of 55.2%. Their won-lost records were

team	won	lost
Royals	79	64
Tigers	80	65

(a) Check the percentage calculations.
(b) Which team was really ahead? By how much?

Exercise 3.10.59. [S] Doubling down on education.

On October 4, 2014 Josh Boak of the Associated Press reported that

> When the Great Recession struck in late 2007 and squeezed most family budgets, the top 10 percent of earners—with incomes averaging $253,146—went in a different direction: They doubled down on their kids' futures.
>
> Their average education spending per child jumped 35 percent to $5,210 a year during the recession compared with the two preceding years. For the remaining 90 percent of households, such spending averaged about a flat $1,000, according to research by Emory University sociologist Sabino Kornrich. [R122]

Use the numbers in the quotation as much as you can to answer the questions that follow. When you need numbers that aren't there, estimate them, with or without the web. Be clear about any assumptions you make. If you do look things up, be sure to cite your sources.

(a) What were the wealthiest households spending per child on education before the increase?
(b) What percentage of their income were the wealthiest households spending on education after the increase?
(c) How did the percentage of their income devoted to education change when the recession started?
(d) What was the average education spending per child for all households?
(e) Estimate the 2007 total national household annual education spending on which this study is based.
(f) What does "double down" mean? Did the wealthiest households double down on education spending?

Exercise 3.10.60. [S] Billions more for nukes.

On November 14, 2014 Secretary of Defense Chuck Hagel announced that

> the Defense Department will boost spending on the nuclear forces by about 10 percent a year for the next five years. . . . That would be a total increase of about $10 billion over the five years. [R123]

(a) What percentage increase will five years at 10 percent per year amount to?
(b) Use your answer to the previous question to estimate 2014 Defense Department spending on nuclear forces.
(c) Compare your answer to the amount spent subsidizing school lunches. (You can do that with a web search or a Fermi problem estimate.)

Exercise 3.10.61. [S] Top 400.

On November 25, 2014 a blogger at *The Washington Post* wrote about an IRS report noting that in 2010 the top 400 households

> . . . or the top 0.0003 percent, for those of you keeping score at home—took home 16 percent of all capital gains. That's right: one out of every six dollars that Americans made selling stocks, bonds, and real estate . . . went to the top-third of the top-thousandth percent of households. [R124]

At `www.irs.gov/pub/irs-soi/10intop400.pdf` you can find out that the total capital gains reported by those 400 households was about $59 million. (That web site seems to be the source for the data in the quote.)

(a) Is 0.0003 one-third of one-thousandth?
(b) Use the data in the quotation to estimate the number of households in the U.S. in 2010. Report only a reasonable number of significant digits in your answer.
(c) Estimate the number of households some other way to show that this answer is the right order of magnitude.
(d) Is 16% the same as "one out of every six"?
(e) How much total capital gains were reported by all households in 2010?
(f) When this blog post appeared tax in the top bracket for ordinary income was 39.6 percent while the tax on capital gains was just 15 percent. How much more in tax would the government have collected if the capital gains had been taxed at the ordinary income rate?

Exercise 3.10.62. [S] Batting averages.

In 1941 Ted Williams's batting average was .406. (We would recommend writing that as 0.406, but baseball doesn't do it that way.) He had 185 hits in 456 official at bats.

According to Wikipedia,

> Before the game on September 28, Williams was batting .39955, which would have been rounded up to a .400 average. Williams, who had the chance to sit out the final, decided to play a doubleheader against the Philadelphia Athletics. Williams explained that he didn't really deserve the .400 average if he did sit out. Williams went 6-for-8 on the day, finishing the baseball season at .406. [R125]

(a) Check the computation of Williams's final batting average.

(b) Check the computation of Williams's batting average before the final day of the season.
(c) Williams had six hits in eight at bats on that last day. Would he have batted .400 for the season if he'd gotten only five? What if only four? Three?

Exercise 3.10.63. [S] Boston city councilors vote for $20,000-a-year raise.

That was the headline in an October, 2014 story in *The Boston Globe*. The article said the new salary would be $107,500. [R126]

(a) What percent increase in salary does this represent?
(b) If the federal minimum wage of $7.25 per hour were increased by the same percentage, how much more per year would someone earn who worked full time at the minimum wage? (Be sure to explain your definition of "full-time".)

Exercise 3.10.64. [S] Dogfish.

The dogfish shark is one of the few fish still abundant in New England coastal waters. In a September 28, 2014 AP story reported in *The Washington Times* you could read that

> Maine fishermen caught a little more than 100,000 pounds of dogfish in 2013 at a total value of $17,945, barely a tenth the price per pound of haddock, and less than 7 percent of the price per pound of cod. The total value of the cod catch was $736,154, while for haddock it was $211,279. [R127]

(a) What was the 2013 price per pound of dogfish?
(b) What were the approximate prices per pound of haddock and of cod?
(c) Estimate the size in pounds of both the haddock and cod catch in Maine in 2013.

Exercise 3.10.65. [U][Section 3.1][Goal 3.3][Goal 7.3] Use first gear!

Figure 3.9 shows a road sign you might see at the bottom of a hill on a road in New Zealand, warning that a steep grade lies ahead.

(a) If the grade is 4 km long, how many meters higher is the top of the hill than the bottom?
(b) If the grade is 2.5 miles long, how many feet higher is the top of the hill than the bottom?
(c) Why are these questions a little easier to answer in the metric system than in the English system?
(d) Is the grade in the figure really about 12%?

Exercise 3.10.66. [U][Goal 1.2][Goal 1.4][Goal 3.3] Writing your dissertation.

Figure 3.9. Steep grade [R128]

Author Joan Bolker's book *Writing Your Dissertation in Fifteen Minutes a Day* was published in 1998. The hundred thousandth copy sold in 2011.

Approximately what percentage of the graduate students writing dissertations in those years bought the book?

Review exercises

Exercise 3.10.67. [A] Some routine percentage calculations.

(a) Calculate 40% of 250.
(b) Calculate 130% of 79.
(c) 85 is what percentage of 140?
(d) 62 is what percentage of 30?
(e) 30% of what number is 211?
(f) 115% of what number is 52?
(g) After a 25% increase a number is 240. What was the number before the increase?
(h) After a 20% decrease a number is 192. What was the number before the decrease?

Exercise 3.10.68. [A] For each problem, calculate the 6.25% sales tax and also the final price. Redo the calculation using the 1+ trick.

(a) A book with sticker price $12.99.
(b) A computer priced at $499.99.
(c) A necklace priced at $32.00.
(d) A DVD selling for $5.50.

4

Inflation

Our everyday experience tells us that pretty much everything costs more each year than the year before. That's known as *inflation*. A dollar today is somehow not the same as a dollar last year. Common sense says it's important to understand that quantitatively in order to understand changes in your salary, the prices you pay and the value of what you own. In this chapter we explore inflation in some familiar contexts. We mine the internet for data about inflation and use the 1+ technique from Chapter 3 to unravel its subtleties.

Chapter goals

Goal 4.1. Use an online inflation calculator.

Goal 4.2. Adjust prices for the effect of inflation.

Goal 4.3. Understand the historic value of money, using the terms current dollars and constant dollars.

Goal 4.4. Reinforce techniques for dealing with percentage changes. Understand the effective change of a price over time.

4.1 Red Sox ticket prices

In Section 3.2 we studied the increase in Red Sox ticket prices from 2003 to 2008 illustrated in Figure 3.1. We found that ordinary ticket prices seemed to have grown by a staggering 23% from \$42.26 to \$52.16.

To understand what that increase means, we need to take inflation into account. How does the 23% increase in the Red Sox ticket price compare to the average increase of pretty much everything from 2003 to 2008?

Fortunately, the federal Bureau of Labor Statistics provides an Inflation Calculator on line at `www.bls.gov/data/inflation_calculator.htm` that you can use to do just that. The screen shot in Figure 4.1 shows that you'd need \$49.45 in 2008 to buy what cost \$42.26 in 2003. That means ordinary ticket prices at Fenway Park increased by \$52.16 − \$49.45 = \$2.71 when measured in "2008 dollars". The relative change was \$52.16/\$49.45 = 1.05480283, an increase of just over 5%—a lot less than 23%.

The \$275 you paid for a premium seat in 2003 would be worth \$321.78 in 2008, so, adjusting for inflation, the price increased by just \$325 − \$321.78 = \$2.22. The relative change was \$325/\$321.78 = 1.01, an increase of just one percent.

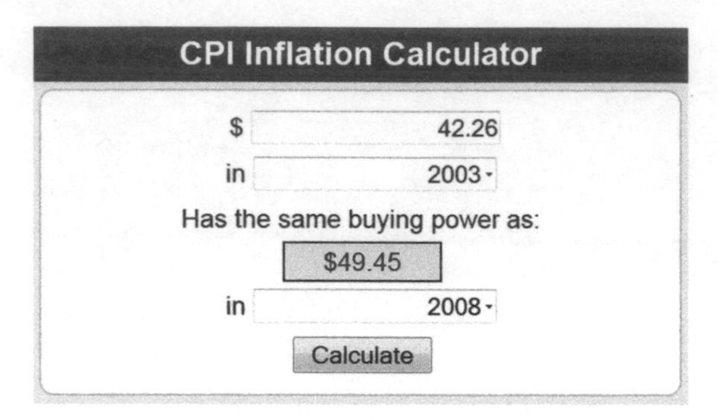

Figure 4.1. Bureau of Labor Statistics inflation calculator [R129]

This calculation shows that the change in Red Sox ticket prices just about matched the average price increase for everything. But that's just an average.

Some costs increase faster than the average inflation rate. For example, health care spending per person increased by 4.6% from 2010 to 2011 according to a report issued by the Health Care Cost Institute. [R130]

Since the inflation rate from 2010 to 2011 was 3.16%, health care costs increased faster than inflation.

4.2 Inflation is a rate

The inflation calculator says you would have to pay $102.85 in 2007 for stuff that cost $100 in 2006.

There are several ways to think about the increase from $100 to $102.85. The absolute change is $2.85. But the relative change is more informative and more traditional:

$$\frac{\$102.85}{\$100} = 1.0285,$$

which we recognize as an increase of 2.85%—the *annual inflation rate.*

Inflation means the "value of a dollar" changes from year to year. That's often confusing. Sometimes it helps to think of adjusting for inflation as currency conversion. To make that point clearer, suppose 2006 dollar bills have faded to yellow, while 2007 dollars were the usual green. The conversion rate is

$$1.0285 \, \frac{\text{green dollars}}{\text{yellow dollar}} = 1.0285 \, \frac{\text{2007 dollars}}{\text{2006 dollar}}$$

or

$$0.9723 \, \frac{\text{yellow dollars}}{\text{green dollar}} = 0.9723 \, \frac{\text{2006 dollars}}{\text{2007 dollar}}.$$

Yellow dollars are worth more than green dollars.

What would be the cost in 2007 of an item that cost $10,000 in 2006? You can enter 10,000 in the inflation calculator, or you can use the fact that you already know the inflation rate is 2.85%. In one step with the 1+ trick you discover that

$$1.0285 \times \$10{,}000 = \$10{,}285.$$

If you ask the inflation calculator to do the work it tells you the answer is \$10,284.82. That's essentially the same 2.85% inflation rate. The pennies add too much precision to the approximate answer.

The 1+ trick lets us combine inflation rates over two years. The inflation rate from 2006 to 2007 was 2.85%. The inflation calculator tells us that the inflation rate from 2007 to 2008 was 3.84%. It's tempting to guess that the inflation rate from 2005 to 2007 was $2.85\% + 3.84\% = 6.69\%$ but we know better. The inflation computation from 2007 to 2008 should start with the already inflated prices from 2007. Here's how to work out the right answer using the trick:

$$1.0285 \times 1.0384 = 1.0679944 \approx 1.0680.$$

Then read off the combined inflation rate as 6.80% by subtracting the 1.

If you're still uncomfortable with the 1+ trick, check the answer the old fashioned way. Stuff that cost \$100 in 2006 would cost $\$100 \times 1.0285 = \$102.85$ in 2007. That stuff would cost $\$102.85 \times 1.0384 = \$106.79944 \approx \$106.80$ in 2008: an inflation rate of 6.80%.

The trailing zero in 6.80 doesn't change the value of the number. Writing 6.8 might seem to be just as good. But that zero at the end does matter. It tells you how precise the figure is 6.8 could be anything between 6.75 and 6.85, while 6.80 must be between 6.795 and 6.805. In this example we want the second decimal place in the answer since there are two decimal places of precision in the data we started with. So don't throw away trailing zeroes.

The inflation calculator confirms our analysis.

The moral of the story (which we've seen before, in Chapter 3):

> Don't add percentages.
> Use the 1+ trick to combine them with a single multiplication.

4.3 More than 100%

The inflation calculator's default starting year (what you see when you first visit the page) is 1980. Using that starting year we see that you would pay about \$262 in 2009 for stuff that cost \$100 back in 1980. That means that prices (on average) have increased by a factor of

$$\frac{\$262}{\$100} = 2.62.$$

They are more than two and a half times what they were. To use the 1+ trick to express the increase as a percentage we have to find the missing "1" that represents the whole:

$$\begin{aligned} 2009 \text{ cost} &= 2.62 \times (1980 \text{ cost}) \\ &= (\mathbf{1} + \mathbf{1.62}) \times (1980 \text{ cost}) \\ &= 1980 \text{ cost} + 1.62 \times (1980 \text{ cost}) \end{aligned}$$

so the percentage change in the 29-year time period is 162%. In the 2.62 there was no visible "1" to ignore. What we need to do is to subtract the 1 from the 2.62. That 1 represents 100% of the 1980 value of the dollar and we are interested in just the percentage increase. Another way of saying this: the inflation rate from 1980 to 2009 is 162%.

Here's another example showing how "adjusting for inflation" changes the impression numbers create.

On Sunday, June 19, 2011 *The Boston Globe* reported that "The United States spends around $30 billion a year on the National Institutes of Health" and later in the article that "NIH funding in 1939 totaled less than $500,000 a year, a sum that supported just one institute. Adjusting for inflation, the budget has since increased nearly 4,000-fold." [R131]

The numbers seem to say that the NIH budget was ($30 billion / $500,000) = 60,000 times as large in 2011 as it was in 1939. But that's not right once you adjust for inflation. Let's check that the author did that correctly. The inflation calculator shows that you'd need just over $8 million in 2011 to buy what cost $500,000 in 1939. Therefore the actual purchasing power of NIH budget grew by a factor of about ($30 billion / $8 million) = 3,750. That's the same order of magnitude as the "nearly 4,000" in the article. It's a lot, but a lot less than 60,000 times as much.

When the change is this large it's better to say "3,750 times as much" than to convert to a percentage increase and say "375,000 percent bigger."

4.4 The Consumer Price Index

The Bureau of Labor Statistics inflation calculator lets you compare prices between any pair of years. The *Consumer Price Index* (CPI) is a single number that captures the same information in a different way. The CPI compares prices with those in 1982. The table at `www.bls.gov/cpi/cpid09av.pdf` says that the annual average CPI for 2009 was 214.537.

When we check that by asking the inflation calculator for the cost in 2009 dollars for what $100 would buy in the base year 1982 we find the answer $222.32 instead. The numbers don't match exactly. That's because

> The Consumer Price Index (CPI-U) is said to be based upon a 1982 Base for ease of thought. But in actuality the BLS set the index to an average for the period from 1982 through 1984 (inclusive) equal to 100. [R132]

To find the inflation rate between two years when you know the CPI for each, the ratio does the job. For example, the 2007 CPI was 207.342. Then

$$\frac{\text{2009 CPI}}{\text{2007 CPI}} = \frac{214.537}{207.342} = 1.0347111 \approx 1.035$$

tells us that the inflation rate from 2007 to 2009 was three and a half percent. In other words, the percentage change in the Consumer Price Index is just the inflation rate.

4.5 How much is your raise worth?

Suppose your salary in 2006 was $40,000 and your contract stated that you would receive a cost of living raise equal to the inflation rate. From 2006 to 2007, the inflation rate was 2.85%, so your 2007 salary would be

$$1.0285 \times \$40,000 = \$41,140.$$

Your salary has increased—you see it in your paycheck each month. But you're not really any better off. Because of inflation, you would need all of the $41,140 in 2007 to buy the same things you bought in 2006, when your salary was $40,000. Since your raise is equal to the inflation rate, your buying power has not changed at all. Effectively, your salary has not increased.

Of course you were clever enough to know this, so you negotiated a 5% raise from 2006 to 2007 so that you would actually be earning more after taking inflation into account. With that raise your 2007 salary would be $42,000.

You don't need a calculator for that arithmetic. 10% of 40 is 4, so 5% is 2, so your new salary will be $40 + 2 = 42$ thousand dollars.

How much has your buying power increased? The inflation rate was 2.85%, so the difference predicts an increase of $5\% - 2.85\% = 2.15\%$. But you should be suspicious. Subtracting percentages is as unreliable as adding them.

To find the actual change in your buying power you should find the buying power of your 5% higher 2007 salary in 2006 dollars, by undoing the inflation. That requires division:

$$1.05/1.0285 = 1.02090423 \approx 1.021.$$

That is, your buying power has increased only by 2.1%. The subtraction estimate was 2.15%—not too far off since the actual percentages involved are small.

This method is pretty subtle (though you should be getting used to it by now). We can check the answer the old fashioned way. Remember that we just figured out that you would need $41,140 in 2007 to maintain the buying power you had in 2006. To see the effective change in your buying power, calculate the relative change:

$$\text{relative change} = \frac{\text{new value}}{\text{reference value}} = \frac{42{,}000}{41{,}140} = 1.0209042295 \approx 1.021.$$

That's a 2.1% increase.

So things aren't always what they seem.

4.6 The minimum wage

The federal minimum wage is the legal minimum employers must offer workers paid by the hour. In 1991, the federal minimum wage was $4.25 per hour. Table 4.3 shows its value at each increase since it was first set in 1938. Note that it changed only twice between 1991 and July 2007, when it increased to $5.85 per hour. If we take inflation into account we see that this was not an effective increase at all. In 2007, you would need $6.50 to buy what you could get for the $4.25 you earned in an hour in 1991. Since the minimum wage increased only to $5.85, employees earning the minimum wage were better off in 1991 than they were in 2007. Figure 4.2 traces the history of the minimum wage in real (2015) dollars using the data in Table 4.3. Note the jumps in the real value in the years when Congress passed an increase, followed by gradual decline as inflation eats away at the gain.

4.7 Exercises

Exercise 4.7.1. [S][R][W][Section 4.1][Goal 4.1] Inflation calculator practice.

Figure 4.2. The minimum wage, in actual and real 2015 dollars [R133]

Table 4.3. Federal hourly minimum wage history [R134]

Month / Year	Minimum Wage	Month / Year	Minimum Wage
October 1938	$0.25	January 1978	$2.65
October 1939	$0.30	January 1979	$2.90
October 1945	$0.40	January 1980	$3.10
January 1950	$0.75	January 1981	$3.35
March 1956	$1.00	April 1990	$3.80
September 1961	$1.15	April 1991	$4.25
September 1963	$1.25	October 1996	$4.75
February 1967	$1.40	September 1997	$5.15
February 1968	$1.60	July, 2007	$5.85
May 1974	$2.00	July, 2008	$6.55
January 1975	$2.10	July, 2009	$7.25
January 1976	$2.30		

Use the inflation calculator to find the inflation rate (as a percent) from 2003 to 2007, using purchase amounts of $1, $100 and $10,000. Explain why the answers differ. Which number is the most appropriate?

Exercise 4.7.2. [S][Section 4.1][Goal 4.1][Goal 4.2] For sale: a $160,000 Apple computer.

> Christie's, the tony auctioneer, is hawking a snazzy computer that it hopes will sell for between $159,800 and $239,700.
>
> ...

> The Apple-1 computer was built and sold by Steve Jobs and Steve Wozniak, Apple's co-founders, in 1976 for $666.66—the strange price was put into effect because Mr. Wozniak liked repetitive numbers. (An inflation calculator determines that price is equivalent to $2560 in today's dollars.) [R135]

(a) The article was written in 2010. Check the inflation calculation used in this article.
(b) Is the expected auction price a bargain?
(c) What did the computer actually sell for at that auction?
(d) (Optional) Why is 666 called "the number of the beast"?

Exercise 4.7.3. [S][Section 4.1][Goal 4.1][Goal 4.2] Good to the last penny.

On the Commonwealth of Massachusetts web page at `www.sec.state.ma.us/trs/trsbok/mod.htm` you can read that

> The Massachusetts State House cornerstone was laid on the Fourth of July, 1795, by Governor Sam Adams and Paul Revere, Grand Master of the Masons. The stone was drawn by fifteen white horses, one for each of the states of the Union at that time. The cost of the original building? $133,333.33
>
> ...
>
> Paul Revere & Sons coppered the dome in 1802 to prevent water leakage. Some seventy years later the dome was gilded with 23 carat gold leaf for the first time. The cost was $2862.50; the most recent gilding, in 1997, cost $300,000. [R136]

(a) What's strange about the number $133,333.33 for the original cost of the State House? How do you think the person who wrote these paragraphs came up with that number?
(b) Adjust the three dollar amounts in this quotation to take inflation into account. Have building and gilding gotten more or less expensive over the years?

[See the back of the book for a hint.]

Exercise 4.7.4. [S][Section 4.1][Goal 4.1] Several ways to skin a cat.

Find another inflation calculator on the internet and compare it to the one provided by the government. Do they give the same answers? If not, why not? Do they cover the same range of years? Are they equally easy to use? Make sure you give a clear citation for the calculator you find.

If it's easy for you, make screen shots comparing your inflation calculator to the one at the Bureau of Labor Statistics.

Exercise 4.7.5. [S][Section 4.1][Goal 4.1][Goal 4.2] When was a dime a nickel?

In 2011 Massachusetts was considering raising the nickel bottle deposit to a dime. On December 29 Chris Lohmann wrote in a letter to *The Boston Globe*

> Let's not forget that a dime now is worth less than a nickel was when the original bottle bill was passed. [R137]

(a) Use the inflation calculator to estimate when the Massachusetts law calling for a nickel bottle deposit was passed.
(b) If you can, check your estimate by finding the actual year.

[See the back of the book for a hint.]

Exercise 4.7.6. [S] [R][Section 4.2][Goal 4.1][Goal 4.4] Twenty years as two decades.

(a) Use the inflation rate calculator to find the inflation rate (as a percent) from 1990 to 2000, from 2000 to 2010 and from 1990 to 2010.
(b) Explain why the inflation rate from 1990 to 2010 is not just the sum of the rate from 1990 to 2000 and the rate from 2000 to 2010.
(c) Use the 1+ trick to compute the 1990 to 2010 rate correctly from the rates for the two previous decades.

Exercise 4.7.7. [S][Section 4.2][Goal 4.1][Goal 4.3] Piecework.

In the 1880s many young women in large cities worked at home sewing clothing. They were paid by the piece, hence the name "piecework" for this type of labor. A pants stitcher would finish a pair of pants, putting in canvas for the pockets and waistband linings, and might be paid 12.5 cents per pair of pants finished. She could generally finish 16 pairs of pants a week, working from 8 am until dark. (You can read more about this on the web at `historymatters.gmu.edu/d/5753/`.)

(a) What was her hourly rate of pay in 1880?
(b) Convert her 1880 wages to current wages.
(c) Compare her hourly rate to the current minimum wage.

Exercise 4.7.8. [S][Section 4.2][Goal 4.1][Goal 4.3][Goal 4.2][Goal 4.4] Hollywood math: bad to worse.

A short article in the January 10, 2011 edition of *The New York Times* discussed the amount spent on DVDs over the past six years:

> [In 2004] consumers spent about $21.8 billion to rent and buy DVDs, Blu-ray discs, digital downloads and other forms of home entertainment . . . [T]he number has fallen every year since, for a total drop of about 13.8 percent, to $18.8 billion in 2010. [R138]

The reporter noted that the actual drop was about double what it seemed to be when the figures were adjusted for inflation: $21.8 billion figure from 2004 would amount to $25.3 billion in current dollars.

(a) Verify the reporter's inflation calculation.
(b) If we adjust for inflation, how dramatic is the drop in spending over the six year period?
(c) Explain how adjusting for inflation doubles the reported drop in spending.
(d) The article also noted that box office revenues rose from $9.3 billion in 2004 to $10.6 billion in 2010. Calculate the percentage increase in box office revenues, then re-calculate the percentage increase using inflation-adjusted dollars. Have box office revenues increased or decreased over this six-year time period?

Exercise 4.7.9. [S][Section 4.2][Goal 4.1][Goal 4.4] Deflation.

What was the inflation rate from 2008 to 2009? [See the back of the book for a hint.]

Exercise 4.7.10. [S][Section 4.2][Goal 4.1][Goal 4.2][Goal 4.4] Holiday Pops.

Adjust the calculations in Exercise 3.10.53 to take into account inflation from 2009 to 2012.

Exercise 4.7.11. [S][Section 4.2][Goal 4.1][Goal 4.2][Goal 4.4] The Jollity Building.

In 1941 A. J. Liebling wrote in *The New Yorker* that Mr. Ormont, the manager of the Jollity Building, was paid $50 a week plus a commission of two percent of the rents. That commission earned him an extra two thousand dollars a year. [R139]

(a) What was the average weekly rent on which Mr. Ormont's commission was based?
(b) What would Mr. Ormont's annual income be today, adjusted for inflation?
(c) Mr. Ormont managed the Jollity Building, a rather seedy low class establishment for "the petty nomads of Broadway—chiefly orchestra leaders, theatrical agents, bookmakers, and miscellaneous promoters." Was he reasonably well paid for this job?
(d) Sometimes we find quantitative reasoning questions in our casual reading (not the daily paper). This one comes from a reprint of "The Jollity Building" in *Just Enough Liebling*, a collection of the author's *New Yorker* pieces, where it was dated 1938. [R140] Redo the previous questions with this date. How much of a difference does the three years make?

Exercise 4.7.12. [S] [C][Section 4.2][Goal 4.2][Goal 4.4] Firefighters' pay.

On June 10, 2010 *The Boston Globe* reported that in the five years between 2006 and 2011 Boston firefighters received raises of 2%, 2.5%, 3%, 3.5%, 2.5% and 1.5%.

(a) Explain why the overall raise for this five year period is not just the sum of the six percentage increases.
(b) What was the firefighters' overall raise for the five year period?
(c) What was the firefighters' overall raise when you take inflation into account?

[See the back of the book for a hint.]

Exercise 4.7.13. [S][Section 4.2][Goal 4.2][Goal 4.4] A raise tied to the inflation rate.

Consider these two different ways to calculate a raise that's "10% over the rate of inflation":

(a) What is your new salary if you add the inflation rate to your 10% raise and use that as your percentage raise?
(b) What will you earn if you first compute your 10% increase and then increase that to take inflation into account?
(c) Which is the best approach (for you—since it's your salary)?

[See the back of the book for a hint.]

Exercise 4.7.14. [S][Section 4.5][Goal 4.2][Goal 4.4] Committee approves 1 percent pay raise. . . .

On June 26, 2013, the Madison, Wisconsin *Star Tribune* reported that most state employees in Wisconsin will get a one percent pay raise—the first since 2008.

> "One percent doesn't even hold the workers even," said Marty Bell, executive director of the Wisconsin State Employees Union. [R141]

(a) What is the effective raise for these state workers?
(b) The reporter noted that the cost of paying the salary increases for University of Wisconsin workers would be $52.4 million over two years. What is the total salary pool of these employees (before the raises)?

(c) Is your answer to the previous question a reasonable estimate for total employee salaries at the University of Wisconsin?

[See the back of the book for a hint.]

Exercise 4.7.15. [U][R][Section 4.5][Goal 4.2][Goal 4.4] When is a raise not a raise?

Your employment contract calls for a 3% annual raise. If the inflation rate is 4.2%, what is your effective "raise"?

Exercise 4.7.16. [S][R][Section 4.5][Goal 4.1][Goal 4.2][Goal 4.3] Private colleges vastly outspent public peers.

The Fort Wayne, IN *Journal Gazette* reported on July 10, 2010 on a study from the Delta Project on Postsecondary Education Costs, Productivity, and Accountability, a Washington-based nonprofit, that said

> Private institutions, on average, laid out $19,520 per student for instruction [in 2007–2008], a 22 percent increase from a decade earlier, . . . Public universities spent $9,732 for each student, up 10 percent in the decade. [R142]

Calculate these increases in spending when inflation is taken into account.

Exercise 4.7.17. [S][Section 4.5][Goal 4.1][Goal 4.2][Goal 4.3] DPS, teachers' union reach accord.

EdNews Colorado reported on June 19, 2010 that Denver school teachers may be about to get their first cost-of-living adjustment in three years.

> Under the terms of the tentative agreement, Denver teachers would receive . . . a 1 percent cost-of-living raise if a proposed $49 million tax increase for operating dollars is . . . approved by voters. . . . In addition, if the increase passes, teachers would receive a .5 percent raise in 2013-14 and a .5 percent raise in 2014-15. [R143]

(a) Assuming all of the increases pass, what is the total cost of living increase that a teacher would receive by the end of the 2015 school year?
(b) If a teacher did not receive any merit increases over the past three years (some did not), what is the effective percentage decrease in that teacher's salary during this time period?

Exercise 4.7.18. [U][S][Section 4.3][Goal 4.1][Goal 4.2] Wages down 14 percent.

On June 6, 2012 the Oregon *Portland Tribune* reported on a study from the Oregon Employment Department and the Bureau of Economic Analysis. The article noted that

> . . . between 2000 and 2011, the average wages for . . . jobs in Columbia County rose by 13 percent. But county wages actually decreased 13.8 percent during that time when inflation is taken into account.
>
> In neighboring Clatsop County, . . . (t)he inflation adjusted wages . . . rose 0.4 percent from 2000 to 2011. [R144]

(a) Use the CPI inflation calculator to make sense of the numbers for workers in Columbia County.
(b) What was the average increase (without taking inflation into account) from 2000 to 2011 for workers in Clatsop County?

Exercise 4.7.19. [S][Section 4.3][Goal 4.1][Goal 4.2][Goal 4.3] Newspaper sales slid to 1984 levels in 2011.

A post to the blog "Reflections of a newsosaur" reported on March 15, 2012 on figures released by the Newspaper Association of America:

> In the poorest showing since 1984, advertising revenues at newspapers last year fell 7.3% to $23.9 billion.
>
> ...
>
> The combined print and digital sales ... for last year are less than half of the all-time sales peak of $49.4 billion achieved as recently as 2005.
>
> ...
>
> The last time sales were this low was 1984, when they totaled $23.5 billion. Adjusted for inflation, the 1984 sum would be worth nearly $50 billion today. [R145]

(a) Use the CPI calculator to verify the claim in the third paragraph.
(b) A comment to the blog post stated, "Shouldn't the headline be 'Newspaper sales slid to half of 1984 level in 2011?' The comparison seems meaningless when not adjusted for inflation." Do you agree? Explain your answer.
(c) The claim in the second paragraph has not been adjusted for inflation. Would the statement change if you adjusted that figure for inflation?
(d) The blog *Carpe Diem* gave updated information on newspaper revenues in a post on September 6, 2012. In that post, the author predicted that print newspaper advertising will be lower in 2012 than in 1950, when adjusted for inflation. [R146]
 In fact total advertising in 2012 was $22.3 million. In 1950, total advertising was $2.07 million (in 1950 dollars). Use the CPI calculator to see if he was right.

Exercise 4.7.20. [S][R][Section 4.5][Goal 4.1][Goal 4.2][Goal 4.3] The MAA.

Annual dues for the Mathematical Association of America were $3 in 1916. They were $169 in 2014.

(a) How much would the $3 dues be in 1916 be in 2014 dollars?
(b) How much have dues gone up (or down) between 1916 and 2014, in 2014 dollars, in absolute and relative terms?

Exercise 4.7.21. [R][S][Section 4.6][Goal 4.1][Goal 4.3][Goal 4.4] Working for the minimum wage.

(a) Figure out the annual income in 2007 dollars for someone who worked at a minimum wage job 40 hours/week for 50 weeks in 1975.
(b) Compare that income to the annual income of someone working the same amount at the minimum wage in 2007.

You'll have to look up the minimum wage for each year—or look back in this chapter to find that information.

Exercise 4.7.22. [S][Section 4.6][Goal 4.2][Goal 4.3] Penny dreadful.

On page 60 of the March 31, 2008 issue of *The New Yorker* David Owen wrote that you'd earn less than the federal minimum wage if it took you longer than 6.15 seconds to pick up a penny. [R147]

(a) Use the information in the quotation to figure out the minimum wage when Owen wrote his article.
(b) Check Owen's arithmetic by comparing your answer to the actual federal minimum wage at that time. (This information is available on the web and in the text.)
(c) How much time could you spend picking up a penny if you wanted to earn the minimum hourly wage today for your work?
(d) What is the origin of the phrase "penny dreadful"?

Exercise 4.7.23. [S][A][Goal 4.1][Goal 4.4] A decade of inflation.

(a) Suppose the annual inflation rate was 3% for 10 years in a row. What was the inflation rate for the decade?
(b) Find the inflation rate for the decade from 1960 to 1970. If the inflation rate had been the same for each year in that decade, what would it have been?

[See the back of the book for a hint.]

Exercise 4.7.24. [S] Running to keep up with inflation.

(a) Use the data in Exercise 3.10.3 to compare the growth in the number of finishers in the New York marathon to the inflation rate for the corresponding years.
(b) Why is this a ridiculous question?

Review exercises

Exercise 4.7.25. [A] For each of the following, use the online CPI Inflation Calculator

(a) Find the buying power in 2010 of $12.50 in 2004.
(b) Find the buying power in 2011 of $42.99 in 2000.
(c) Find the buying power in 2008 of $20.00 in 2010.
(d) Find the buying power in 2010 of $100 in 1992.

Exercise 4.7.26. [A] Rewrite each sentence, expressing the change as a percentage.

(a) The cost of gasoline doubled during this time period.
(b) The value of the painting tripled over the past decade.
(c) The price of milk is two-thirds of what it cost last year.
(d) The CEO's salary is ten times the salary of the lowest-paid worker in the company.
(e) I cut my bills by a third last year.
(f) His salary is one and a half times more than her salary.
(g) Our budget this year is half of what it was last year.

5

Average Values

We start by remembering that to compute an average you add the values and divide by the count. We quickly move on to weighted averages, which are more common and more useful. They're a little harder to understand, but worth the effort. They help explain some interesting apparent paradoxes.

Chapter goals

Goal 5.1. Compute means using weighted averages.

Goal 5.2. Investigate what it takes to change a weighted average.

Goal 5.3. Understand paradoxes resulting from weighted averages.

Goal 5.4. Study the Consumer Price Index.

5.1 Average test score

Suppose a student has taken ten quizzes and earned scores of

$$90, 90, 80, 90, 60, 90, 90, 70, 90, 80.$$

To keep the computations simple and focus on the ideas, we've made up this short unrealistic example. Later we will return to our prejudice in favor of real data.

To find her average score you add the ten scores and divide by ten:

$$\frac{90+90+80+90+60+90+90+70+90+80}{10} = \frac{830}{10} = 83.$$

There are only four different values in this list: 60, 70, 80 and 90. But her average score isn't just $(60+70+80+90)/4 = 75$. A correct calculation must take into account the fact that she had lots of grades of 90 but just one 60. Here is a way to do that explicitly:

$$\frac{1\times 60+1\times 70+2\times 80+6\times 90}{10} = \frac{830}{10} = 83.$$

To work with the fraction of times each score occurs rather than the number of times just divide the 10 in the denominator into each of the terms in the numerator:

$$\frac{1}{10}\times 60+\frac{1}{10}\times 70+\frac{2}{10}\times 80+\frac{6}{10}\times 90 = 83.$$

Viewed that way, we see exactly how each of the four different quiz grades contributes to the average with its proper weight.

We can rewrite that weighted average showing the weights as decimal fractions

$$0.1 \times 60 + 0.1 \times 70 + 0.2 \times 80 + 0.6 \times 90 = 83$$

or percentages:

$$10\% \times 60 + 10\% \times 70 + 20\% \times 80 + 60\% \times 90 = 83.$$

In each case all the tests are accounted for so the weights expressed as fractions or as decimals sum to 1. As percents they sum to 100%.

The same strategy finds the average value of a card when (as in blackjack) the value of the face cards (Jack, Queen and King) is 10. Imagine that you choose a card at random, write down its value, return it to the deck, shuffle, and do it again—many times. What will the average value be? It's sure to be greater than the simple average of the numbers 1 to 10 (which is 5.5) since there are more cards with value 10 than any other. This is a weighted average, where the weight of each value is its probability. There are 4 chances in 52 that you see (say) a four, and 16 chances in 52 that you see a card with a value of 10.The average value is

$$\frac{4}{52} \times 1 + \frac{4}{52} \times 2 + \cdots + \frac{4}{52} \times 9 + \frac{16}{52} \times 10 = \frac{1 + 2 + \cdots + 9 + (4 \times 10)}{13}$$
$$\approx 6.54.$$

The first step in the computation puts all the fractions over a common denominator and cancels a 4. We could have started there by thinking about the cards one suit at a time.

5.2 Grade point average

The UMass Boston registrar posts the rules used to compute student grade point averages at the web site www.umb.edu/registrar/grades_transcripts/grading_system. Here is what that web page said in 2015.

> **Grading System**
> Each letter grade has a grade point equivalent. List your grades in a column, then each grade point equivalent next to the letter grade. Multiply each grade point equivalent by the number of credits for each class. Total all products and divide by the total number of credits. The answer will be your grade point average for that semester. [R148]

The site then displays Table 5.1 showing a sample computation for a student who took 12 courses for a total of 33 credits and earned one grade of each kind.

The "average" in "grade point average" is a weighted average of the grades, with credits as the weights.

Suppose a student has taken four courses worth 4, 3, 3 and 2 credits and earned grades of B+, A, B+ and C-, respectively. Since she has earned a total of 12 credits, her GPA is

$$\frac{4 \times 3.3 + 3 \times 4 + 3 \times 3.3 + 2 \times 1.7}{12} = \frac{38.5}{12} = 3.21.$$

That's a solid B average in spite of the C-.

Table 5.1. Computing a grade point average [R149]

Grade	Grade Point	Equivalent Credits	Quality Point
A	4.00	3	12.00
A−	3.70	3	11.10
B+	3.30	3	9.90
B	3.00	2	6.00
B−	2.70	3	8.10
C+	2.30	3	6.90
C	2.00	1	2.00
C−	1.70	2	3.40
D+	1.30	3	3.90
D	1.00	4	4.00
D−	0.70	3	2.10
F or IF	0.00	3	0.00
*TOTAL		33	69.40

* Example: Divide total quality points (69.40) by total number of registered credits (33) = 2.103 (grade point average).

To see the weights more clearly, rewrite the computation showing the fraction of credits for each course:

$$\frac{4}{12} \times 3.3 + \frac{3}{12} \times 4 + \frac{3}{12} \times 3.3 + \frac{2}{12} \times 1.7 = 3.21.$$

The Registrar's web site tells you to multiply the grade equivalent by the number of credits. We do the multiplication in the other order—fraction of credits times grade equivalent—to make the weights more visible.

5.3 Improving averages

Suppose a student at the end of her junior year has a GPA of 2.8 for the 90 credits of courses she has taken so far. What GPA must she earn for the 30 credits she will take as a senior so that she can graduate with a GPA of 3.0?

The ordinary average of 2.8 and 3.2 is 3.0, so she might think at first that a 3.2 as a senior will do the trick. We know that can't be right, since there are more credits in her first three years than in her last. She will need more than a 3.2 to bring her GPA up to 3.0. We need to figure out how much more.

If her senior year GPA is G then her combined GPA is the weighted average

$$\frac{90 \times 2.8 + 30\,G}{120}. \tag{5.1}$$

What value of G will make the arithmetic in this expression come out at least 3.0?

We'll answer this question three ways. Each method has its advantages and disadvantages.

If you remember even a little bit of algebra you can solve the equation

$$\frac{90 \times 2.8 + 30\,G}{120} = 3.0$$

for the unknown G. Multiply both sides by 120 to get

$$90 \times 2.8 + 30\,G = 360$$

so

$$30\,G = 360 - 90 \times 2.8 = 360 - 252 = 108$$

so

$$G = \frac{108}{30} = 3.6.$$

Once you see the answer you can see why it is right. The 3.0 GPA she wants will be just one fourth of the way from the 2.8 GPA she has so far to what she needs as a senior, since she has already earned three fourths of her credits—the 2.8 carries 75% of the weight. That leads to another way to solve the problem. If you noticed the one fourth at the start you could do it in your head: G must be three times as far from 3.0 as 2.8 is, so it must be 3.6.

Finally, you can answer the question even if you've forgotten your algebra and don't see the answer right away. Just guess, check your guess, and guess again as long as you have to.

Try a first guess of $G = 3.0$ in (5.1):

$$\frac{90 \times 2.8 + 30 \times 3.0}{120} = 2.85,$$

so 3.0 is too small. How about 4.0 (straight A work as a senior)?

$$\frac{90 \times 2.8 + 30 \times 4.0}{120} = 3.1,$$

which is more than she needs. The answer is somewhere between 3.0 and 4.0. Try 3.5:

$$\frac{90 \times 2.8 + 30 \times 3.5}{120} = 2.975.$$

That's almost enough. Guess 3.6 next:

$$\frac{90 \times 2.8 + 30 \times 3.6}{120} = 3.0.$$

Bingo! Got it!

Guess-and-check isn't as efficient as algebra, but it's easy to remember, and it works in places where algebra won't help. That makes it a better life skill.

5.4 The Consumer Price Index

The CPI calculator says that the inflation rate from 2006 to 2007 was 2.85%. That does not mean every price increased by that percent. The inflation rate is an average of the changes in costs of various items. The Bureau of Labor Statistics surveys the population to find out how

Table 5.2. Consumer Price Index market basket

Category	Weight (%)	2006–2007 change (%)
Food and beverages	15.009	4.0
Housing	44.377	3.0
Apparel	3.697	−1.5
Transportation	16.030	1.2
Medical care	5.780	4.8
Recreation	5.387	−0.4
Education and communication	6.455	2.7
Other goods and services	3.265	2.6
All items	100	2.58

Components in the Consumer Price Index Urban consumers, northeast region (`www.bls.gov/cpi/cpid07av.pdf`)

much we spend on various goods and services to discover what weights to use when averaging the changes in prices:

> The CPI market basket is developed from detailed expenditure information provided by families and individuals on what they actually bought. For the current [2010] CPI, this information was collected from the Consumer Expenditure Surveys for 2005 and 2006. In each of those years, about 7,000 families from around the country provided information each quarter on their spending habits in the interview survey. To collect information on frequently purchased items, such as food and personal care products, another 7,000 families in each of these years kept diaries listing everything they bought during a 2-week period.
>
> Over the 2 year period, then, expenditure information came from approximately 28,000 weekly diaries and 60,000 quarterly interviews used to determine the importance, or weight, of the more than 200 item categories in the CPI index structure. [R150]

Table 5.2 shows the weights and the percentage changes for the several categories for urban consumers in the northeast for the period 2006–2007. In fact, each of these changes is a weighted average of subcategories within each category. For example, food and beverages are broken down into those consumed at home and those consumed away from home.

The percentages in the second column add to 100, as they should. It's interesting to note from the third column that average prices in Apparel and Recreation actually decreased. The 2.58 in the last row, last column is the weighted average of the changes in each category, calculated this way:

$$\begin{aligned} 0.15009 \times 4.0 + 0.44377 \times 3.0 - 0.03697 \times 1.5 + \cdots \\ + 0.03265 \times 2.6 = 2.583642 \\ \approx 2.58. \end{aligned}$$

That figure for the 2006–2007 inflation rate for urban consumers in the northeast doesn't exactly match the nationwide average of 2.85% from the inflation calculator. That's because

Table 5.3. Vehicle sales: second quarter 2008

Model	Average price ($K)	Percent of market
car	20.2	55
truck	32.2	45

the nationwide average is itself a weighted average of the regional averages, and prices in the northeast seem to have increased less than those elsewhere.

5.5 New car prices fall . . .

On September 5, 2008 the business section in *The Columbus Dispatch* carried an Associated Press story headlined "New-vehicle prices plunge, report says". The article said that in the second quarter the average cost of a new vehicle was $25,632, 2.3 percent less than a year earlier. [R151]

Since average costs (the Consumer Price Index) increased in that year, we were puzzled. We were pretty sure car prices had gone up too. When we read further we found that

> . . . Truck-based vehicles such as pickup trucks, minivans, and SUVs accounted for less than half of all sales in the second quarter for the first time since 2001. . .

Aha! That means vehicle prices could all have risen even while the average fell! We've made up some numbers to show how the arithmetic might work. Suppose truck-based vehicles accounted for 45% of the sales in 2008—that's "less than half," and that the average prices for truck-based vehicles and cars were $32.2K and $20.2K respectively, as shown in Table 5.3.

Then the average price for a vehicle would be

$$0.55 \times \$20.2\text{K} + 0.45 \times \$32.2\text{K} = \$25{,}600 = \$25.6\text{K},$$

which is close to the reported average of $25,632.

Now imagine what the numbers might have been a year earlier, when, perhaps, truck-based vehicles outsold cars 55% to 45%, as in Table 5.4.

The average price for a vehicle would have been

$$0.45 \times \$19.6\text{K} + 0.55 \times \$31.6\text{K} = \$26{,}200 \approx \$26.2\text{K}.$$

With these assumptions, the average price fell by $600, a relative decrease of $\$600/\$26,200 = 0.0229 \approx 2.3\%$, as the article reports. But average prices for cars and trucks separately both increased.

This is really a warning about averaging averages. The average vehicle cost is an average of the average car cost and the average truck cost. The weights in that second average matter.

Table 5.4. Vehicle sales: second quarter 2007

Model	Average price ($K)	Percent of market
car	19.6	45
truck	31.6	55

5.6 An averaging paradox

The professor's average class size is smaller than the student's average class size.

How can that be? Sometimes the best way to understand a paradox is to imagine a simple extreme case rather than trying to untangle complex real data.

Suppose a small department (with just one professor) offers just two classes. One is a large lecture, with 100 students. The other is a seminar on a topic so narrow that no students sign up. Then the average class size (from the professor's point of view) is $(100 + 0)/2 = 50$ students, while each student's average class size is 100.

5.7 Exercises

Exercise 5.7.1. [S][A][Section 5.1][Goal 5.1] Can she earn a B?

Suppose a student's final grade in a biology course is determined using the following weights:

quizzes are worth 5%
exam 1 is worth 20%
exam 2 is worth 20%
lab reports are worth 15%
research paper is worth 15%
final exam is worth 25%

Just before the final, she has earned the following grades (all out of 100):

Lab report grades:	75, 90, 85, 69, 70, 75, 80, 75
Quiz grades:	85, 80, 0, 60, 70, 80, 80, 75
Exam 1:	80
Exam 2:	70
Paper:	85

(a) What is her lab report average?
(b) What is her quiz average?
(c) What is her course average just before the final?
(d) For a B she needs a course average of at least 82%. What is the lowest grade she can get on the final and achieve that goal?

Exercise 5.7.2. [S][Section 5.1][Goal 5.1] Fundraising.

On October 20, 2011 the Elizabeth Warren campaign provided *The Boston Globe* with the data we used to draw Figure 5.5. The first bar in the chart shows the total dollar contributions to her campaign, broken down according to where the money came from (Massachusetts vs. out of state). The second bar shows the total number of donors, broken down the same way.

(a) What was the average donation (per donor)?
(b) What was the average donation from Massachusetts?
(c) What was the average donation from outside Massachusetts?
(d) Check that calculating appropriate weighted average of your answers in parts (b) and (c) gives the answer you found in (a).

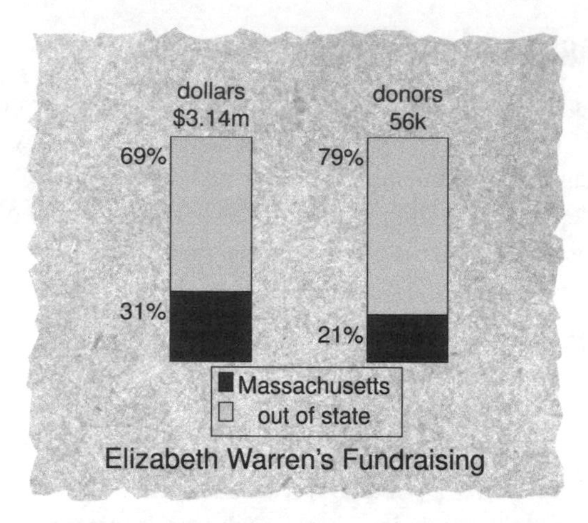

Figure 5.5. Where did the money come from? [R152]

Exercise 5.7.3. [Section 5.2][Goal 5.1] A grade point average that matters.

Check that your own GPA has been computed correctly using the rules in effect at your school.

If you are in your first semester and don't yet have a GPA, imagine the grades you expect at the end of the semester and figure out what your GPA would be.

If you are in your last semester then this exercise will take you a long time. Instead you may check your GPA for just one semester, or for the courses in your major.

If all the courses you took carry the same number of credits, check that you get the correct GPA if you compute the average the old-fashioned way, without using any weights.

Exercise 5.7.4. [W][S][Section 5.3][Goal 5.2] Ways to raise your GPA.

A UMass Boston student has earned 55 credits toward his degree but has a GPA of just 1.80. The registrar has informed him that he is on probation and will be suspended unless his overall GPA is at least 2.0 after one more semester.

(a) If he takes 12 credits, what is the minimum GPA he must earn in that semester to avoid being suspended?
(b) Answer the same question if he takes 9 credits.
(c) What if he took just 6 credits?
(d) (Optional) What would you do if you found yourself in a similar situation—take fewer courses in hopes of doing really well in them, or take more courses so that you could afford to do not quite so well in each?

[See the back of the book for a hint.]

Exercise 5.7.5. [S][Section 5.3][Goal 5.1][Goal 5.2] Gaming the system.

In the *Chess Notes* column in *The Boston Globe* on Tuesday, March 10, 2008 Harold Dondis and Patrick Wolff wrote that

> ... the sensation of the [Amateur East] tourney was the team with the highest score, GGGg (no relation to the song Gigi but standing for the three Grandmasters and one future Grandmaster.) The players were our Eugene Perelshteyn, Roman Dzindzichasvili, Zviad Izoria, all Grandmasters, and 5-year-old Stephen Fanning who rounded out the team. Was this a valid lineup? Well, yes, it was. The rules of the Amateur provide that the average rating of a team could not exceed 2200. GGGg's three Grandmasters were well above that, but Stephen Fanning ... had a current rating of 178, which brought the average rating to 2017.
>
> The Grandmasters of GGGg ... delivered wins, while their fourth board, who would have won the prize as the cutest chess player (if the sponsors had had the foresight to establish such a prize) struggled to make legal moves, sometimes failing to do so. Naturally, there followed an extensive debate as to whether the victorious ensemble had gamed the system. [R153]

(a) If the three grandmasters had the same rating, what would it have been?
(b) Did GGGg game the system?
(c) How might the tourney organizers change the rules to prevent this kind of team from winning?

Exercise 5.7.6. [S][Section 5.3][Goal 5.1][Goal 5.2] Good day, sunshine.
On June 23, 2009 *The Boston Globe* reported that

> June 2009 in Boston might turn out to be the dimmest on record. So far in the month, the sun was shining only 32% of the time. The record low was in 1903, when the sun shone only 25% of the time. [R154]

(a) What percentage of sunshine for the remaining days of the month would make 2009 at least a tie for the dimmest June?
(b) Did that happen?

Exercise 5.7.7. [S][Section 5.3][Goal 5.1][Goal 5.2] Five million unemployed.
In *The Hightower Lowdown* (Volume 12, Number 5, May 2010) you could read

> - **5 MILLION PEOPLE** (about 10% of the workforce are out of work).
> - **UNEMPLOYMENT IS HEAVILY SKEWED BY CLASS.** Among the **wealthiest 10%** of American families (incomes above $150,000), only **3% are unemployed**—a jobless rate that rises as you go down the income scale. Among the **bottom 10%, more than 30% are out of work.** [R155]

What average unemployment rate for the middle 80% of families fits with the given values for the top and bottom 10% to work out to the overall (weighted) average unemployment rate of 10%?

Exercise 5.7.8. [S][A][W][Section 5.3][Goal 5.1][Goal 5.3] Who wins?

Alice and Bob are both students at ESU. In September they start a friendly competition. In June they compare transcripts. Alice had a higher GPA for both the fall and spring semesters. Bob had a higher GPA for the full year.

(a) Explain how this can happen, by imagining their transcripts — number of credits and GPA for each, for the two semesters and for the full year, as in this table:

	fall credits	fall GPA	spring credits	spring GPA	year GPA
Alice					
Bob					

(b) Who wins?

[See the back of the book for a hint.]

Exercise 5.7.9. [S][Section 5.4][Goal 5.4][Goal 5.1][Goal 5.2] Your rate may vary.

(a) Suppose a student in 2007 estimated that her expenses were distributed among the CPI categories as listed in Table 5.6. (She had no housing expenses since she lived at home.) Calculate the inflation rate she would have experienced relative to the year before.
(b) Compare the answer in (a) to the northeast urban average of 2.58% and explain why the rate the student experienced was higher or lower.
(c) Estimate the weights in your life now for the various categories used to compute the CPI in Table 5.2.

Then figure out how much your cost of living would have increased from 2006–2007 if you'd been living then with the same lifestyle.

[See the back of the book for a hint.]

Table 5.6. How one student spent her income

Category	Weight (%)	2006–2007 change (%)
Food and beverages	15	4.0
Housing	0	3.0
Apparel	15	−1.5
Transportation	25	1.2
Medical care	5	4.8
Recreation	10	−0.4
Education and communication	15	2.7
Other goods and services	15	2.6
All items	100	

Exercise 5.7.10. [S][C][Section 5.4][Goal 5.4][Goal 5.1][Goal 5.2] Regional differences in the CPI.

We saw in Section 5.4 that the average 2006–2007 inflation rate for the northeast urban consumer was 2.58% while the national average was 2.85%.

(a) Estimate the fraction of the population of the United States that counts as urban in the Northeast.

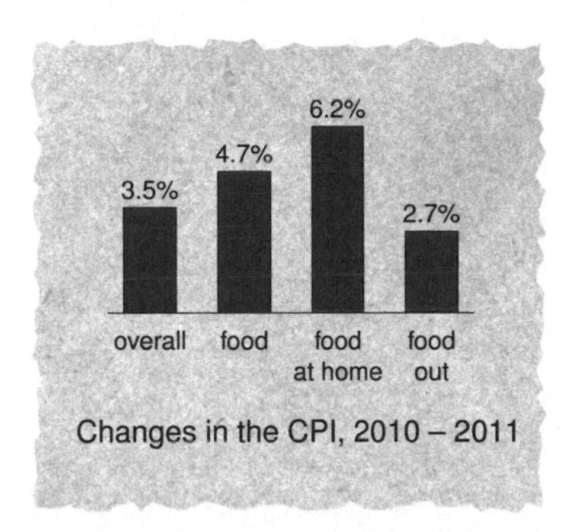

Figure 5.7. The cost of everything and the cost of food [R156]

(b) Use your estimate to estimate the average inflation rate for the rest of the country.

[See the back of the book for a hint.]

Exercise 5.7.11. [S][Section 5.4][Goal 5.4][Goal 5.1][Goal 5.2] Eat out or in?

The overall change in the Consumer Price Index is a weighted average of changes in various categories. On November 19, 2011 *The Boston Globe* reported the data displayed in Figure 5.7.

(a) The change in the CPI for food is a weighted average of the change in the cost of food at home and the change in the cost of food away from home. Find the weights.
(b) Is it reasonable to say that about 40% of food expenses are for food away from home?
(c) The overall change in the CPI is a weighted average of the change for food and the change for everything else. Do you have enough information to find the weights?

Exercise 5.7.12. [S][C][Section 5.5][Goal 5.1] Projected cost of F-35 program up to $382B.

On June 2, 2010 *USA Today* reported a story from Bloomberg News that said that the total price tag for the project to build the new F-35 Joint Strike Fighter would be 65% higher than the original $232 billion estimate. The cost now is estimated to be $382 billion.

The total cost of each plane would be $112.4 million, up 81% from the original $62 million. The cost to produce each plane would be $92.4 million, about 85% higher than the original estimate of %50 million. [R157]

(a) Check the arithmetic to see that the three percentage increases reported are correct.
(b) Can you explain how the project cost can have increased by just 65 percent when the overall total per plane increased by 81 percent and the construction cost alone increased by almost 85 percent?
(c) How would the description of the increase in project cost change if you took into account inflation between 2002 and 2010?

Exercise 5.7.13. [S][Section 5.6][Goal 5.3][Goal 5.1] Your car is more crowded than you think.

Table 5.8 reports results from a 1969 Personal Transportation Survey on "home-to-work" trips in metropolitan areas.

Table 5.8. Rush hour car occupancy [R158]

Number of riders	Percentage of cars
1	73.5
2	18.2
3	4.7
4	1.9
5	1.1
6	0.5
7	0.1

(a) The survey stated that the average car occupancy was 1.4 people. Check that calculation.
(b) Show that the average number of riders in the car of a typical commuter is 1.9.
(c) (Optional and tricky) Suppose you could persuade enough people who drive alone to switch to five person car pools in order to increase the average number of riders per car from 1.4 to 2.

What would the percentage of single occupant cars be then? What percentage of the people would be driving alone? How many people would there be in the car of a typical commuter?

[See the back of the book for a hint.]

Exercise 5.7.14. [S][Section 5.6][Goal 5.1][Goal 5.3] Why your friends have more friends than you do.

Imagine a small social network—perhaps Facebook when it was just starting out—with 100 people. One of them is friends with the other 99, but none of those 99 is a friend of any of the others.

(a) What is the average number of friends in this network?
(b) Explain why the statement "your friends have more friends than you do" is true for almost everybody in this network.

This network is just a toy one for making easy computations. But the paradox is true for real social networks, whenever there are some people with many friends and some with few. It's surely true for Facebook.

For more on this interesting paradox, see the article "Why Your Friends Have More Friends Than You Do" by Scott L. Feld, *American Journal of Sociology*, Vol. 96, No. 6 (May, 1991), pp. 1464–1477, written long before Facebook.

Review exercises

Exercise 5.7.15. [A] Find the average of each set of numbers.

(a) 40, 20, 30, 50, 40, 40, 55, 45, 60, 30
(b) 0, 0, 0, 0, 5, 10, 10, 10, 10
(c) 5, 5, 5, 5, 5, 5, 5, 5, 5, 5

(d) 0, 0, 0, 100, 100, 100
(e) 0.2, 0.4, 0.33, 0.45, 0.2, 0.1, 0.1, 0.1, 0.2, 0.4
(f) 5, 3, 6, 1, 4, 4, 4, 7, 6, 6, 4, 5, 1
(g) 86, 72, 86, 90, 91, 86, 75, 88, 42, 89, 90

Exercise 5.7.16. [A] Find each weighted average.

(a) Three test scores were 75, five test scores were 88 and two test scores were 90.
(b) Over the past month, I took three 5-mile runs, four 7-mile runs and ten 3-mile runs.
(c) Last week, the store sold 25 DVDs at $14.99 each, 13 DVDs at $12.99 each and 19 DVDs at $7.99 each.

Exercise 5.7.17. [A] Find the grade point average for each student, using the chart given in the book.

(a) Bob earned an A in a 3-credit course, a B− in a 3-credit course, a B+ in a 4-credit course and an A− in a 3-credit course.
(b) Mary earned a C+ in a 3-credit course, an A− in a 4-credit course, a B+ in a 2-credit course, a B+ in a 3-credit course and a B in a 3-credit course.
(c) Alice earned a D+ in a 2-credit course, an A in a 3-credit course, an A− in a 4-credit course and a B+ in a 4-credit course.

Exercise 5.7.18. [A] Find the semester and cumulative GPA for each student, using the chart in the book.

(a) Mike has a 2.9 cumulative GPA and 45 credits. This semester he took three 4-credit courses and earned grades of B−, B+ and B+. He also took a 3-credit course and earned a C grade.
(b) Hilda has a 2.5 cumulative GPA and 16 credits. This semester she took a 3-credit course and earned a C+, a 4-credit course and earned a B− and a 5-credit course and earned an A−.

6

Income Distribution—Excel, Charts, and Statistics

This chapter covers a lot of ground—two new kinds of average (median and mode) and ways to understand numbers when they come in large quantities rather than just a few at a time: bar charts, histograms, percentiles and the bell curve. To do that we introduce spreadsheets as a tool.

Chapter goals

Goal 6.1. Work with mean, median and mode of a dataset.

Goal 6.2. Introduce the normal distribution, with its mean and standard deviation.

Goal 6.3. Understand how skewed distributions lead to inequalities among mean, median and mode.

Goal 6.4. Make routine calculations in Excel.

Goal 6.5. Use Excel to ask and answer "what-if" questions.

Goal 6.6. Create bar charts and other types of charts in Excel.

Goal 6.7. Use histograms to group and explore data.

Goal 6.8. Calculate averages for grouped data.

Goal 6.9. Understand what a percentile is and how to interpret percentile information.

Goal 6.10. Understand the basics of descriptive statistics, including bell curve, bimodal data and margin of error.

6.1 Salaries at Wing Aero

Table 6.1 shows the distribution of workers' salaries at Wing Aero, a small hypothetical company. In keeping with our *Common Sense Mathematics* philosophy we should work with real data. But most companies keep this kind of information private. Any similarity between our imagined Wing Aero and any real company is purely coincidental. In Exercise 6.14.18 we'll apply the lessons we learn to look at income distribution in society at large.

The company has about 30 employees. We want to understand the salary distribution. A natural place to begin is with the average. But adding thirty numbers by hand (even with a

Table 6.1. Wing Aero salary distribution

Employee	Salary (thousands of $)	Employee	Salary (thousands of $)
CEO	299	Supervisor	43
CTO	250	Supervisor	51
CIO	250	Supervisor	38
CFO	290	Supervisor	33
Manager	77	Supervisor	42
Manager	123	Supervisor	49
Manager	84	Worker	25
Manager	63	Worker	19
Manager	68	Worker	41
Manager	49	Worker	17
Manager	82	Worker	26
Manager	87	Worker	25
Supervisor	42	Worker	21
Supervisor	37	Worker	28
Supervisor	29	Worker	27

calculator) is tedious and error-prone. A spreadsheet computer program can do the arithmetic faster and more accurately. In this text the spreadsheet we use is Microsoft's Excel.

We've organized this chapter as a spreadsheet tutorial—you can follow it step by step in Excel as you read it.

See Section 6.3 for some general software tips and information about alternatives to Excel.

If you're online you can save typing time by downloading the Wing Aero spreadsheet from `WingAero.xlsx`. That spreadsheet and all the others you'll need live at `commonsensemathematics.net/spreadsheets`. If you build it for yourself, use column `A` for the employees and column `B` for the salaries. Put the labels in row `7` and the data in rows `8:37`, not side by side as in the table. You should see Figure 6.2.

What do you notice?

The employees are listed in decreasing order of importance (or prestige), but only approximately in decreasing order of salary. Some supervisors earn more than some managers, and some workers more than some supervisors. We can make those discrepancies visible by sorting the data.

Select the rectangular block of data in rows `8` through `37`, columns `A` and `B`. Be sure to select both columns so they will be sorted together. Choose the `Sort` dialog box from the `Data` tab, select sorting by Salary, `Largest to Smallest`, as in Figure 6.3.

Often Excel offers you more than one way to do a job. This is one way to sort, in Excel 2013. There are others. Other versions of Excel may use different menus. But the ability to sort will be available in any spreadsheet program you use.

If you sort the data again alphabetically (by Employee, `A to Z`) the table returns (nearly) to its original state, because the employee categories were alphabetical at the start. But each category is now sorted by salary.

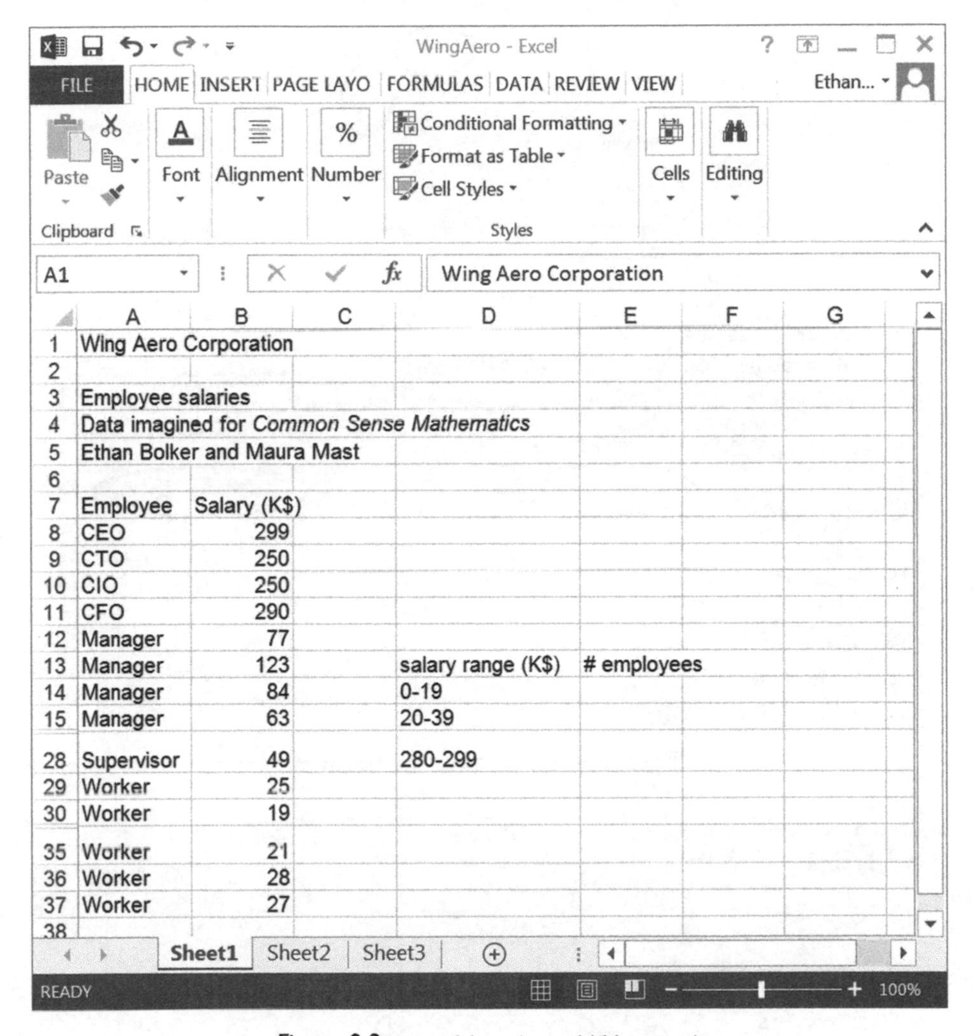

Figure 6.2. spreadsheet (some hidden rows)

To find the average salary at Wing Aero we tell Excel to add the entries in column `B` and divide by the number of employees.

Enter the label `Total` in cell `A38`. Then go to cell `B38`. Make sure the `Formula Bar` is visible. (Use the `View` menu to find it if it's not.) In the formula box type an equals sign =, to tell Excel you want it to do some arithmetic, and then the name of the operation,

```
=SUM(
```

since you are about to add up some numbers. The open parenthesis asks Excel to prompt you for information. It suggests

```
SUM(number1, [number2], ...)
```

as in Figure 6.4.

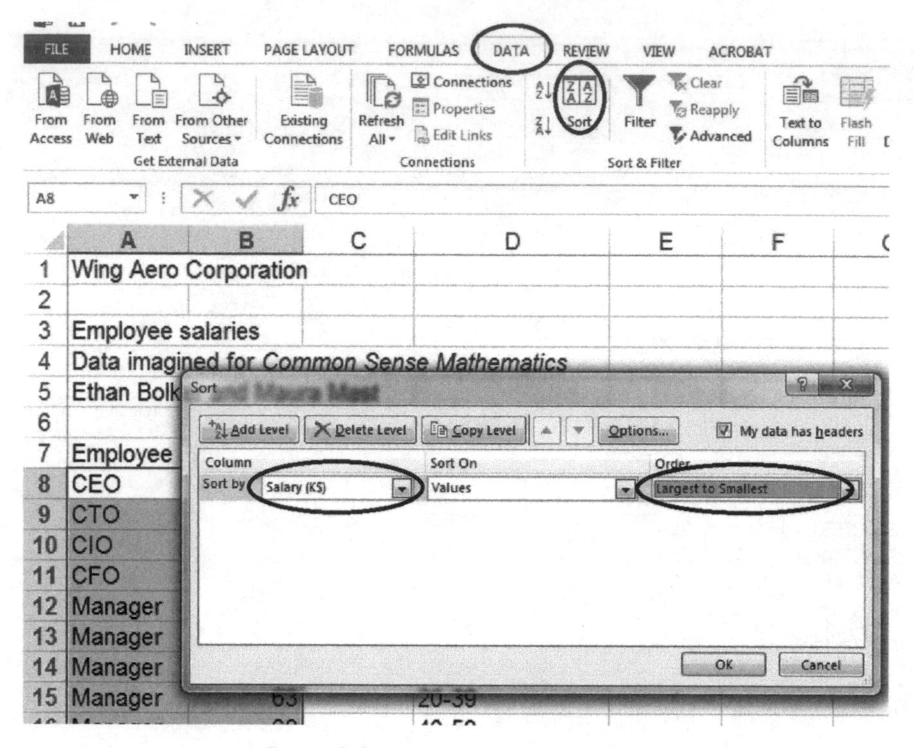

Figure 6.3. Sorting Wing Aero salaries

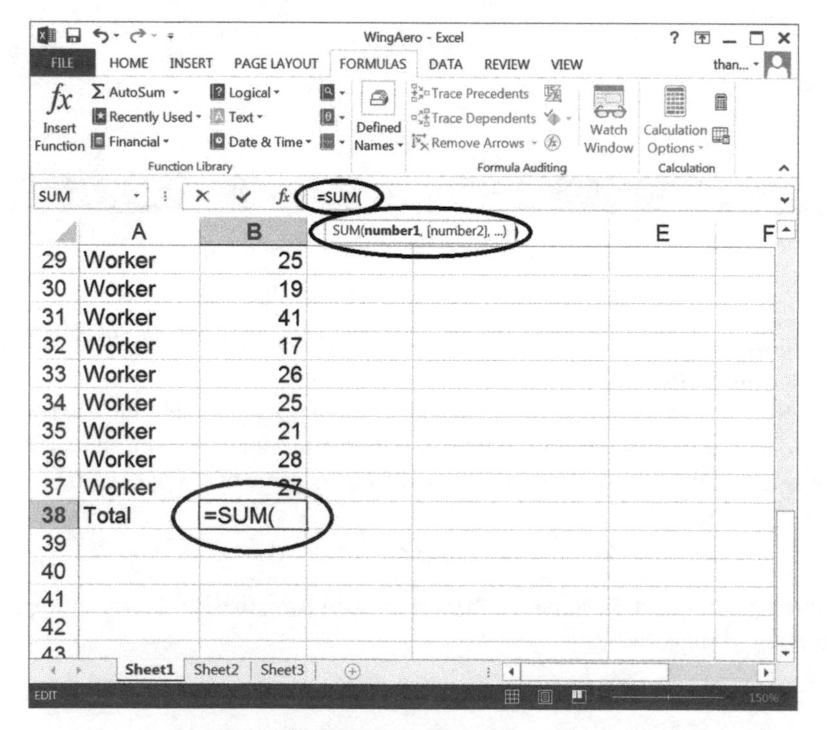

Figure 6.4. Summing Wing Aero salaries

Select cells B8:B37, close the parentheses and type enter or click the check icon on the Formula Bar. You should see Total 2315 in cells A38 and B38. Wing Aero's total annual payroll is $2315K—about $2.3 million.

To find the average salary we must divide the total $2315K payroll by the number of employees. Rows 8 through 37 contain employee records so there are $37 - 8 + 1 = 30$ employees. But it's better to ask Excel to count the rows for you. Type the label Count in cell A39 and begin formula =COUNT(in cell B39. Finish the formula by selecting cells B8:B37 or by typing the addresses of those cells and closing the parentheses. You should see

```
Count    30
```

Finally, type the label Average in cell A40 and put formula =B38/B39 in cell B40. You should see

```
Average    77.16667
```

so the average annual salary at Wing Aero is about $77,000. Excel rounded the exact 77.1666 ... to 77.16667 when it ran out of space. We rounded to two significant digits because that's all the precision we have in almost all the data. In Exercise 6.14.3 you'll learn how to tell Excel to round for you.

Excel's built-in functions SUM and COUNT are separately useful, which is part of why we showed them to you, but if it's the average you want Excel can find it in one step.

Enter =AVERAGE(B8:B37) in cell B41, click the check icon and Excel tells you again that the average is 77.16667. Put "computed using SUM/COUNT" in cell C40 and "computed using AVERAGE function" in cell C41.

6.2 What if?

Suppose that the CEO convinced the Board of Directors to double his salary, to $598K (even though the company lost money). To see how that would affect the payroll statistics, go to cell B8 and change the 299 there to 598. Excel automatically updates all your computations, increasing the total annual payroll to $2,614 thousand and the average annual salary by about $10,000 to more than $87,000.

We can learn several useful lessons from our work so far.

- Excel is faster and more accurate at arithmetic than we are.
- Excel is an excellent tool for answering what-if questions, because it automatically updates its computations when the data change.
- The average we calculated with SUM/COUNT and then again with AVERAGE is a terrible way to summarize the salary structure at Wing Aero. The CEO's 100% raise increases the average salary by about $10,000, or 13%—but he's the only employee who's actually better off!

After you've tried out these changes, restore the original values by clicking the Undo icon on the toolbar. This very useful feature allows you to undo the last few changes you've made. Use it to fix mistakes or to get back to where you were before you asked a "what-if" question.

6.3 Using software

Why use a spreadsheet at all? There are several reasons.

- to carry out large tedious calculations rapidly and correctly.
- to answer "what-if?" questions without having to redo arithmetic.
- to draw charts.
- to learn a tool you may use long after you've finished this course.

Here are some tips for working with Excel, and with software packages in general.

- To figure out something new, you can use the application's built in help, search the web, ask a friend or teacher, or just play around. Which you try first depends on your personality.
- Many applications provide several ways to do the same task. That means you may get different advice or instructions from different sources. Choose a way that suits your style.
- Use the menus for things you do rarely, but learn the keyboard shortcuts for things you do often—in particular, `control-C` for copy and `control-V` for paste can save time.
- Learn about `undo`. **Save your work often**.
- When you are about to make significant changes to a spreadsheet or a document make a copy of what you have, so that you can return to it if you change your mind about what you should have done. In the Wing Aero study we created several, calling them `WingAero1.xlsx`, `WingAero2.xlsx` and so on.
- Create backup copies of important documents often—off your computer. Use a thumb drive (flash drive), a CD, or a web service.
- Some things work the same way in different applications (browsers, Word, Excel)—for example, selecting with the mouse, cut and paste with keyboard shortcuts.
- In many software applications placing the mouse over a feature you are interested in and right clicking often lets you view and change the properties of that feature. "Do the right thing" is a good mnemonic.
- Applications often try to guess what you intend to do. That can be good or bad. We'll see soon that Excel can adjust cell references automatically—that's usually, but not always, what you want. Word processors may try to fix your spelling—perhaps correctly, perhaps not.

Why Excel? It's the most commonly available spreadsheet, so it's the one we use. But there are others available. In particular, Open Office (`www.openoffice.org/`) offers free spreadsheet and word processing software..

With Google Sheets (`www.google.com/sheets/`) you can create spreadsheets in the cloud. That software is powerful enough to do the arithmetic we need for *Common Sense Mathematics* but it has far fewer chart formatting features than full fledged programs. Excel on tablet computers lacks those features too.

6.4 Median

In Section 6.1 we found that the average salary at Wing Aero was $77,000. This is a pretty good annual salary. If you saw that in a job advertisement you'd think it was a pretty good company to work for. Maybe, maybe not. In Section 6.2 we saw how it's skewed by the CEO's earnings.

When his (or her) salary increased from $299,000 to $598,000, the average salary jumped by $10,000 to $87,000. But no one else's salary changed!

The $77,000 "average" is misleading in other ways too. Most of the employees—26 out of 30—have salaries less than the average. That contradicts what we like to think "average" means. To find a salary that's "in the middle", sort the 30 line table again so that salaries are increasing. Since the table starts in row `8` and has 30 entries, rows `22` and `23` are the middle rows and the entries in cells `B22` and `B23` are the middle salaries. That means half the employees make $42,000 or less (the entry in cell `B22`) and half make $43,000 or more. So we might want to say that the "average" salary is $42,500. There's a name for this kind of "average"—it's the *median*. The first "average" we computed above is called the *mean*.

On the spreadsheet, change the `Average` label in cell `A41` to `Mean`.

Then put

```
Median    42.5    computed by finding middle of sorted list
```

into cells `A45`, `B45` and `C45`.

In some ways the median is a fairer "average" than the mean for describing the Wing Aero salary structure. It tells you more about the way salaries are distributed. In particular, the median salary isn't affected by the CEO's big raise. Try changing that salary again in Excel: the mean changes, as it did before, but the median stays the same.

Excel knows how to compute medians. Enter `=MEDIAN(B8:B37)` in cell `B46` and check that you get the same value: 42.5. Enter "`computed using MEDIAN function`" in `C46`.

Finding the median with the `MEDIAN` function is better than finding the middle of the sorted list yourself because it works even when the data aren't sorted. Suppose the supervisor making $43K gets a raise to $50,000. Enter that new value as `50` in the spreadsheet. Then see that Excel has recalculated the median in cell `B46`: it's now 45.5.

There's a third kind of average, the *mode*. We'll return to that after we've summarized the salary data in a different way.

6.5 Bar charts

Often pictures are better than numbers when we wish to convey information convincingly. A *bar chart* is one such picture.

We use bar charts when we have data that fall naturally into categories. The height of each bar represents the value for that category. When you want general understanding rather than numerical detail it's easier to compare the heights of bars visually (in both relative and absolute terms) than the values of numbers.

In order to understand the Wing Aero income distribution better we will use Excel to draw two bar charts with four columns each, one showing the number of employees in each of the four job categories executive, manager, supervisor and worker, the other the total earnings in each category. We will use the original Wing Aero salary data from Section 6.1.

You can find the data from Table 6.5 in the range `D7:F11` in the copy of the Wing Aero salary distribution spreadsheet at `WingAeroBarCharts.xlsx`. Excel calculated the values in columns `E` and `F` using the `COUNT` and `SUM` functions.

Figure 6.6 shows those calculations. We created it by selecting `Show Formulas` from the `Formulas` tab.

Table 6.5. Wing Aero salary distribution by job category

Job	Number	Total salary ($K)
Executive	4	1089
Manager	8	633
Supervisor	9	364
Worker	9	229

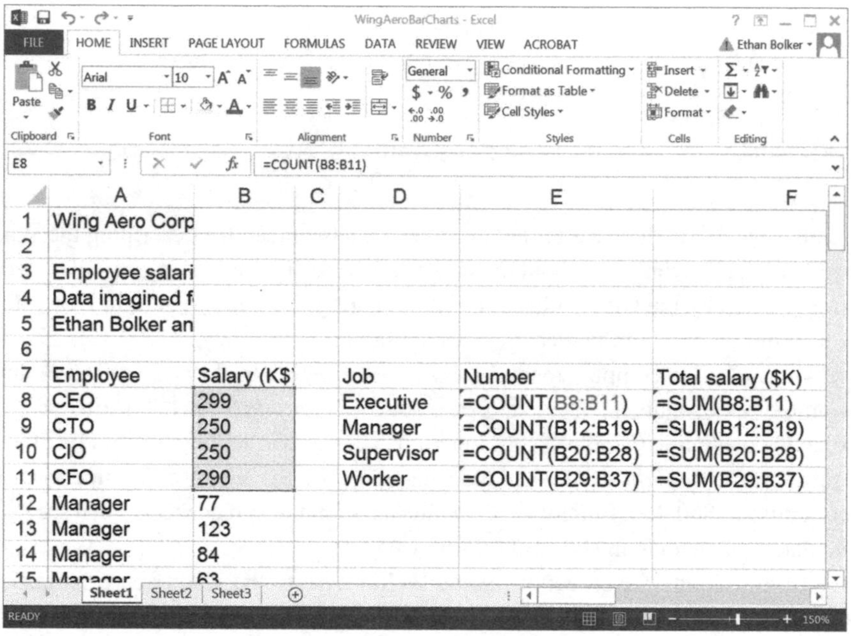

Figure 6.6. Showing the formulas used in a spreadsheet

Figure 6.7 shows the first two pictures from that spreadsheet.

These side by side bar charts dramatically demonstrate that the executives make up a small part of the workforce but enjoy a large part of the salary expenses!

Delete the charts from your copy of the spreadsheet, so that you can learn how to build them for yourself. For the first one, select the data in columns `D` and `E`, rows `8` through `12`—the range `D8:E12`. Then ask for a new chart by selecting `Column` and `2D Column` from the `Insert` tab, as in Figure 6.8.

Use the mouse to move the chart so that it does not hide any of the data. Then add the appropriate labels and change the colors so that the result is suitable for black and white printing. There are several ways to go about these tasks; experiment until you find ones that work for you. Section 6.3 has a tip about how the right mouse button can help with these tasks.

To build the second picture the same way you must select the data in columns `D` and `F` without selecting column `E`. To do that, select rows `8:12` in column `D`. Then hold down the control (PC) or apple or command (Mac) key and use the mouse to select those rows in column `F`.

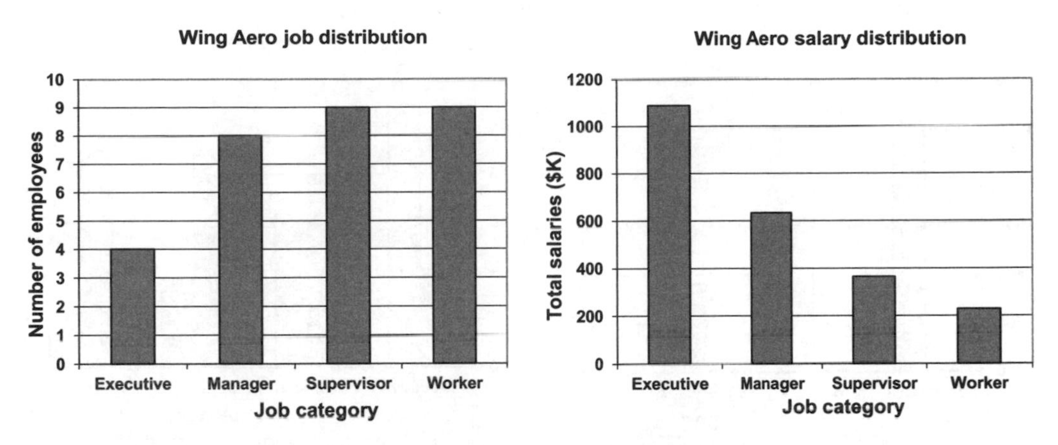

Figure 6.7. Wing Aero employee information by category

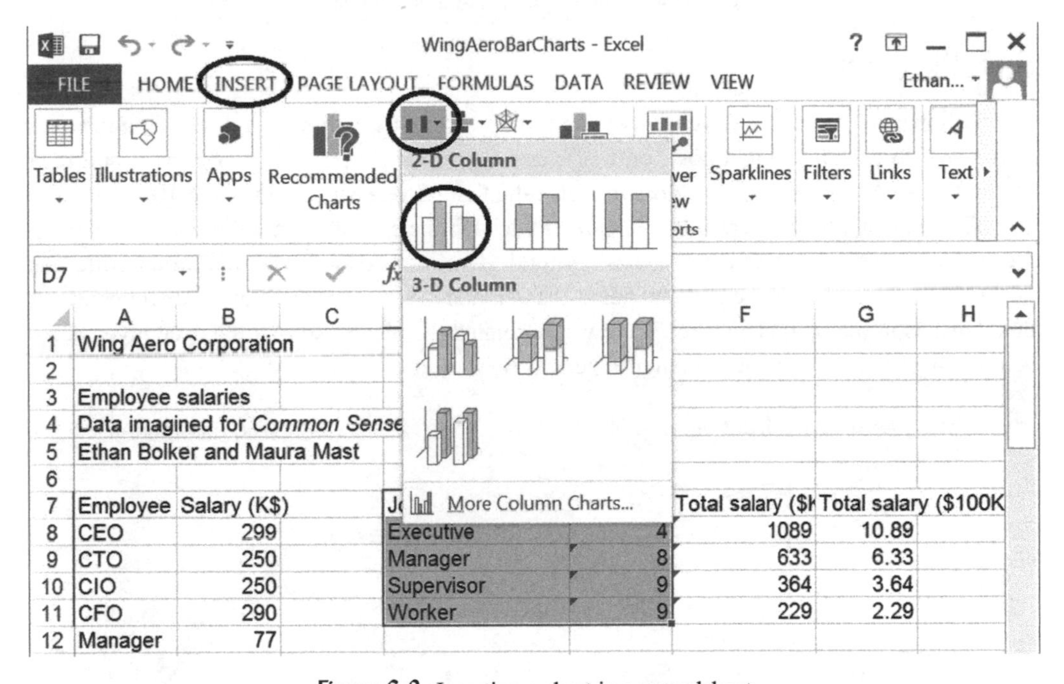

Figure 6.8. Inserting a chart in a spreadsheet

It would be even better to combine the two pictures. We can do that in Excel by building a column chart for the full range `D8:F12`. The result (after adding titles and fixing colors) is the chart on the left in Figure 6.9.

The problem with that picture is that the bars for the employee numbers are nearly invisible, because data values for the numbers vary from 4 to 9 while those for the total salaries vary from \$200K to \$1200K. We can fix that by reporting the salary totals in hundreds of thousands of dollars rather than just in thousands of dollars (that is, by changing the units for measuring salary). Column `G` contains those numbers; we used it to draw the second picture. There you see clearly the opposing trends in the categories: total wages decrease as the number of employees increases.

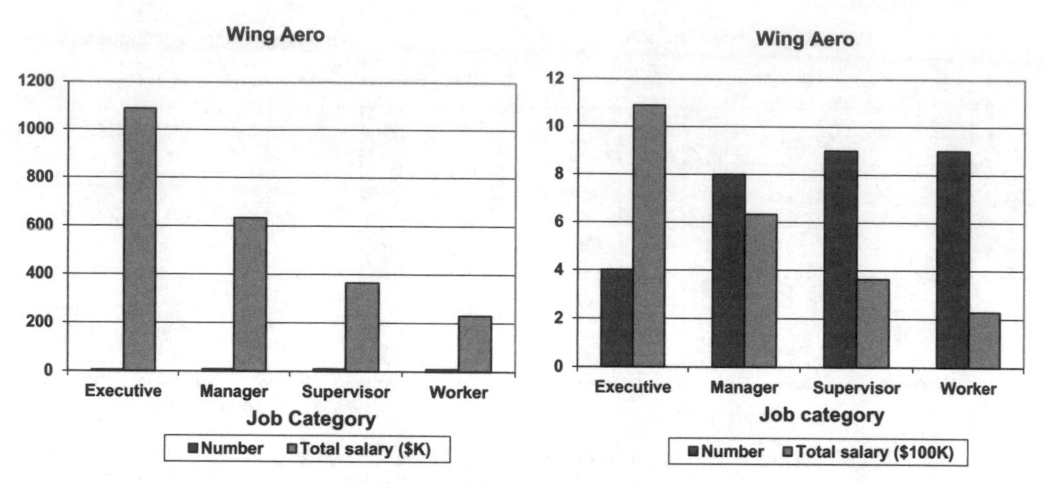

Figure 6.9. Wing Aero: side by side bar charts

6.6 Pie charts

Excel allows you to change the chart type on the fly. Select the column graph showing the total salary by job category. Right click on that chart. Select `Change Chart Type...` and then the first `Pie`. Adjust labels and colors to create the first of the charts in Figure 6.10.

The pie chart shows clearly in yet another way that the executives are the winners at Wing Aero. They take home nearly half the salary total. If you wanted to make that look a little less dramatic, you could omit the percentages and ask Excel to show a three dimensional version of the chart, as in the second picture. There we've rotated the picture so that the executive wedge is at the back, so it looks even smaller in perspective.

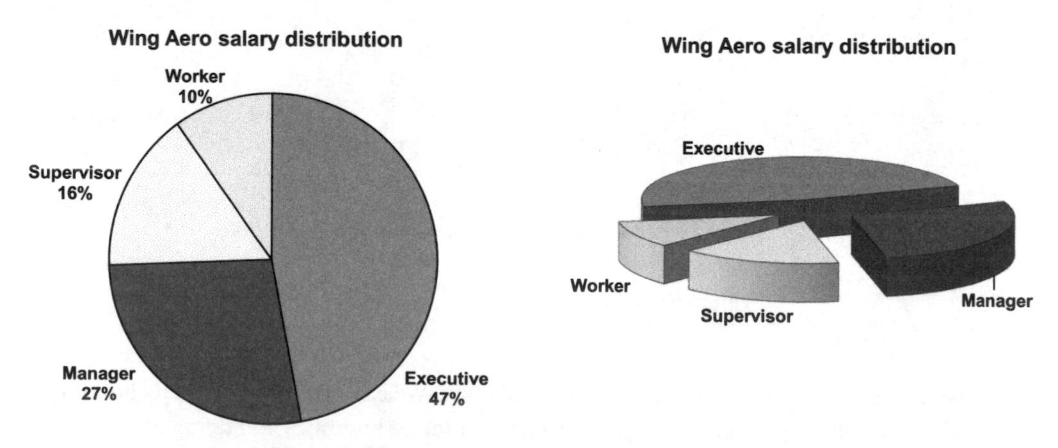

Figure 6.10. Wing Aero salary distribution pie charts

6.7 Histograms

Wing Aero is small enough so that we can see the whole salary table at a glance. But if there were 1000 employees that wouldn't be possible. To understand the numbers we'd have to summarize them. The four categories in the previous section might not provide enough detail.

Another useful way to summarize the data is to divide the salaries into ranges and count the number of employees whose salary falls in each range. Then we can use the ranges as the categories in a bar chart. In this example we'll use $20K ranges. That means we think of two employees who make $29K and $33K as having approximately the same salary, since each falls in the $20K–$39K category.

We need to count the number of employees making less than $20K, then the number making between $20K and $39K, and so on. That's easy when the data are sorted in increasing order. Table 6.11 shows what we found.

Table 6.11. Wing Aero salary distribution by salary range

Salary range ($K)	Number of employees
0–19	2
20–39	10
40–59	7
60–79	3
80–99	3
100–119	0
120–139	1
140–159	0
160–179	0
180–199	0
200–219	0
220–239	0
240–259	2
260–279	0
280–299	2

To save you typing, we've listed the categories in cells `D15:D29` in `WingAero.xlsx`. You should check our work and enter the data in column `E`.

To see that you haven't missed anyone, `SUM` the range `E15:E29` to make sure the answer is 30, the known number of employees. (The sum isn't a perfect check. Although the total is correct, we might have put some employees into the wrong categories.)

We can use the data to draw a *histogram*—a bar chart where the categories on the *x*-axis specify data ranges and the *y*-axis counts or percentages for each range. Figure 6.13 shows the result.

Start as usual by building a column chart from the data in cells `D14:E29`. In a histogram it's conventional to make adjacent bars touch. To do that in Excel, right click on one of the columns and select `Format Data Series ...` from the menu. Then adjust the `Gap Width` using the slider shown in Figure 6.12. While you're there, figure out how to change the colors of the bars, and fix the labels. You can see the full spreadsheet with all the computations we've done so far at `WingAeroHistogram.xlsx`.

It's worth taking some time to study this histogram, which shows how the data are distributed. Most of the salaries are less than $100K and there's a large gap in salaries between

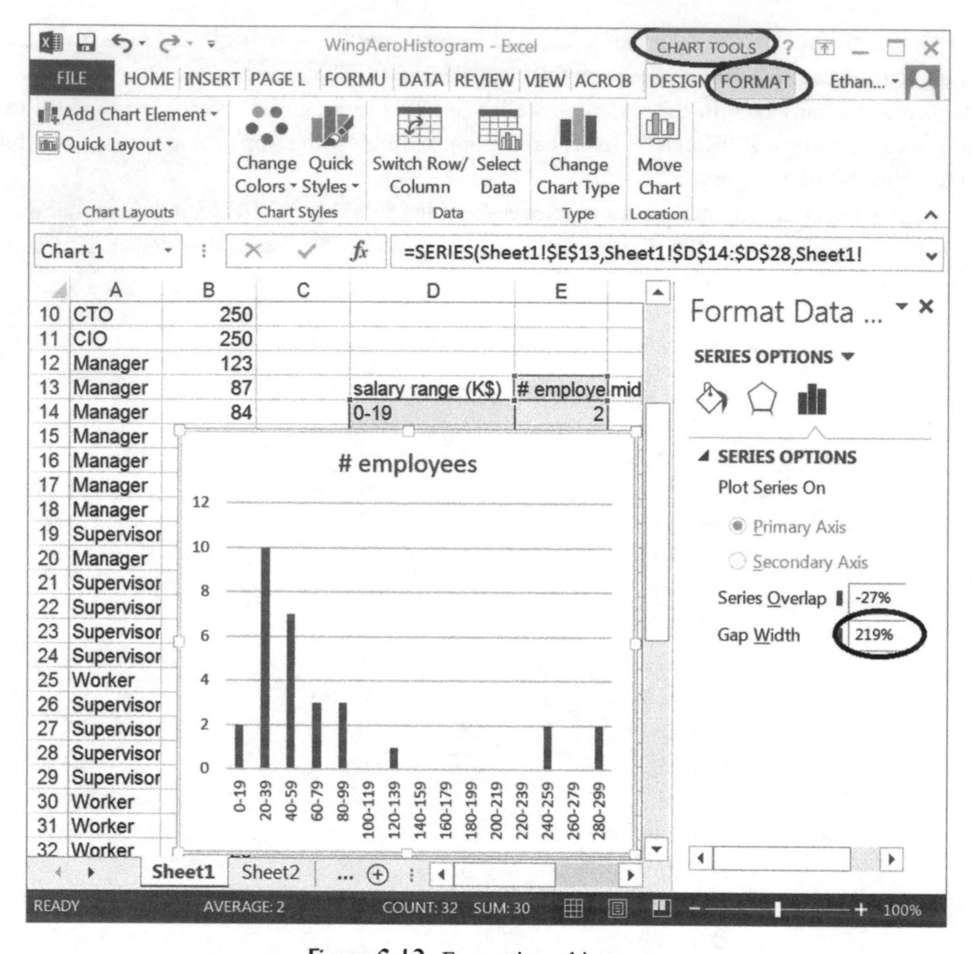

Figure 6.12. Formatting a histogram

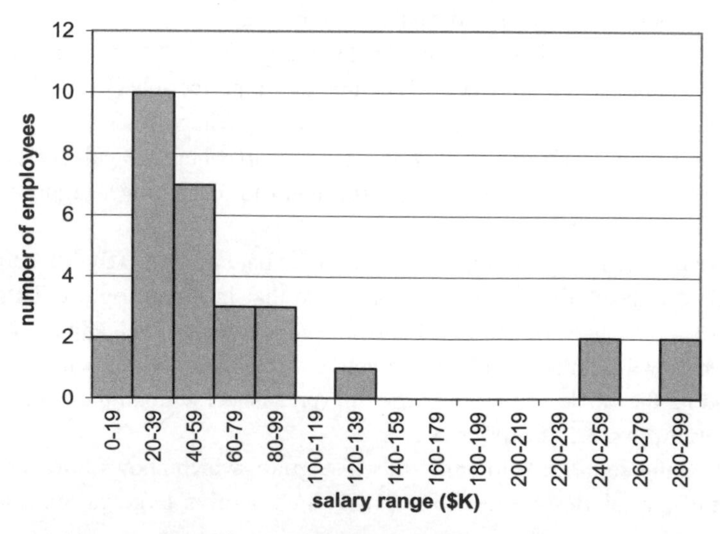

Figure 6.13. Wing Aero salary histogram

$139K and the executives who make more than $200K. Although this information is all in the table, the histogram makes it visible and dramatic.

We chose $20K for the size of the salary ranges so that we would have enough categories to show what was going on but few enough to make the graph understandable. In Exercise 6.14.14 you can explore what happens with different choices.

One of the themes of this chapter has been to use Excel for tedious repetitive calculations. So you might wonder whether Excel could build Table 6.11 for the grouped data if we told it the ranges we were interested in. Then the numbers would automatically update when we asked "what-if" questions. The good news is that Excel can do this job. The bad news is that its histogram building tools are rather clumsy. We will content ourselves with doing the counting by hand. If you're ambitious, try Excel help or search the internet for "excel histogram" to find out how to group data for an Excel histogram.

6.8 Mean, median, mode

There's a third kind of "average"—the mode—that's sometimes informative. The *mode* is the most common value. The histogram from the previous section shows that there are more employees with salaries in the $20K–$40K range than any other. So the mode is about $30K. It's the category with the highest bar.

The mode is most useful for data aggregated into ranges, as in a histogram. In the raw Wing Aero data there is no well-defined mode. Each of the values $250K, $49K, $42K and $25K appears twice. In Excel the function `MODE(B8:B37)` reports $250K when column B is sorted in descending order and $25K when the column is sorted the other way.

Each of "mean," "median" and "mode" can legitimately be called an "average." That ambiguity makes it easy to lie with statistics without actually lying. The CEO at Wing Aero may brag that workers at his company earn an average $77K per year, while the union argues that the average salary is $30K per year.

A cynic would advise you to use the "average" that tells the story your way and hope your listener won't know the difference.

When distributions are symmetric, the mean, median and mode are in the same place. The Wing Aero salary distribution isn't symmetric, it's skewed to the right. That's the fancy way to say that the bulk of the data cluster toward the left of the histogram with a long tail off to the right. For data that's right skewed, as this is:

$$\begin{array}{ccccc} \text{mode} & < & \text{median} & < & \text{mean} \\ 30 & < & 42.5 & < & 77.1 \end{array} .$$

In a histogram the mode is the peak, the median splits the area in half and the mean is where the graph would balance if it were a cardboard cutout.

If you're learning about these several kinds of averages for the first time you may want mnemonics to help you remember which is which. The "med" in median suggests correctly that it's in the middle. You can remember mode because there are "mo' of them than anything else."

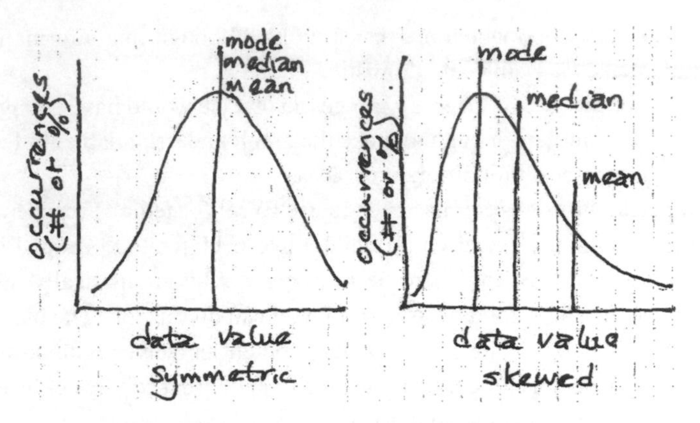

Figure 6.14. Symmetric and skewed distributions

6.9 Computing averages from histograms

Often all we know about data is a summary like that presented in a histogram. We'll see here how to estimate the mode, median and mean from that information. We will use the histogram in Figure 6.13 as an example. Since we know the correct averages we can see how good our estimates are.

We've already found the mode. It's the highest bar in the histogram—the range with the largest entry: \$20,000—\$39,000. Note carefully—the mode is the range, not the height of the bar. We could report that range, or report the mode as the middle of the range: about \$30,000. The raw data do not have a mode that makes sense, so there's nothing to compare this estimate to.

To estimate the median salary from the histogram we need to find the salary such that half the employees make less and half more. The first bar tells us that two make less than \$19K. Adding the number of employees in the first two ranges tells us that $2 + 10 = 12$ make less than \$39K. Looking at the next range we see that $2 + 10 + 7 = 12 + 7 = 19$ make less than \$59K. Since there are 30 employees, it's the 15th and 16th whose salaries are closest to the median. They are clearly pretty much in the middle of the third category, so we can estimate the median as the midpoint of that category, say \$50,000. The correct value (from the raw data) is \$42,500.

Since there are just eight ranges with data and the median occurred in the third one, we didn't need Excel to do the arithmetic. In a more complicated example things might not be so easy, so let's see how to make Excel do the work. Label column `H` in cell `H15` as "`count so far`". Then copy the value from `E15` to `H15` by typing `=E15` in H15. Then enter `=H15+E16` in `H16`, to add the number of employees counted so far (in `H15`) to the number in this range (in `E16`).

Now select that formula and copy it to `H17:H29`. A miracle has happened! Cell `H29` contains the value 30. If you click on that cell you can see that it's the result of the formula `=H28+E29`. Excel read your mind, and automatically updated the row references to cells in columns `H` and `E` with each copy down the column. It knew you wanted to add the value in the cell above to the value in the cell three over to the left. The last entry should be 30 since all the employees make less than \$299K. Now it's easy to see that the count so far passes the midpoint of 15 in the middle of the third range.

Estimating the mean salary is the hardest. Since the only Wing Aero data we have is what we used to draw the histogram we can't expect to find it exactly. All we can say about the 10 employees in the \$20K–\$39K range is that they earn somewhere in the neighborhood of \$30K. So our best guess is to assume they all earn exactly that. If we make a similar assumption for each of the other ranges then our estimate for the mean is the weighted average

$$\frac{2 \times \$10\text{K} + 10 \times \$30\text{K} + \cdots + 2 \times \$290\text{K}}{\text{total number of employees}}.$$

That's too much arithmetic to do by hand so we'll use Excel.

Put the label `mid range` in cell `F13`. We want to use cells `F14:F28` to hold the values 10, 30, ..., 290 that are the middles of the ranges in cells `D14:D28`. There's a quick trick for that. Enter the 10 in cell `F14` and enter the formula

```
=F14+20
```

in cell `F15`. Excel will display `30` there. That's because it reads the formula as

add 20 to the contents of cell `F14`.

Now we want to add 20 each time you move down a row. To do that, copy the formula in `F15` and paste it into cells `F16:F28`. (This takes advantage again of Excel's correct guess about what we are trying to do.)

Next label column `G` by typing `range total` in cell `G13`. Then put the formula

```
=E14*F14
```

in cell `G14`. That asks Excel to multiply the numbers in cells `E14` and `F14`. You should see 20. That's the first number to add in the weighted average computation we're working on.

When you copy that formula to cells `G15:G28` you should see 580 at the end of the list. That's the miracle yet again.

To compute the weighted average you must sum the values in column `G`. Since cell `E29` contains the sum of the values in column `E`, just copy the formula from that cell to cell `G29`. Excel will automatically change the column reference, turning the formula `=SUM(E14:E28)` into `=SUM(G14:G28)`. The sum is 2360. To find the mean, enter the formula `=G20/E29` in cell `G30`. Excel shows you 78.66667. Label that value as the mean.

Our estimate of the mean salary from the histogram is \$79,000. We shouldn't report more precision than that, since we made many approximations along the way. Do note that the estimate is not very far from the true mean of \$77,167 computed from the raw data.

6.10 Percentiles

On June 7, 2011 we visited the website `www.infantchart.com/`. There we collected the data displayed in Table 6.15. Reading down the first column, you can see that a one year old male baby weighing 21.5 pounds would be in the 30.9th *percentile*. That means that he would weigh more than 30.9 percent of the babies, less than 69.1 percent. A baby weighing 22.5 pounds would be in the 46.5th percentile, so 46.5% – 30.9% = 15.6% of the year old male babies weigh between 21.5 and 22.5 pounds.

Table 6.15. Year old male baby weights [R159]

Weight (pounds)	Percentile	Difference
15.5	0.1	0.1
16.5	0.2	0.1
17.5	0.8	0.6
18.5	3.0	2.2
19.5	8.2	5.2
20.5	17.5	9.3
21.5	30.9	13.4
22.5	46.5	15.6
23.5	61.8	15.3
24.5	74.9	13.1
25.5	84.8	9.9
26.5	91.4	6.6
27.5	95.4	4.0
28.5	97.7	2.3
29.5	98.9	1.2
30.5	99.5	0.6
31.5	99.8	0.3
32.5	99.9	0.1
33.5	100.0	0.1

To find the mode, we should look for the largest *difference* in percentiles. That turns out to be the 15.6% we just found, so the mode is about 22 pounds.

With guess-and-check on the website we found that a weight of 22.75 pounds was as close as we could get to the 50th percentile, so that's the median. Some other commonly used percentiles are the

- tenth: any baby weighing less than about 19.2 pounds is in the tenth percentile,
- ninetieth: about 90% of the babies weigh less than about 26.5 pounds,
- the quartiles—the 25th, 50th and 75th percentiles. The 50th percentile is, of course, the median.

Using the techniques from Section 6.9 we computed the mean as a weighted average. It is about 22.9 pounds.

Therefore

$$\begin{array}{ccccc} \text{mode} & < & \text{median} & < & \text{mean} \\ 22 & < & 22.75 & < & 22.9 \end{array}.$$

That suggests that the distribution is just little bit skewed to the right.

6.11 The bell curve

The baby weight data in Table 6.15 are just what we need to construct the histogram in Figure 6.16.

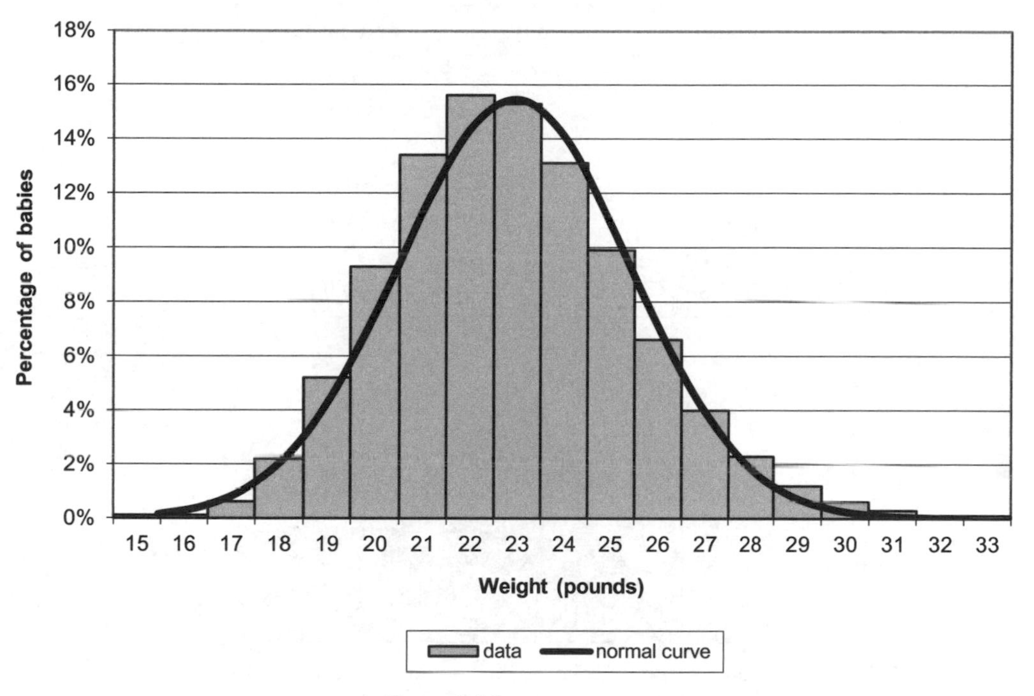

Figure 6.16. Baby weights

There you can see the modc (the highest bar) and you can see the little bit of right skew.

Many factors contribute to a baby's weight at one year: birth weight, heredity, nutrition,.... Whenever many small effects combine to give a total the distribution of values forms a *bell curve*. The one shown in the figure is a mathematically correct bell curve that approximates the real data. The mathematically correct name for the bell curve is *normal distribution*.

A normal distribution is always symmetrical. Its mean, median and mode are in the same place. To construct the one in the figure we need one more number besides the mean—a measure of how fast the curve spreads out. That measure is called the *standard deviation*. It's usually written with the Greek letter sigma: σ; the mean is usually written with the Greek letter mu: μ. For the baby weight data the standard deviation is about 2.6 pounds.

There's a nice way to use the standard deviation to describe how a bell curve spreads out.

- 2/3 of the values are less than one standard deviation away from the mean,
- 95% of the values are less than two standard deviations away,
- 99.7% are less than three standard deviations away.

Figure 6.18 illustrates these percentages. Table 6.17 summarizes them in terms of percentiles.

Since the baby body weight distribution is very close to normal, with mean μ about 22.9 pounds and standard deviation σ about 2.6 pounds, we know about 2/3 of the babies weigh between $\mu - \sigma = 22.9 - 2.6 = 20.3$ and $\mu + \sigma = 22.9 + 2.6 = 25.5$ pounds. Approximately

Table 6.17. Percentiles for the normal curve, mean μ, standard deviation σ

value	percentile
$\mu - 3\sigma$	0.1
$\mu - 2\sigma$	2.3
$\mu - \sigma$	15.9
μ	50.0
$\mu + \sigma$	84.1
$\mu + 2\sigma$	97.7
$\mu + 3\sigma$	99.9

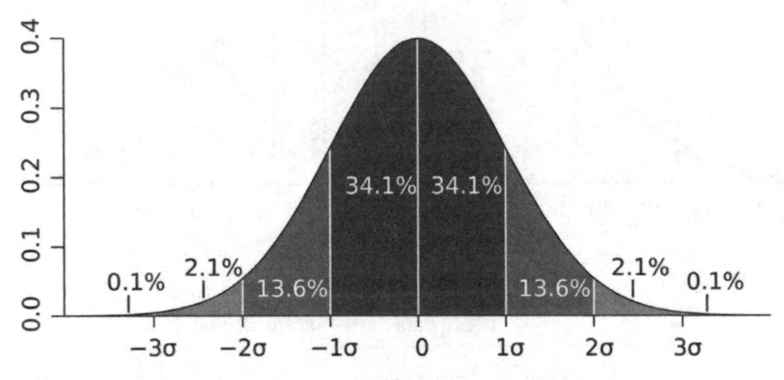

Figure 6.18. How the normal distribution spreads out [R160]

95% weigh between 17.7 and 28.1 pounds. 2.5% weigh more than 28.1 pounds and 2.5% weigh less than 17.1 pounds.

Figure 6.19 shows three bell curves with the same mean $\mu = 22.9$, but three different standard deviations, $\sigma = 1.3$, 2.6 and 5.2. The middle one matches the baby weight data.

The mathematics needed to calculate standard deviations and draw bell curves is more than we will present here. You will learn about it if you take an introductory course in statistics. All you should remember now is the rough relationship between standard deviation and spread we sketched above and that the official name for the bell curve is *normal distribution*. If you want to see how we asked Excel to calculate the standard deviation for the baby weights and draw Figures 6.16 and 6.19, you can look at the spreadsheet `babyweights.xlsx`.

6.12 Margin of error

The Pew Research Center conducted a study in July of 2012 that asked about support for President Obama's tax position. Their report said (in part)

> By two-to-one (44% to 22%), the public says that raising taxes on incomes above \$250,000 would help the economy rather than hurt it, while 24% say this would not make a difference.
>
> [The poll reached 1,015 adults and has a margin of sampling error of plus or minus 3.6 percentage points.] [R161]

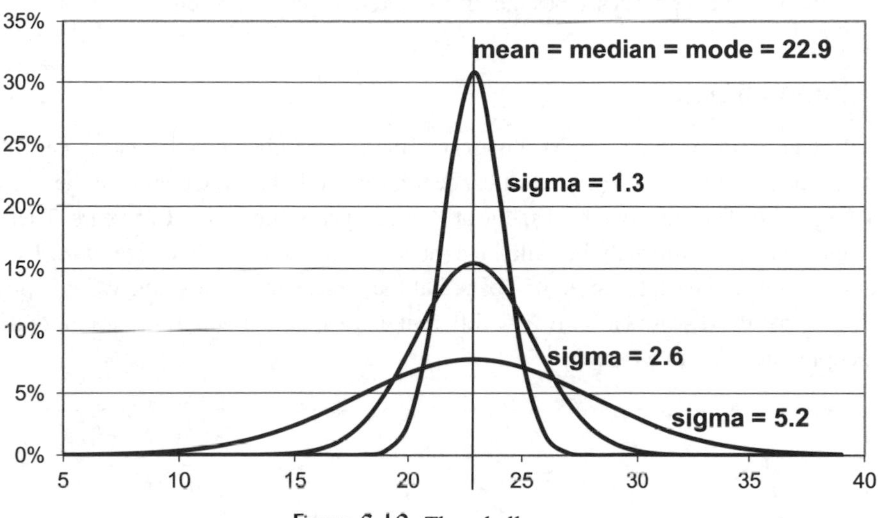

Figure 6.19. Three bell curves

The first paragraph quoted above reports the results of a survey. The second tells you something about how reliable those results are. It's clear that the smaller the margin of error the more you can trust the results. Understanding the margin of error quantitatively—seeing what the number actually means—is much more complicated. A statistics course would cover that carefully; we can't here. Since the term occurs so frequently, it's worth learning the beginning of the story. Even that is a little hard to understand, so pay close attention.

The survey was conducted in order to discover the number of people who thought the tax increase would benefit the economy. If everyone in the country offered their opinion then we would know that number exactly. If we gave the survey to just three or four people we could hardly conclude anything. The people at the Pew Research Center decided to survey the opinions of a *sample* of the population—1,015 people chosen at random. Of the particular people surveyed, $0.44 \times 1{,}015 = 447$ people thought the tax increase would benefit the economy. If they'd surveyed a different group of 1,015 people, they would probably see a different number, so a different percent.

The 3.6 percentage point margin of error says that if they carried out the survey many times with different samples of 1,015 people, 95% of those surveys would report an answer that was within 3.6 percentage points of the true value.

There's no way to know whether this particular sample is one of the 95%, or one of the others. About five of every 100 surveys you see in the news are likely to be bad ones where the margin of error surrounding the reported answer doesn't include the true value. Survey designers can reduce the margin of error by asking more people (increasing the sample size). But there will always be five surveys in every hundred where the true value differs from the survey value by more than the margin of error.

The report doesn't explicitly mention 95%. That's just built into the mathematical formula that computes the margin of error from the sample size. Even that conclusion may be too optimistic. The margin of error computation only works if the sample is chosen in a fair way, so that everyone is equally likely to be included. If they asked 1,015 people at random from an area where most people were Democrats (or Republicans) or rich (or poor) the result would be even less reliable. The report describes the efforts taken to get a representative sample.

6.13 Bimodal data

The black data points in Figure 6.20 shows how the rate of diagnosis (in cases per 100,000 people) of Hodgkin lymphoma (a kind of cancer) for white females depends on the age of the woman diagnosed. There are two peaks, one at about 20 years, the other at 75 years. There is no single value that can legitimately be called the mode—this is a *bimodal distribution*. The mean and the median would each be about 45 years, but they make no sense at all. What is probably happening is that the disease has two very different causes, one of which occurs more often in young people, the other in old people.

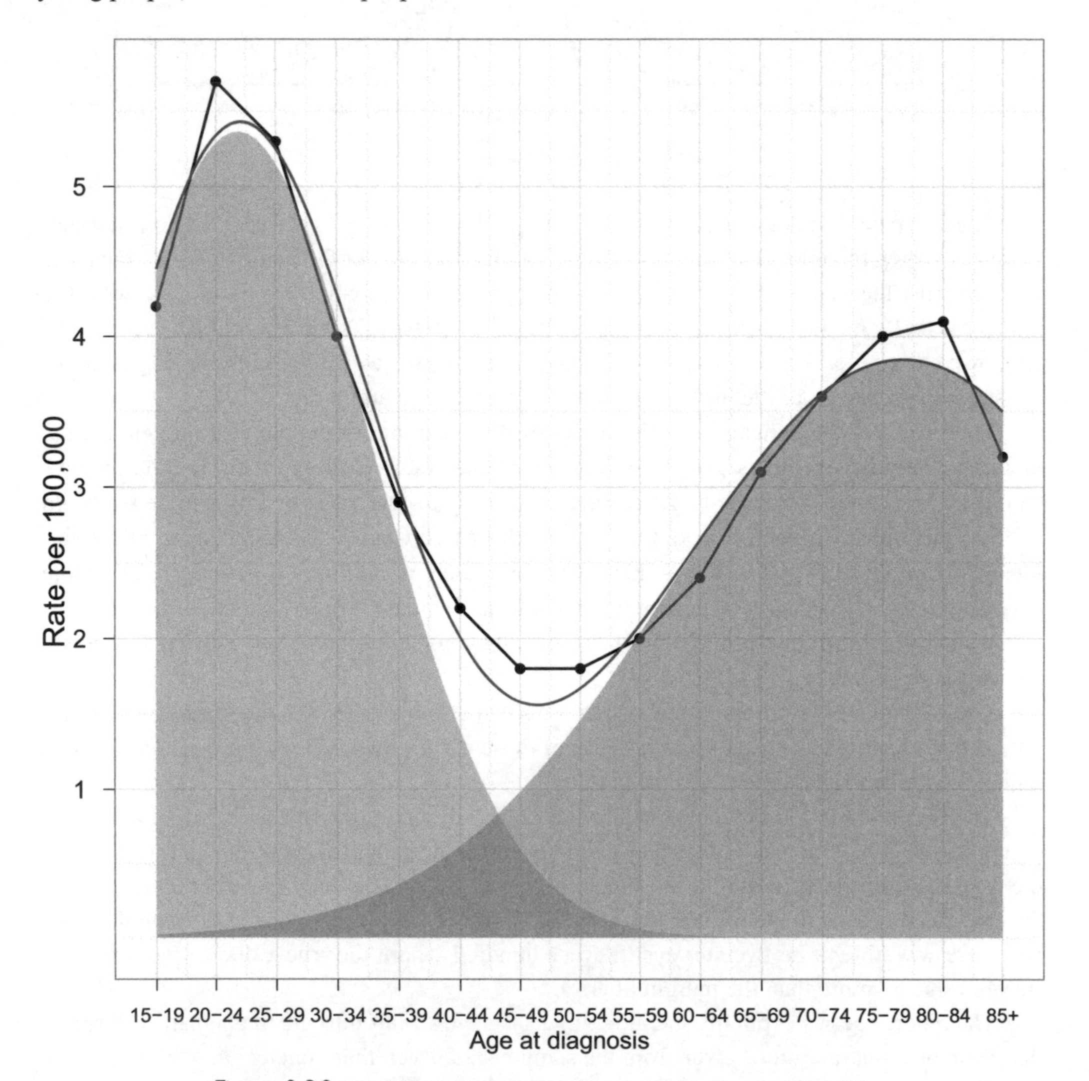

Figure 6.20. Hodgkin lymphoma incidence (white females) [R162]

We can understand this distribution as a combination of two normal distributions. The left bell curve for the early onset of Hodgkin lymphoma has a mean of about 24 years with a standard deviation of 11 years. The right bell curve for late onset has a mean of about 79 years with a standard deviation of 19 years—it spreads out more slowly. The smooth curve is the sum of the two normal distributions. It matches the data very well.

6.14 Exercises

Excel is just a tool. It doesn't answer questions, it provides numbers and pictures you use to answer questions. So keep on writing complete sentences explaining the meaning and context of the numbers you report. The numbers in cells in a spreadsheet are no more use by themselves than the numbers in a calculator display.

If an exercise asks for hard copy of a spreadsheet, format it well. Use `Print Preview` to make sure that charts fit on one page and don't cover important numbers. Label the data columns, cells containing important computations, axes and legends in charts. Numbers are useless when they can't be understood.

Exercise 6.14.1. [S][R][Section 6.1] CXO.

What do "CEO", "CTO", "CIO" and "CFO" stand for in the Wing Aero salary table in Section 6.1?

Exercise 6.14.2. [S][Goal 6.1][Goal 6.5][Section 6.2] What if...?

Open up the original Wing Aero spreadsheet and use Excel to calculate the mean and median, as we did in Section 6.1 and Section 6.4.

(a) Suppose all the managers get a $10K raise. Change their salaries and see how Excel updates the mean and median calculation.
(b) Go back to the original spreadsheet (use the `Undo` button if you can). Experiment with changing salaries (of any of the workers—your choice) so that the mean and the median increase.
(c) Reset back to the original spreadsheet. How would you change salaries so that the mean decreases but the median stays the same?
(d) Reset back to the original spreadsheet again. How would you change salaries so that the mean stays the same but the median decreases?

Exercise 6.14.3. [S][R][Section 6.3] [Goal 6.4] Formatting in Excel.

Format the cells containing averages in the Wing Aero spreadsheet so that the numbers displayed for the various averages are rounded to the nearest thousand dollars (no decimal places).

Exercise 6.14.4. [S][Section 6.4][Goal 6.1][Goal 6.4] [Goal 6.6] Practice finding the median.

Open up the original Wing Aero spreadsheet from `WingAero.xlsx`.

(a) Use Excel to find the median salaries for each category of employees (workers, managers, etc.).
(b) Show the data in a properly labeled bar chart.
(c) Do you think the median is a representative "average" for each category? Explain.

Exercise 6.14.5. [S][Section 6.5][Goal 6.6][Goal 6.1] Averaging averages.

(a) Find the average (mean) salary at Wing Aero for each of the four categories of employees (use the original data set, from Section 6.1).
(b) Show the data in a properly labeled bar chart.
(c) Compute the weighted average of these averages to check that you get the correct mean for the whole payroll.

[See the back of the book for a hint.]

Exercise 6.14.6. [S][Section 6.5][Goal 6.1] [Goal 6.6] Cash-strapped T proposes 23 percent fare increase.

On March 29, 2012 the Massachusetts Bay Transportation Authority (MBTA) provided the fare data in Table 6.21. Riders can pay with a stored-value Charlie Card or with a Charlie Ticket bought on the spot.

Table 6.21. MBTA fare increases [R163]

Fare Category	Current	Proposed
	Charlie Card	
Bus	$1.25	$1.50
Subway	$1.70	$2.00
Senior Bus	$0.40	$0.75
Senior Subway	$0.60	$1.00
Student Bus	$0.60	$0.75
Student Subway	$0.85	$1.00
	Charlie Ticket	
Bus	$1.50	$2.00
Subway	$2.00	$2.50

(a) What are the relative and absolute changes in the Charlie Card bus fare?
You don't have to do this in your head, without a calculator, but you should be able to.
(b) Create a spreadsheet for these data, with eight rows (one for each of the eight categories) and three columns, for the category name, the existing fare and the proposed fare. Create a properly labeled bar graph to display the data.
(c) Label the next two columns appropriately to hold the relative and absolute changes. Fill those columns with Excel formulas to compute the correct values. Do not compute the values elsewhere and enter them in the spreadsheet as numbers.
(d) Imagine that you are addressing a public meeting about these fare increases. How would you argue that an unfair burden is being placed on people who pay using a Charlie Ticket? How would you argue that senior citizens are most hard hit?
(e) Find the mean of the relative percent increases. Explain why the answer is not the 23 percent quoted in the headline.
(f) What extra information would you need to check that the correct mean is 23 percent?

(g) (Optional) Find out why the stored value card is called a Charlie Card.

(h) (Optional, and difficult) Create a column for the percentage of MBTA revenue for each of the categories and fill in some values that sum to 100% and give a weighted average fare increase of about 23%.

There's an important and subtle distinction here between weights as a percentage of revenue and weights as a percentage of trips.

Exercise 6.14.7. [S][Section 6.6][Goal 6.6] Why not pie charts?

Do some internet research to discover why bar charts are usually better than pie charts. Write the reasons in your own words (don't just cut and paste). Identify the sources of your information and comment on why you think those sources are reliable.

Exercise 6.14.8. [A][U][Section 6.6][Goal 6.6] Misleading pie charts.

Table 6.22 shows some student enrollment data at a college. It's clearly incomplete: Juniors and Seniors are missing.

Table 6.22. Freshman and sophomore enrollments

Class	Percentage
Freshman	40
Sophomore	25

Flashy pie charts are common in the media, but staid and boring bar charts are usually more informative and less deceiving. Because Excel makes it so easy to switch to a pie from a bar, designers may be tempted by the glitz. You should not succumb to that temptation. Here's an exercise where you can find out why.

(a) Construct a bar chart displaying these data, with columns for Freshman and Sophomore enrollments.

(b) Change the chart type in Excel.

(c) Explain what is wrong with the chart Excel built for you.

Exercise 6.14.9. [S][W][Section 6.8][Goal 6.5] [Goal 6.1] What if?

Add five workers each earning $18K to the original Wing Aero payroll by inserting some rows in the table. Comment on what happens to the mean, median and modal incomes.

Note that Excel automatically recomputes these averages, but not the charts for which you created data by hand.

When you've done this exercise, undo your changes and check that the three kinds of averages revert to their old values.

Exercise 6.14.10. [S][Section 6.8][Goal 6.3][Goal 6.7] Population pyramids.

At `www.census.gov/population/international/data/idb/informationGateway.php` you can choose a country and a year, then ask for a kind of bar chart known as a *population pyramid*. You can also download the data used to build the chart.

(a) Construct population pyramids for the United States and for Sudan for the year 2010.

(b) Estimate the number of people in the United States in 2010 between 0 and 9 years old. Do the same for Sudan. What fractions of the populations do these numbers represent?

(c) Find the modal age range for the United States population. Do the same for Sudan.
(d) Find the age range for the United States that has the smallest population. Do the same for Sudan.
(e) Compare the population distributions of the United States and Sudan. Write several sentences that highlight aspects of each distribution that you think are quantitatively significant. Use the results of the previous part of the problem.

Exercise 6.14.11. [S][Section 6.8][Goal 6.1] Lake Wobegon.

(a) Look back at the Wing Aero data and answer the following questions:
  - What percentage of the employees make more than the mean salary?
  - What percentage of the employees make more than the median salary?
  - What percentage of the employees make more than the mode salary?
(b) In his radio show *A Prairie Home Companion* host Garrison Keillor regularly tells his audience about Lake Wobegon, where "all the children are above average".
  Is that possible, with any of the meanings of "average"?

Exercise 6.14.12. [S][C][Section 6.8][Goal 6.1] Working for Walmart.

On December 2, 2009 Bloomberg News reported on a settlement in which Walmart agreed to pay $40 million to up to 87,500 employees because the company had failed to pay overtime, allow rest and meal breaks and, in addition, manipulated time cards.

Eligible present and former employees would receive $400 to $2,500 each—on average $734.

The lawyers for the employees asked for fees of $15.2 million from the $40 million. [R164]

(a) Compute the mean compensation, assuming that there are 87,500 eligible employees and that the lawyers have taken their cut.
(b) Compare your answer to the reported minimum compensation of $400 and the reported $734 the average worker will receive.
(c) Draft a letter to the editor or the reporter, politely pointing out that both the reported "averages" made no sense, and asking for more detail or a correction.

Exercise 6.14.13. [R][A][S][Section 6.8][Goal 6.1] Is it discrimination?

Table 6.23 shows the salary structure of two departments in a hypothetical university.

(a) What is the average (mean) salary of the professors? Of the women professors? Of the men?
(b) Answer the same questions for the median.
(c) Answer the same questions for the mode.

Table 6.23. Salary structure at a university

	Physics		English	
	professors	salary	professors	salary
Women	1	$100K	8	$50K
Men	9	$90K	2	$40K

(d) Write a few sentences to convince someone that men in this university are paid better than women. Then write a few sentences to convince someonc of just the opposite. Explain the contradiction.

These calculations are so straightforward that they're easier with pencil and paper (maybe a calculator) than with Excel.

Exercise 6.14.14. [S][Section 6.9] [Goal 6.8][Goal 6.7] Choosing data ranges for a histogram.

When you change the widths of the intervals in a histogram you get a (slightly) different picture of the data.

(a) Redo the Wing Aero histograms in the text using salary ranges of size $10K and then of size $50K.

[See the back of the book for a hint.]

(b) In each case use the techniques from Section 6.9 to estimate the mean, median and mode.
(c) Discuss the advantages and disadvantages of these possible choices for the salary range, comparing them to our choice of $20K.

Exercise 6.14.15. [S][Section 5.1][Goal 5.1] Texting teens.

We drew the chart in Figure 6.24 using data from *The Boston Globe* on April 15, 2012.

(a) Estimate the mode, median and mean number of text messages sent by teenagers each day.
(b) Estimate the mode, median and mean number of text messages sent by teenagers each day.
(c) In total, approximately how many text messages are sent by the 23 million American teens each day?
(d) The percentages don't add up to 100%. Why might that have happened?
(e) If you asked a random teenager how many text messages she sent yesterday what are the chances (what is the probability) that it was more than 50? More than 100? More than 25?

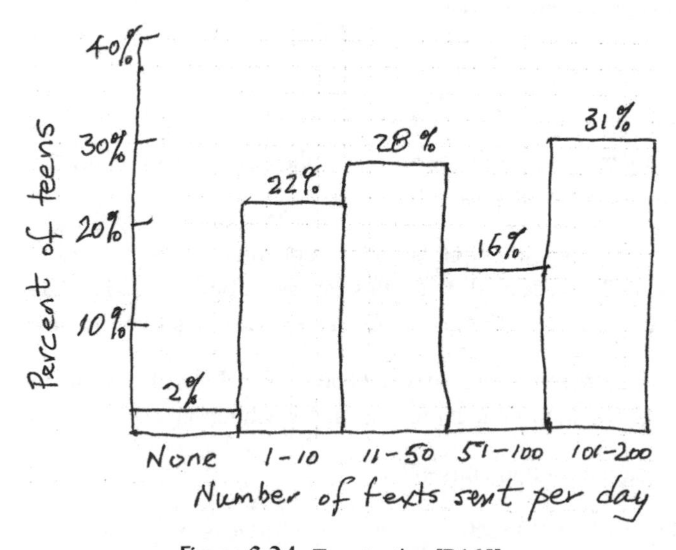

Figure 6.24. Teen texting [R165]

(f) What percent of teenagers text more than the median amount?
(g) Does the figure display a histogram?
(h) Create an Excel chart that reproduces the figure.

Exercise 6.14.16. [S][Section 6.9][Goal 6.8] [Goal 6.7] Websites are often confusing.

Jakob Nielsen evaluated the usability of voter information websites for the 2008 election for each of the fifty states and the District of Columbia. You can read his analysis at `www.nngroup.com/articles/aspects-of-design-quality/`. His article includes the histogram in Figure 6.25.

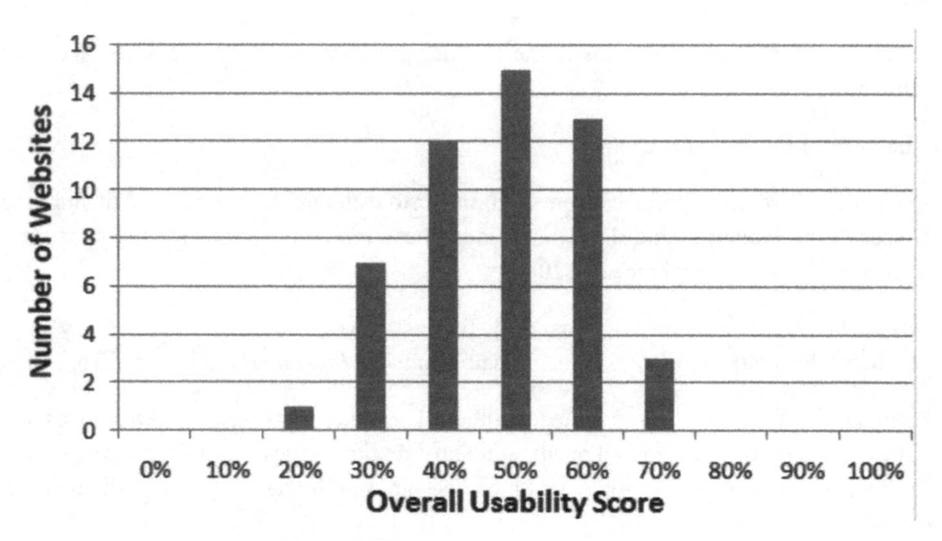

Figure 6.25. Website usability [R166]

(a) What is the modal usability score for these 51 home pages?
(b) Reproduce this histogram in Excel.

[See the back of the book for a hint.]

(c) Estimate the median usability score for these 51 home pages.
(d) How many of them have a usability score less than the median score?
(e) Estimate their mean usability score.

Exercise 6.14.17. [U][Section 6.4][Goal 6.1] Doublethink.

In January 2012 one could read this at `www.businessinsider.com/where-the-one-percent-live-the-15-richest-counties-in-america-2012-2`:

> Living in Arlington isn't cheap, so you'd better be making at least the median household income to live in this county just outside Washington, D.C. [R167]

How does this statement contradict itself?

Exercise 6.14.18. [S][C][W][Section 6.9][Goal 6.8][Goal 6.7] Household income in the United States.

The histogram in Figure 6.26 shows the percentages of households in income groups \$5,000 increments apart, except for the two farthest right columns.

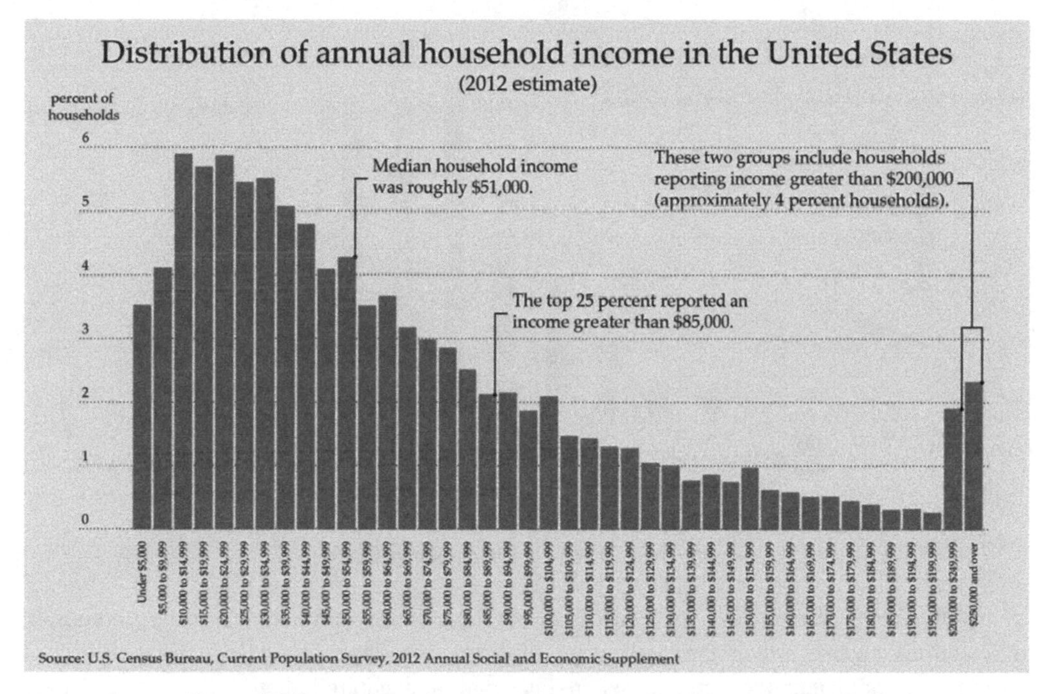

Figure 6.26. United States household income—2012 Estimate [R168]

We've put the data (and a copy of the figure) in the spreadsheet `Households2012.xlsx`.

(a) Check the quantitative assertions in the text in the Wikipedia chart.

(b) Build a histogram in Excel that comes as close as possible to matching the one from Wikipedia. Create the same chart and axis titles. Change the grid lines. Put in the comments as text boxes. Match the fonts.

(c) Do the percentages sum to 100%? If not, what might explain the discrepancy?

(d) To estimate the mean household income you will need an estimate of the mean for the households with incomes greater than \$250,000. There's no top to this range, so you can't use the middle of the range.

What value for the mean for the last category makes the mean for the whole population equal to the median?

(e) Search for an estimate of the mean household income for the whole population. What mean for the last category results in this overall mean?

Exercise 6.14.19. [S][Section 6.9] [Goal 6.8][Goal 6.7] Fight for the Senate.

A graph like the one in Figure 6.27 appeared in Nate Silver's Five Thirty Eight column in *The New York Times* on October 31, 2012. The x-axis displays the number of seats held by each party: the Tie in the middle is 50 Democrats, 50 Republicans. The +10 Dem corresponds to 55 Democrats, 45 Republicans.

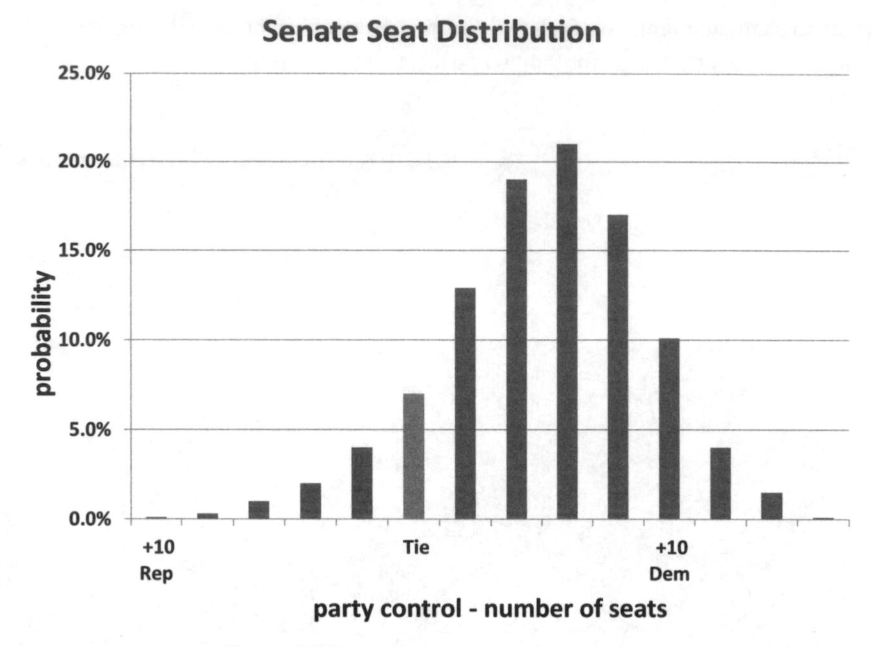

Figure 6.27. The fight for the Senate [R169]

Nate Silver constructed this histogram by imagining (simulating) many thousands of elections and recording the percentage of time each Democratic/Republican split occurred. We estimated the percentages in his chart and entered them in the spreadsheet `Oct31SenateProjection.xlsx` so you don't have to type them yourself. (We rounded the really tiny percentages to zero.) Use Excel whenever it's most convenient for you.

(a) What is the most likely number of Democratic senators?
(b) What number of Democratic senators represents the mode of this distribution?
(c) What is the probability that there are more than 50 Democratic senators?
(d) What number of Democratic senators is the median of this distribution?
(e) If you had the complete list of all Nate Silver's imagined elections and sorted it by the number of Democratic senators, how many Democratic Senators would there be in the middle election on that list?
(f) Use Excel to compute the (weighted) average number of Democratic senators for these imagined elections.
(g) What actually happened in the election?

Exercise 6.14.20. [S] [W][Section 6.9] [Goal 6.1][Goal 6.8] What cities pay for fire protection.

On Monday, March 30, 2009 *The Boston Globe* published an article comparing the amount various cities spent on Fire and EMS services. Figure 6.28 is a screenshot of a spreadsheet where we entered some of the data, along with population figures. You can download that spreadsheet from `FireSpending.xlsx`. Use the data to answer these questions.

(i) Population
  (a) What is the mean population of the twelve cities for which data are presented?
  (b) What is the median population of the twelve cities for which data are presented?

Spending on Fire/EMS services

Study from the Boston Globe, March 30, 2009
2007 population figures from Wikipedia

City	Population	Fire/EMS spending per resident	Fire/EMS personnel per 1000 residents
Boston	608,352	$452.15	3.4
San Francisco	799,183	$315.81	2.2
Columbus, OH	747,755	$255.70	2.1
Seattle	594,210	$247.75	1.8
Baltimore	637,455	$225.98	2.7
Memphis	674,028	$220.22	2.5
Detroit	916,952	$201.54	1.6
Nashville	590,807	$194.43	1.9
Philadelphia	1,449,634	$187.63	1.6
Jacksonville	805,605	$179.99	1.5
New York	8,274,527	$157.56	1.7
Los Angeles	3,834,340	$137.80	0.9

Figure 6.28. Fire protection spending [R170]

(c) Create Table 6.29 in Excel. Fill in the second column there. Then create a properly labeled histogram for the data.
(d) Use your histogram to estimate the mode population for these cities.
(e) What percent of the U.S. population lives in these twelve cities?

(ii) Fire/EMS spending per person
(a) What is the mean amount spent for Fire/EMS services per person in these twelve cities?
(b) Estimate the median amount spent for Fire/EMS services per person in these twelve cities.
(c) Estimate the mode amount spent for Fire/EMS services per person in these twelve cities.

(iii) What do firefighters earn?
There is enough information in the spreadsheet to calculate the average (mean) earnings of Fire/EMS personnel in each of the twelve cities. Do that, in a fresh column in your spreadsheet.

Table 6.29. City populations

Population range	Number of cities
500K–600K	
600K–700K	
700K–800K	
800K–900K	
900K–1000K	
1000K–2000K	
2000K–3000K	
3000K–4000K	
>4000K	

(a) In which city do Fire/EMS personnel have the highest average salary? How much is it?
(b) In which city do Fire/EMS personnel have the lowest average salary? How much is it?
(c) Where does Boston rank in the list of Fire/EMS personnel salaries?
(d) Explain how Boston can be at the top of the list in Fire/EMS expenses per resident although it does not pay the highest salaries.

(iv) Correction the next day!
On Tuesday, the next day, *The Boston Globe* published a correction, which said that Boston's fire department expenses were $285 per resident in the last fiscal year. [R171]

Look at the answers to the questions above and indicate which have changed (and how) and which stayed the same.

[See the back of the book for a hint.]

Exercise 6.14.21. [S][Section 6.9] [Goal 6.8] College presidents' pay.

Figure 6.30 is a histogram showing the total compensation for the 100 best paid presidents of public universities. The data are from an article in the April 3, 2011 issue of *The Chronicle of Higher Education* (`chronicle.com/article/Presidents-Defend-Their/126971`).

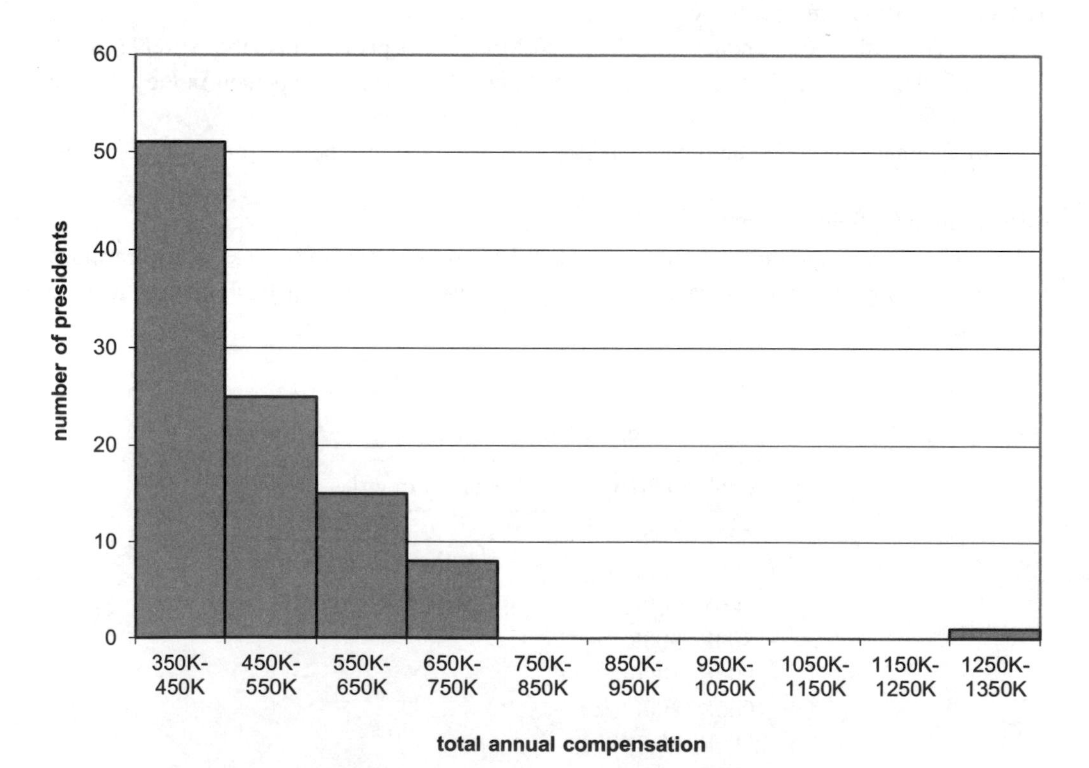

Figure 6.30. College presidents' pay [R172]

To save you staring at the picture, the number of presidents in each of the ranges (reading from left to right) is 51, 25, 15, 8 and then 1 in the last range.

You may use Excel for this exercise if you wish, but you don't have to.

(a) What is the mode of this distribution?
(b) Estimate the median compensation.
(c) Estimate the mean compensation.
(d) Write two arguments one of which indicates that the average president is paid appropriately, one that presidential pay is too high.

Exercise 6.14.22. [A][S][W][Section 6.9] [Goal 6.8][Goal 6.7][Goal 6.10] Xorlon fleegs.

Table 6.31 shows the distribution of weights of a sample of fleegs from the planet Xorlon, where weight is measured in frams.

We made up the numbers in this artificial problem so that the arithmetic would be easy. We made up the words, too. "frams" is one Prof. Bolker could have used in a Scrabble game with his wife—for 57 points on a triple word score.

Table 6.31. Xorlon fleegs

weight range (frams)	percent of sample
20–40	10
40–60	10
60–80	20
80–100	10
100–120	50

(a) Sketch a neat properly labeled histogram that displays the data.
(b) Create a properly labeled histogram in Excel that matches the one you just drew. You can start by downloading `XorlonFleegs.xlsx`.

Answer the following questions. You may work in Excel or with a calculator or do mental arithmetic.

(c) Estimate the mode fleeg weight.
(d) Estimate the median fleeg weight.
(e) Estimate the mean fleeg weight.
(f) Estimate the percentage of fleegs that weigh more than the median.
(g) Estimate the percentage of fleegs that weigh more than the mean.
(h) Estimate the percentage of fleegs that weigh more than the mode.

Exercise 6.14.23. [S][Section 6.9] [Goal 6.10] Erdős numbers.

Paul Erdős (1913–1996) was the most prolific mathematician of the twentieth century. He was famous (in mathematical circles) for the way he worked—he travelled from school to school, writing joint papers with mathematicians at each.

From Wikipedia:

> An Erdős number describes a person's degree of separation from Erdős himself, based on their collaboration with him, or with another who has their own Erdős number. Erdős alone was assigned the Erdős number of 0 (for being himself), while his immediate collaborators could claim an Erdős number of 1, their collaborators have Erdős number at most 2, and so on. [R173]

You can think of the Erdős Number Project as a description of the social network of mathematicians. Its home page is `www.oakland.edu/enp`; there you can find out that

> ... the median Erdős number is 5; the mean is 4.65, and the standard deviation is 1.21.

Table 6.32 shows the raw data.

Table 6.32. Erdős numbers [R174]

Erdős number	mathematicians
0	1
1	504
2	6,593
3	33,605
4	83,642
5	87,760
6	40,014
7	11,591
8	3,146
9	819
10	244
11	68
12	23
13	5

(a) Use Excel to draw a histogram for the distribution of Erdős numbers.
(b) What is the mode of this distribution?
(c) Verify the claims for the median and mean.
(d) Verify the claim for the standard deviation.
(e) Professor Bolker (one of the authors of this book) has an Erdős number of 2 (he wrote a paper with Patrick O'Neil, who wrote a paper with Erdős). Why is Professor Mast's Erdős number at most 3? How might it be less than 3?
(f) How many mathematicians have a finite Erdős number?
(g) There are some mathematicians whose Erdős number is infinite. How can that be?

Exercise 6.14.24. [S][A][Section 6.9] [Goal 6.8] Ruritania.

Find the mean, median and mode age for male residents of Ruritania using the histogram or the data in the spreadsheet `Ruritania.xlsx`. (Ruritania is a fictional country in central

Europe which forms the setting for *The Prisoner of Zenda*, a fantasy novel written by Anthony Hope.)

Exercise 6.14.25. [A][S][Section 6.10][Goal 6.9] Wing Aero percentiles.

(a) How many Wing Aero employees are in the bottom tenth percentile in salary?
(b) What is the salary cutoff for the bottom tenth percentile?
(c) Answer the same questions for the top tenth percentile.

Exercise 6.14.26. [U][Section 6.12][Goal 6.2] [Goal 6.10] Sick-leave proposal.

An article in the *Orlando Sentinel* on August 6, 2012 discussed a ballot initiative that would require employers with 15 or more workers to provide paid time off for employees for illness-related issues. The article polled voters to gauge support for placing this question on the ballot for the November election, and noted that among likely voters, 67 percent supported the initiative while 26 percent opposed it.

The article reported that the poll surveyed 500 people and had a margin of error of 4.4 percentage points. [R175]

(a) Why don't the percentages from this poll add up to 100%?
(b) Explain why this statement is not true: "67% of the residents likely to vote in November support the measure."
(c) Explain what the 4.4 percentage point margin of error means for this poll.

Exercise 6.14.27. [S] The Boston Marathon. Table 6.33 contains data for the numbers of men and women who finished the 2012 Boston marathon, grouped by finishing times. For example, 26 men and one woman finished with a time between two and two and a half hours. (That one woman was a wheelchair racer.)

Table 6.33. The 2012 Boston marathon

Finishing time	Men	Women
2:00–2:30	26	1
2:30–3:00	444	27
3:00–3:30	1,844	260
3:30–4:00	3,389	1,714
4:00–4:30	2,819	2,833
4:30–5:00	1,861	1,966
5:00–5:30	1,068	1,013
5:30–6:00	607	609
6:00–6:30	323	339
6:30–7:00	160	162

We've entered the data in the spreadsheet `Marathon2012.xlsx`.

Answer the following questions. Do as much of the arithmetic in Excel as possible.

(a) Sketch a neat histogram for this data.
(b) Draw your histogram with Excel. Does it match your sketch?
(c) How many men finished the marathon? How many women?

(d) Use the data to estimate the mode, median and mean for the men's finishing times. Mark these times on the handwritten histogram sketch.
(e) Suppose my friend ran the marathon and finished ahead of half the men. What was his finishing time (approximately)?
(f) About what percentage of the women finished ahead of half the men?

[See the back of the book for a hint.]

Exercise 6.14.28. [U] Income growth.

On April 26, 2014 *The Boston Globe* reported that 2012 per capita income in Massachusetts grew to $49,354, up 3.2% from 2008, after adjusting for inflation.

(a) How much was Massachusetts per capita income in 2008, in 2012 dollars?
(b) How much was Massachusetts per capita income in 2008, in 2008 dollars?
(c) This income figure is an average. Is it a mean, a median or a mode? Explain how you know.
(d) Estimate the total 2012 income for Massachusetts.

[See the back of the book for a hint.]

Exercise 6.14.29. [S] Scrabble.
The Wikipedia page `en.wikipedia.org/wiki/Scrabble_letter_distributions` shows the point value of each of the 100 Scrabble tiles.

(a) Draw a bar chart illustrating the number of tiles with each of the point values from 0 to 10. The x-axis labels should be

0 1 2 3 4 5 6 7 8 9 10

The heights of the bars should correspond to the number of tiles with each value.

This is a difficult chart to create in Excel. Before you try, draw it by hand, so you know what you want the end result to look like. What Excel shows you first is likely to be far from your goal.
(b) What are the mode, median and mean point values? Show them on your (hand written) chart.
(c) What percentage of the tiles are worth more than 1 point?
(d) What percentage of the tiles are worth less than the median number of points?
(e) Answer these questions for Scrabble in some other language (your choice) and discuss the differences between that language and English.

Exercise 6.14.30. [U][Goal 6.9] Who's gifted?

The blog `giftedissues.davidsongifted.org/BB/ubbthreads.php/topics/152941/Re_Innumeracy_in_Gifted_Educat.html` offers two very interesting quotations: [R176]

From *The Everything Parent's Guide to Raising a Gifted Child* by Sarah Robbins (p. 125):

> Unfortunately, highly gifted children (those in the 95th percentile) only occur in approximately 1 out of 1,000 preschoolers, and profoundly gifted children (those in the 99.9th percentile) are as rare as 1 in 10,000 preschoolers.

From *Giftedness 101* by Linda Silverman (p. 87):

> In our mushrooming populace, over 3 million Americans and approximately 70 million global citizens are highly gifted or beyond (99.9th percentile).

What is wrong with the arithmetic?

Exercise 6.14.31. [S] Many flights arrive early!

The spreadsheet at `ArrivalDelays.xlsx` contains data on how many minutes late American Airlines flights to Boston's Logan airport were in January, 2014.

1. What does a "negative delay" mean?
2. Later you'll be asked to draw a histogram of this data in Excel. Sketch a neat approximate version first, with proper titles and reasonable scales for both axes and a proper title for the whole chart. You don't need to draw all the bars!
3. Draw your histogram with Excel. Does it match your sketch?
4. How many flights were counted in this data?
5. What percentage of the flights arrived on time?
6. Use the data to estimate the mode, median and mean arrival delay. Show these values on your histogram sketch.
7. Flights that are more than two hours late are *outliers*—the delay is probably not American Airlines' fault. Estimate the mode, median and mean arrival delays if you don't include the outliers.

Exercise 6.14.32. [S] Quotes in *Common Sense Mathematics*.

The Excel file `CSMquotes.xlsx` contains data on the number of words in quotes used in an early draft of this text.

(a) Create a properly labeled histogram displaying the data. You may sketch the histogram with pencil and paper, or use Excel.
(b) Calculate the total number of quotes.
(c) Estimate the total number of words in the quotes.
(d) Estimate mode, median and mean quote sizes, and mark them on your histogram.
(e) Explain why the mean is the largest of the three averages.
(f) Estimate the total number of words in the text.
(g) Estimate the percentage of words in the text that are in quotes.

Exercise 6.14.33. [S][W] Ricky's tacos.

A story in *The Boston Globe* on February 7, 2015 stated that

> Food prices over the past year have increased at four times the rate of overall inflation, with fresh products, such as meat, vegetables, and dairy, soaring even faster. Ground beef prices, for example, are up about 20 percent from a year ago. Shoppers at local grocery stores have felt the sharp rise in prices, but for Ricky Reyes, owner of the taqueria on Dorchester Avenue, costlier ingredients mean it is getting harder to keep the price of his signature beef taco down. [R177]

Use Table 6.34 to answer the following questions. We've entered the data in the spreadsheet `tacos.xlsx`.

(a) Is the *Globe* correct about the percent increase in the cost of beef?

Table 6.34. Taco costs

Ingredient	2013 cost per pound	2014 cost per pound	percent increase in cost	percent of taco filling
Beef	$3.46	$4.16		45
Tomato	$1.73	$2.19		20
Cheese	$5.39	$5.44		20
Lettuce	$0.99	$1.11		15

(b) Fill in the column showing the percent increase in cost of each of the ingredients.
(c) Find the cost of a pound of taco filling in 2013 and 2014. Then find the percent increase in the cost of the filling.
(d) One way for Mr. Reyes to reduce the cost increase would be to change the percentages of meat and cheese, keeping the lettuce and tomato the same. What would the percent of each be if he wanted to keep the increase in a pound of filling to just 10%?
(e) Do you think customers would notice if Mr. Reyes changed the recipe using your answer to (d)?

[See the back of the book for a hint.]

Exercise 6.14.34. [S] State populations.

Figure 6.35 shows the U.S. population distribution among the 50 states based on the 2010 U.S Census.

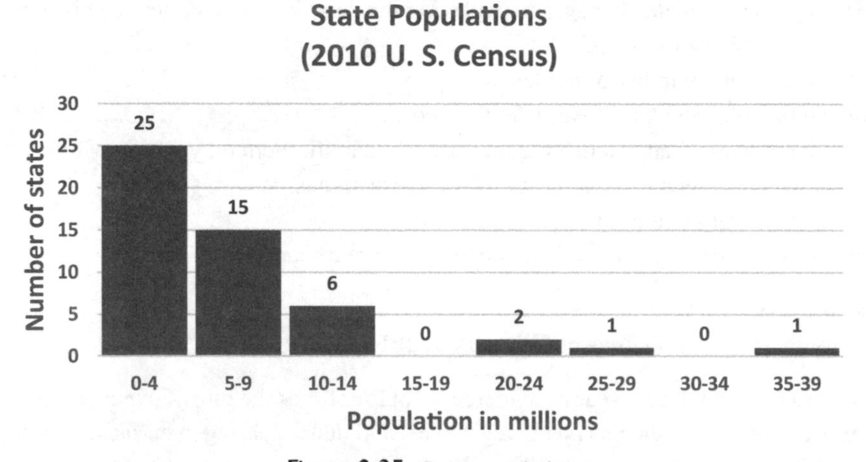

Figure 6.35. State populations

(a) Recreate these data in Excel as both a table and a histogram.
(b) What is the modal population of states?
(c) What is the median population of states?
(d) What is the mean population of states? How did you find it?
(e) How good an estimate is the mean you calculated in (d)?

Review exercises

Exercise 6.14.35. [A] [R][Section 6.1][Goal 6.1][Goal 6.4]

Create an Excel spreadsheet and put the following numbers in the first column.

14 15 22 50 0 33 16 18 23 40 47

(a) Use Excel to find the mean, median and mode of these numbers

(b) Change the first number from 14 to 23. How do the Excel calculations change?

(c) Click the "undo" button and confirm that Excel reverts back to the original set of numbers.

(d) Change the last four numbers to 0 (so that the data now read

14 15 22 50 0 33 16 0 0 0 0

How do the different averages change? Explain how the data are skewed.

7

Electricity Bills and Income Taxes—Linear Functions

We use an electricity bill as a hook on which to hang an introduction to functions in general and linear functions in particular, in algebra and in Excel. Then we apply what we've learned to study taxes—sales, income and Social Security. You'll also find here a general discussion of energy and power.

Chapter goals

Goal 7.1. Understand direct proportion as a linear equation with intercept 0.

Goal 7.2. Study situations governed by linear equations.

Goal 7.3. Stress the meaning and units of the slope and intercept.

Goal 7.4. View functions as tables, graphs and formulas.

Goal 7.5. Construct flexible spreadsheets to model linear equations.

Goal 7.6. Understand the piecewise linear income tax computations.

Goal 7.7. Sort out the confusing distinction between energy and power.

7.1 Rates

We began Chapter 2 with a discussion of the relationship

$$\text{distance} = \text{rate} \times \text{time},$$

or, with units,

$$\text{miles traveled} = (\text{rate, in miles/hour}) \times (\text{hours on the road}). \tag{7.1}$$

There we worked with particular numbers; now we want to look at that relationship a little more generally. If the rate is fixed (say 60 miles/hour), then we can find the distance traveled whenever we know the time. To say that with some algebra, write D for distance and T for time. Then

$$D = 60 \times T,$$

or, more generally,

$$D = r \times T,$$

where r is the rate of travel, in miles/hour.

That formula says that distance traveled is *proportional* to travel time. The rate in miles per hour is the *proportionality constant*. If you drive for twice as long you go twice as far. If you drive ten times as long you go ten times as far.

In Section 2.2 we introduced the units (gallons per hundred miles) as a useful way to measure automobile fuel economy: the amount of gas you use is proportional to the distance you drive. Let F be the amount of gasoline you use, in gallons, to drive D hundred miles. Then

$$F = r \times D$$

where the proportionality constant r is the fuel use rate, with units gallons per hundred miles. If you drive 10 times as far you use 10 times as much gas. You don't use any gas at all just sitting in the driveway (unless you're idling to warm up the car).

The proportionality constant is always a rate: it appears with units. In these examples the units are miles per hour and gallons per 100 miles.

The unit pricing discussion in Section 2.4 provides more examples of proportionality.

Finally, sales tax is computed as a proportion. If you spend twice as much you pay twice as much tax. The tax rate is the proportionality constant. If it's 5% then you pay (five dollars of tax) per (hundred dollars of purchase).

7.2 Reading your electricity bill

The more electricity you use at home, the more you pay. But the relationship isn't quite proportional. You don't pay twice as much to use twice as much. Figure 7.1 shows a simple sample electricity bill from `www.squashedfrogs.co.uk`; there's a local copy at `TamworthElectricBill.docx`. [R178]

This bill explains itself. We'll study it before we look at a real one. It comes from England, so the costs are expressed in pounds and pence rather than dollars and cents, and it comes once a quarter (every three months) rather than once a month, but you can ignore that while you read it—from the bottom up.

The last line is the total bill, computed as

`Cost of electricity + fixed charge` .

Checking the arithmetic:

$$£34.62 + £9.49. = £44.11.$$

The third line from the bottom explains the £34.62:

`Number of units × cost per unit` .

That's our old friend proportionality. The previous line gives the proportionality constant: the cost per unit as 7.35 pence per unit. Later on in the document we're told that a unit is just a kilowatt-hour, abbreviated "kwh". (It's too bad the bill talks about "units" instead of just "kwh" since for us "units" has a more general meaning.)

Electricity Bills

How to read your electricity bill

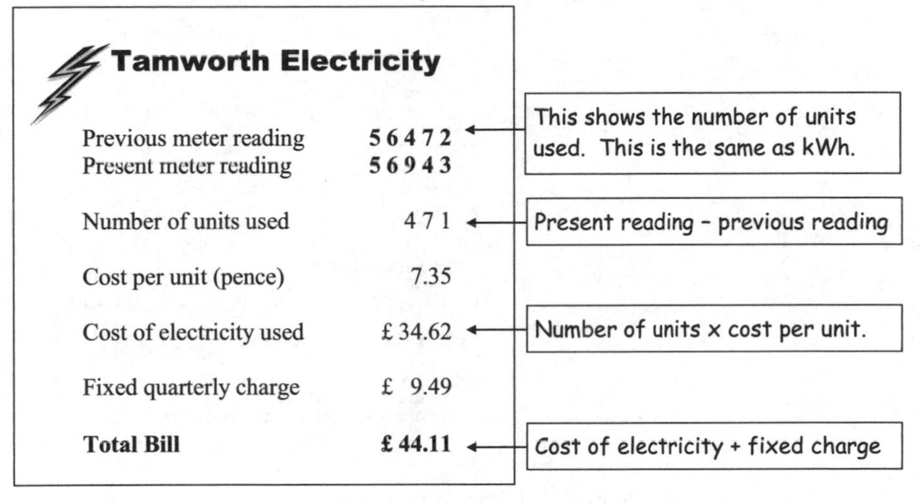

Tamworth Electricity

Previous meter reading	**56472**
Present meter reading	**56943**
Number of units used	471
Cost per unit (pence)	7.35
Cost of electricity used	£ 34.62
Fixed quarterly charge	£ 9.49
Total Bill	**£ 44.11**

Figure 7.1. Tamworth electricity bill

There are 100 pence in a pound, so the cost per unit is

$$0.0735 \frac{£}{\text{kwh}}.$$

(The computation would have been much more complicated before February 15, 1971—the day England converted from pounds/shillings/pence to decimal currency. See `news.bbc.co.uk/onthisday/hi/dates/stories/february/15/newsid_2543000/2543665.stm`.)

In the current quarter this customer used 471 kwh of electricity—the difference between the meter reading before and after the quarter.

Here's all the arithmetic, with units:

$$£44.11 = 0.0735 \frac{£}{\cancel{\text{kwh}}} \times 471\,\cancel{\text{kwh}} + £9.49.$$

That English bill is easy to read. Figure 7.2 shows a real one that's a little more complex, from NStar, in Boston.

We can identify the same two components. The fixed charge is the \$6.43 labelled "Customer Charge." It's the part of the \$145.26 total that does not depend on the amount of electricity used—in this case, 813 kwh. The six lines on the bill that do depend on that contribute

$$\begin{aligned}&(0.04432 + 0.01039 + 0.00468\\&\quad + 0.00050 + 0.00250 + 0.10838) \times 813\\&= 0.17077 \times 813\\&= 138.83601\end{aligned}$$

Service Provided to

E D BOLKER
10 CHESTER ST
NEWTON HLD MA 02461

Account Summary	
Previous Bill	143.04
Payment - Thank You	-143.04
Total Cost Electricity	145.26
Amount Due	$145.26

Electricity Used

Rate A1-Residential Non-Heating
Meter 1764836

Nov 16, 2007 Actual Read	33289
Oct 18, 2007 Actual Read	- 32476
29 Day Billed Use	813

1764836	KWH
11/16	813
10/18	800
09/18	814
08/16	855
07/18	1037
06/19	738
05/18	491
04/20	835
03/20	847
02/20	1358
01/16	989
12/18	1118
11/15	939

Cost of Electricity

Delivery Services			
Customer Charge			6.43
Distribution	.04432 X	813 KWH	36.03
Transition *	.01039 X	813 KWH	8.45
Transmission	.00468 X	813 KWH	3.80
Renewable Energy	.00050 X	813 KWH	0.41
Energy Conservation	.00250 X	813 KWH	2.03
Delivery Services Total			57.15
Supplier Services			
Generation Charge			
Basic Svc Fixed	.10838 X	813 KWH	88.11
Total Cost of Electricity			**145.26**

*PART OF WHAT WE COLLECT IN THE TRANSITION CHARGE IS OWNED BY EACH OF BEC FUNDING LLC AND BEC FUNDING II LLC

Figure 7.2. NStar electricity bill (2007)

to the total bill, which is

$$\$145.26 = 0.17077 \frac{\$}{\cancel{\text{kwh}}} \times 813 \, \cancel{\text{kwh}} + \$6.43.$$

Note that NStar rounded $138.83601 down to $138.83 rather than up to the nearest penny. We should be grateful for small favors.

7.3 Linear functions

So far *Common Sense Mathematics* has called for hardly any algebra. Now a little bit will come in handy.

Suppose you buy your electricity from NStar as in the example above and want to study how your bill changes when you use different amounts of electricity. The monthly $6.43 Customer Charge does not change. The rest of your bill is proportional to the amount of electricity. The proportionality constant is 0.17077 $/kwh in the sample bill. We will assume that it does not change, although in fact it does change slightly when the electric company changes its rates.

If in a given month you use E kwh of electricity your total bill B can be computed with the formula

$$B\$ = 0.17077 \frac{\$}{\cancel{\text{kwh}}} \times E \, \cancel{\text{kwh}} + \$6.43.$$

That formula captures how the dollar amount of your electricity bill depends on the amount of electricity you use, measured in kwh. The first term is the part that's proportional to the amount of electricity used. The second term (the amount $6.43) is fixed. It represents the electric company's fixed costs: things like generating the bill and mailing it to you and maintaining the power lines on the street in front of your house. Those are expenses they must cover even if you're on vacation and have turned off all the appliances.

You probably encountered a similar formula once in an algebra class—it may look more familiar without the units

$$B = 0.17077 \times E + 6.43.$$

It may look even more familiar if we call the variables by the traditional names x (for the independent variable) and y (for the dependent) instead of E and B:

$$y = 0.17077x + 6.43.$$

This is a *linear function*, which standard algebra texts write in *slope-intercept* form

$$y = mx + b.$$

In this example the *slope m* is $0.17077 \frac{\$}{\text{kwh}}$ and the *intercept b* is \$6.43. For the English bill the slope is $0.0735 \frac{£}{\text{kwh}}$ and the intercept is £9.49.

There are many everyday examples where a linear equation describes how a total is computed by adding a fixed amount to a varying part that's a proportion. The fixed amount is the intercept. The proportionality constant is the slope.

- The most familiar examples are the ones where the intercept is 0: all the ones in Section 7.1.
- When renting a truck, the amount you pay is

  $$(\text{rate in dollars/mile}) \times (\text{miles driven}) + (\text{fixed charge}).$$

- Your monthly cell phone bill might be

  $$(\text{rate in dollars per minute}) \times (\text{number of minutes}) + (\text{fixed fee}).$$

  (A real cell phone bill will probably be more complicated, perhaps with separate charges for phone minutes, text messages and data transfer, perhaps with some of each kind of use built in to the fixed fee.)
- If you work as a salesperson and your commission is 15% of total sales your total wages are

  $$0.15 \times (\text{total sales}) + (\text{your base salary}).$$

The pattern is

$$\text{total} = (\text{rate}) \times (\text{amount of some quantity}) + (\text{fixed constant}).$$

In each case the slope is the rate and the intercept is the fixed constant. The units of a slope are always those of a rate. In the truck example, the slope is the rate in dollars per mile; in the cell phone example, the units of the slope are dollars per minute; in the salesperson example, the slope is 0.15 dollars of commission per dollar of total sales.

Think of the intercept as an initial or starting value; it's what happens when the input is zero. It has proper units too—in each of these examples that unit is dollars. If you rent a truck but don't drive it anywhere, you still pay the fixed charge. If you make no cell phone calls you still pay the fixed fee. If you don't sell anything in a month, your commission is \$0 but you still earn your base salary.

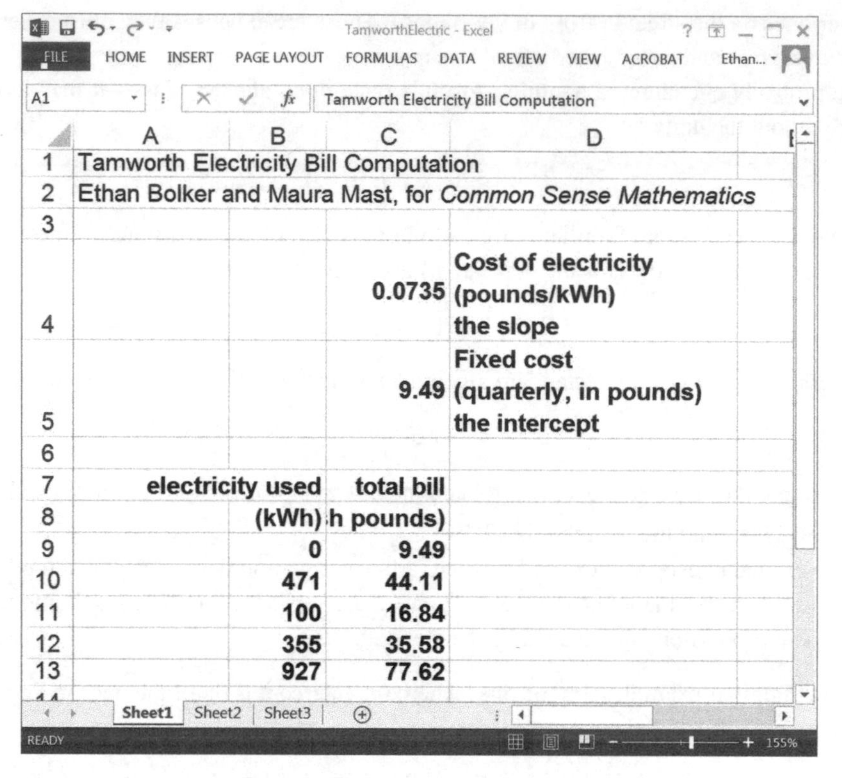

	A	B	C	D
1	Tamworth Electricity Bill Computation			
2	Ethan Bolker and Maura Mast, for *Common Sense Mathematics*			
3				
4			0.0735	Cost of electricity (pounds/kWh) the slope
5			9.49	Fixed cost (quarterly, in pounds) the intercept
6				
7		electricity used	total bill	
8		(kWh)	h pounds)	
9		0	9.49	
10		471	44.11	
11		100	16.84	
12		355	35.58	
13		927	77.62	

Figure 7.3. Tamworth electricity bill

7.4 Linear functions in Excel

In Section 7.2 we saw that the amount you pay for electricity in a month is a linear function of the amount you use. In this section we'll use Excel to calculate electricity bills and to draw a picture of the results.

Figure 7.3 is a screen shot of the Excel spreadsheet `TamworthElectric.xlsx`. We put the slope 0.735 in cell `C4`, with its units £/kwh in cell `D4`. We put the intercept 9.49 in cell `C5` and the units (£) in cell `D5`.

Then we entered column labels in cells `B7:C8` and a few values in rows `9` through `13` in column `B`. Finally, we asked Excel to calculate the electricity bills in column `C`. To do that, we started with the formula

```
=C4*B9+C5
```

in cell `C9`. The = sign tells Excel to multiply the numbers in cells `C4` and `B9` and add the number in `C5`. The result is `9.49`, as expected.

The next step is to copy that formula from cell `C9` to cell `C10`. With any luck Excel will guess what we want to do, change `B9` to `B10`, compute

```
=C4*B10+C5
```

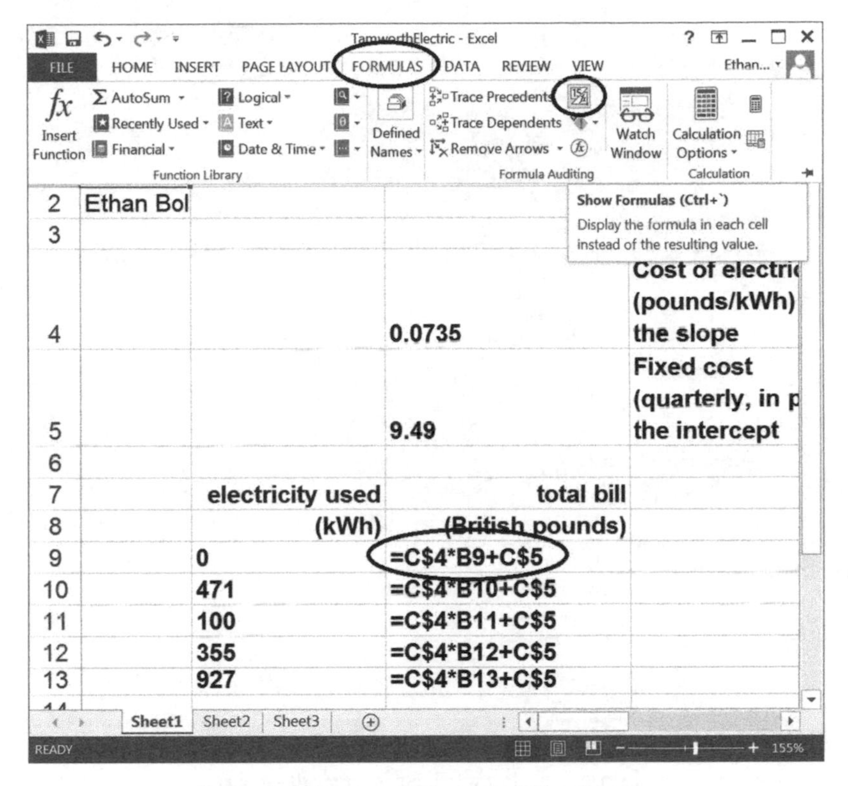

Figure 7.4. Tamworth electricity bill—formulas

and display `44.11`. But that's not what happens! Excel shows `4469.79` instead! If you look at the contents of cell `C10` you will find

```
=C5*B10+C6
```

so Excel added 1 to the row numbers for cells `C4` and `C5` as well as for `B9`. There's nothing in cell `C6`. Excel treats that as a zero and adds it to 471 × 9.49 to get 4469.79.

That's not what we want. Changing `B9` to `B10` is right, but we want Excel to leave the references to cells `C5` and `C6` alone. The trick that makes that happen is to put a `$` in front of the `5` and the `6`. This is not something you could have figured out. There's no particular reason why this trick should work. Just remember it. The right formula to use in

cell `C9` is

```
=C$4*B9+C$5 .
```

When we copy that formula from `C9` to cells `C10:C13` we get Figure 7.3.

Figure 7.4 is a screen shot of the same spreadsheet—after we asked Excel to show the formulas for each cell instead of the values.

In Chapter 6 we learned how to use Excel to draw bar charts and histograms so that we could visualize data organized into categories. The x-axis displayed category names, with corresponding values on the y-axis. That won't work for the data in Figure 7.3, since there both the

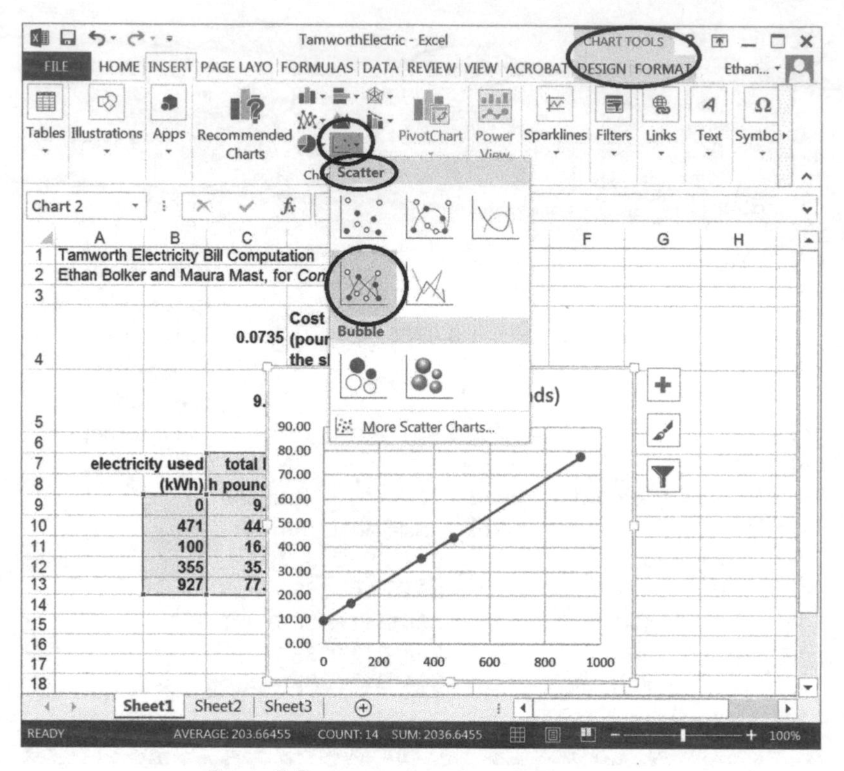

Figure 7.5. Tamworth electricity bill—chart

x- and y-axes have numerical values. Instead, after selecting cells `B9:C13` we must ask Excel for a chart of type `XY (Scatter)`. Figure 7.5 shows the result.

The graph is a straight line—that's why the function is called "linear". The slope tells us how steep the line is and the intercept tells us where it crosses the vertical axis—in this case at the value £9.49, the total bill when you use no electricity at all.

Excel will let you change the type of a chart once it's built. If you change the chart in Figure 7.5 to a `Line Chart` Excel will use the data in the column `B` as category labels rather than as the numbers of kilowatt-hours. It will space them evenly along the x-axis, whatever their values, and draw the nonsense you see in Figure 7.6.

If you change the chart type to scatter you get two scatters, one for each column. You can get Figure 7.5 only if you start with a scatter plot. If you select the two columns of data and build a line chart first, things are even worse. Excel thinks each row is a category for which you have two pieces of data. It labels the categories 1, 2, ... and shows a line for each. Try this and see what happens.

7.5 Which truck to rent?

Table 7.7 shows the cost of renting a truck for one day in Boston in March of 2015. Four companies offer equivalent models. Which one should you choose?

It's clear from the numbers that for a very few miles Budget will be cheapest, since it's tied with U-Haul for the lowest fixed cost and charges less per mile. For a really long move Watertown will be best since there is no mileage charge. The Excel chart in Figure 7.8 tells the

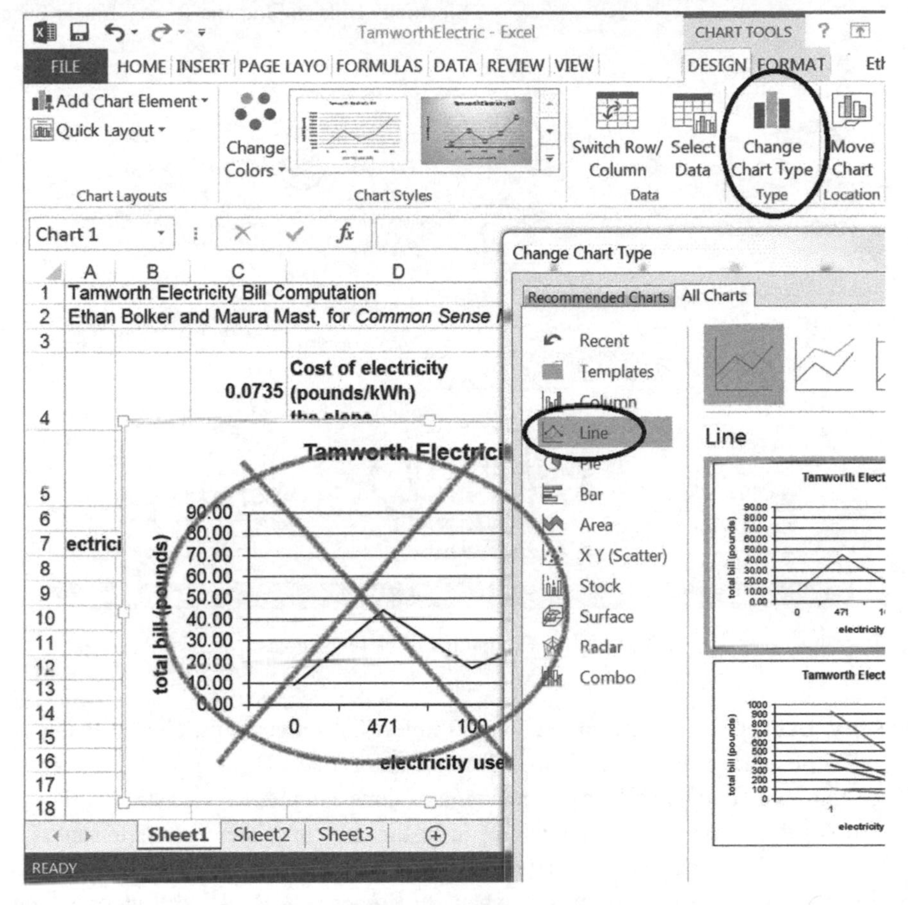

Figure 7.6. How NOT to draw a line

Table 7.7. One day truck rental costs

	Watertown	U-Haul	Budget	Enterprise
fixed cost	79	29.95	29.95	59.99
$/mile	0	1.39	0.99	0.59

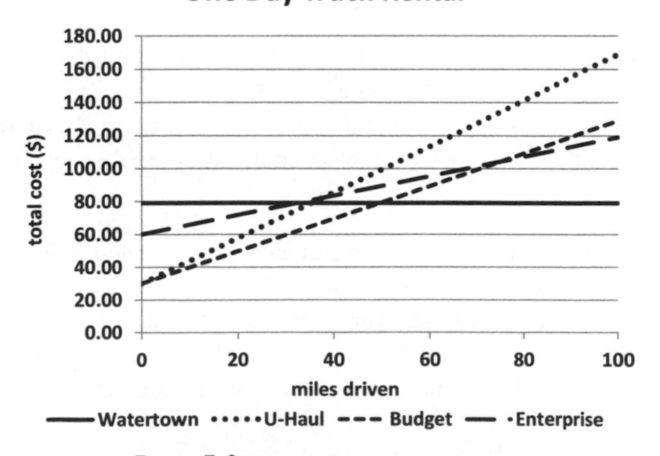

Figure 7.8. Comparing truck rental costs

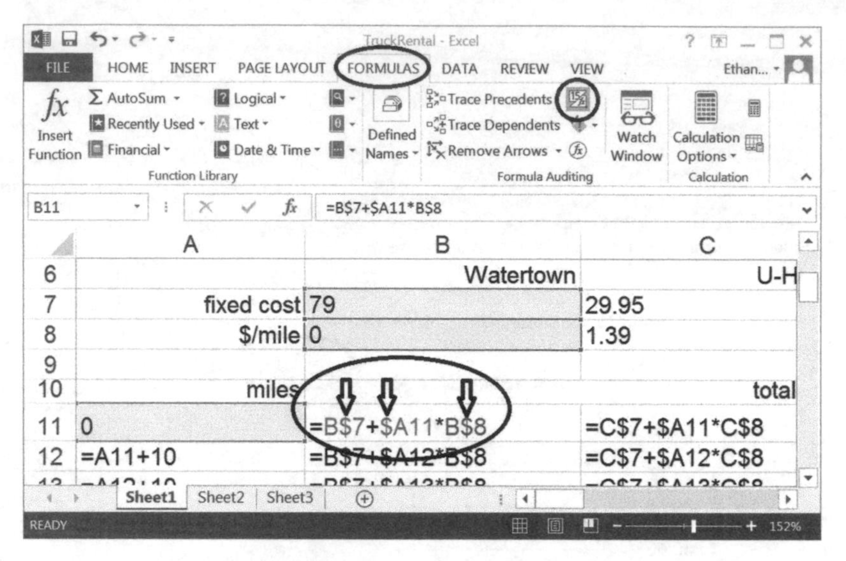

Figure 7.9. Computing truck rental costs

whole story. Budget is cheapest up to about 50 miles. For longer trips, choose Watertown. It never makes financial sense to rent from U-Haul or Enterprise.

The figure is a good reminder of the meaning of slope and intercept for straight line graphs. A line crosses the vertical axis at the intercept. In this example intercepts represent the fixed costs. The units of the intercept are dollars—the units on the y-axis. The slope of a line measures how steep it is. That's particularly visible when you compare U-Haul and Budget, which have the same intercept but different slopes—1.39 dollars/mile and 0.99 dollars/mile. The units of the slope are always (units on y-axis)/(units on x-axis), The line for Watertown is horizontal since its slope is 0 dollars/mile.

The spreadsheet with that figure is at `TruckRental.xlsx`. Figure 7.9 shows how we arranged the formulas in the spreadsheet to compute the total cost for each company. Column `A` lists the mileages we're considering, from 0 in cell `A11` to 100. The formula `=A11+10` in cell `A12` fills column `A` when we copy it to cells `A13:A21`. The formula in cell `B11` is

```
=B$7+$A11*B$8 .
```

It uses three `$` signs to keep Excel from adjusting references for rows `7` and `8` and for column `A`. That allowed us to copy it to all of the range `B11:E21`.

The problem we've just solved is typical of situations where you have to decide among options, some with small startup cost but a high ongoing rate, others the reverse. Here are some examples:

- Insulate your house (high initial investment compared to doing nothing) in order to pay less for heat in the winter (lower rate for use).
- Buy a hybrid instead of a conventional car (higher initial cost, lower rate of fuel consumption).
- Buy energy efficient light bulbs (more expensive to start with, but they use less electricity to run),

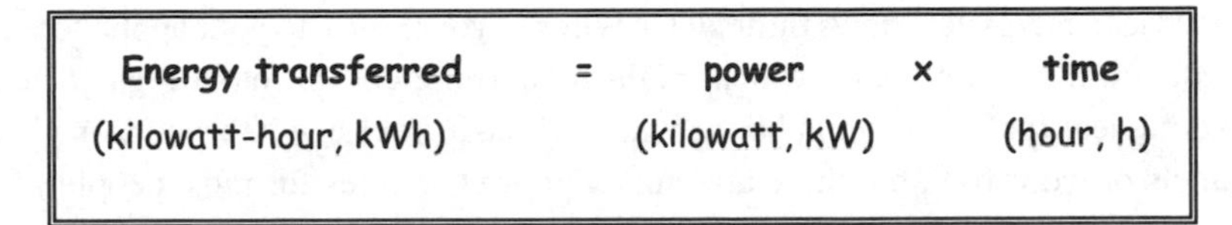

Figure 7.10. How much electrical energy?

- Select a phone plan with unlimited text messaging (more expensive than pay-as-you-text, but the slope ($ per text) is zero).

Each of these can be thought of as looking at linear equations to see where their graphs cross. You can do that by building a table of values or by drawing the graphs (in Excel, or with pencil and paper) or by writing down the equations and solving them with remembered algebra, or by guess-and-check.

But you can't rely on just this mathematics to make a decision. There are always other important things to think about. How long does the more expensive purchase last? Can you afford the initial high payment? If so, what else might you rather do with that money? Do you need to take depreciation or inflation into account?

7.6 Energy and power

How much electrical energy does a 100 watt light bulb use? That depends on how long it's on. When it's switched off, it doesn't use any at all. If it's on for two hours it must use twice as much as it does in one hour. Figure 7.10 (from the second page of the Tamworth bill) shows the proportion lurking there.

The second line in that figure displays the units for the quantities in the first line. Time is measured in hours, of course. Power is measured in kilowatts. Energy is measured in kilowatt-hours: the product of the units for power and for time.

That tells us right away that energy and power are not the same thing. Comparing the figure to Equation 7.1, you can see that energy is like distance—a thing that's consumed or traveled, while power is like speed—the rate at which energy is used or distance covered.

If you turn a 100 watt light bulb on for 2 hours the formula in Figure 7.10 tells you how much electrical energy you use:

$$100 \text{ watts} \times 2 \text{ hours} = 200 \text{ watt-hours} \tag{7.2}$$

$$= 0.2 \text{ kilowatt-hours} \tag{7.3}$$

(7.2) is just multiplication. (7.3) changes watt-hours to kilowatt-hours.

How much does it cost to leave a 40 watt bulb on all the time in your basement for a year? There are about 9,000 hours in a year. That is 9 kilo-hours, so you'll use about

$$40 \text{ watts} \times 9 \text{ kilo-hours} = 360 \text{ kilowatt-hours.}$$

If you pay $0.10 per kwh for electricity it will cost you about $36 per year to guarantee that you don't fall down the basement steps in the dark.

The electrical energy that flows through the wires in your house to your appliances probably comes to you from a power plant, which might be burning coal or natural gas or extracting energy from nuclear fuel. (You might have a wind turbine in your neighborhood, or a hot spring, or solar panels on your roof, but these are unlikely power sources for most people.) So power plants produce energy, not power. The power of a power plant is the rate at which it can produce energy, so "energy plant" would be a better name than "power plant".

The website for Chicago's Cook Nuclear Plant says that

> The 1,048 net megawatt (MW) Unit 1 and 1,107 net MW Unit 2 combined produce enough electricity for more than one and one half million average homes. [R179]

Let's check this. The combined total power is 2,155 megawatts. This is the rate at which that plant produces electrical energy when it is running at full power. (When it's not running it's still just as powerful, but not producing any energy.) When it's running, how many average homes could it produce electricity for?

If the Cook plant ran all year (about 9,000 hours) it would produce

$$\begin{aligned} 2,155\,\text{megawatts} \times 9{,}000\,\text{hours} &\approx 18{,}000{,}000\,\text{megawatt} - \text{hours} \\ &= 18{,}000{,}000{,}000\,\text{kilowatt} - \text{hours} \end{aligned}$$

of electrical energy. Googling "average household electricity usage" finds

> 6,000 kwh per household per year for 3 residents average per household

from `www.physics.uci.edu/~silverma/actions/HouseholdEnergy.html`. The source is a physics professor's website, so it's probably reliable. At 6,000 kwh per household per year Cook could power 3 million homes. The quotation claims half that, so it's clearly in the right ballpark. The 6,000 kwh per household per year is a southern California average—households in northern Illinois might well use more electricity.

Energy comes in many forms besides electric. The Cook plant converts the energy in its nuclear fuel to electricity. Driving a car uses the energy stored in the gasoline. Running a marathon uses the energy in the food you eat. Each form of energy has its own units. We've seen that electrical energy is measured in kilowatt-hours. If you cook on a gas stove, the energy in the gas is measured in *therms*. The energy in the oil that heats your house is measured in *British Thermal Units* or Btus. The energy in the food you eat is measured in *calories*. Physicists measure energy in *ergs* or *joules*; you rarely see those units in everyday life. You can look up conversion factors for these units—for example, the energy in a barrel (42 gallons) of oil is about 5.8 million Btu, which is equivalent to 1700 kilowatt-hours. So it would take about a fifth of a barrel to keep that 40 watt light bulb burning for a year.

Converting among the units for energy is just like converting among the units for length (meters, feet, yard, miles, ...). You can use a table, an online calculator like the one at the National Institute of Standards and Technology (`physics.nist.gov/cuu/Constants/energy.html`) or the Google calculator.

Possibly the most interesting energy conversion is the one that Einstein discovered in 1905: mass and energy are the same thing, measured in different units. The conversion factor is the square of the speed of light—hence the famous equation

$$e = mc^2.$$

To see that at work, look again at the yearly energy output of the Cook plant. The Google calculator tells us that

$$\boxed{18\,000\,000\,000 \text{ kilowatt-hours} = 6.48 \times 10^{16} \text{ joules}}\ .$$

The National Institute of Standards and Technology website says that corresponds to a mass of about 0.72 kg, which is 720 grams. That means just about 1.6 pounds of matter must be converted to energy to power millions of Chicago homes for a year. The Google calculator does not know Einstein's equation, so it wouldn't convert kwh to grams directly!

7.7 Taxes: sales, Social Security, income

Taxes are a part of life (the only other certainty is death), so it's only common sense to learn how they work. In Section 3.6 and Section 7.1 we studied sales taxes. They are collected by cities and states and are computed as a percentage of the purchase price. In this section we'll explain two important federal taxes that depend on your income, not on how you spend it.

The first of these is the FICA tax. "FICA" is the acronym for the "Federal Insurance Contributions Act". Those taxes pay for Social Security and Medicare. In 2014 the tax rate was 6.2% for Social Security and 1.45% for Medicare. The Social Security tax is collected only on the first \$117,000 of your earnings. Up to that income level the combined rate is 7.65%. (The actual rules are a little more complicated. First, the tax applies only to wages. Other income (like stock dividends or interest) are not subject to this tax. Second, the real rates are twice the quoted amounts, but your employer is required to pay half. If you're self-employed you pay it all.)

Here are some sample computations.

- If you earn \$1,000 your FICA tax is $0.0765 \times \$1,000 = \$76.50$.
- If you earn \$50,000 your FICA tax is $0.0765 \times \$50,000 = \$3,825$.
- If you earn \$117,000 your FICA tax is $0.0745 \times \$117,000 = \$8,950.50$.
- If you earn \$500,000 your FICA tax is

$$\text{Social Security} + \text{Medicare} = 0.062 \times \$117,000 + 0.0145 \times \$500,000 = \$14,504.$$

Once you make more than \$117,000 the amount of Social Security tax remains constant; the Medicare tax continues at the rate of 1.45%. That means the percentage of your earnings collected for FICA taxes decreases as your earnings increase. That percentage is called the *effective tax rate*. Up to \$117,000 the effective rate is 7.45%. For \$500,000 the effective rate is just $\$14,504/\$500,000 = 2.9\%$. For higher incomes, the effective rate is even smaller. Because they decreases as earnings increase, Social Security taxes are *regressive*. Once you reach the Social Security maximum your effective tax rate decreases. So the more you earn, the smaller a percentage you pay.

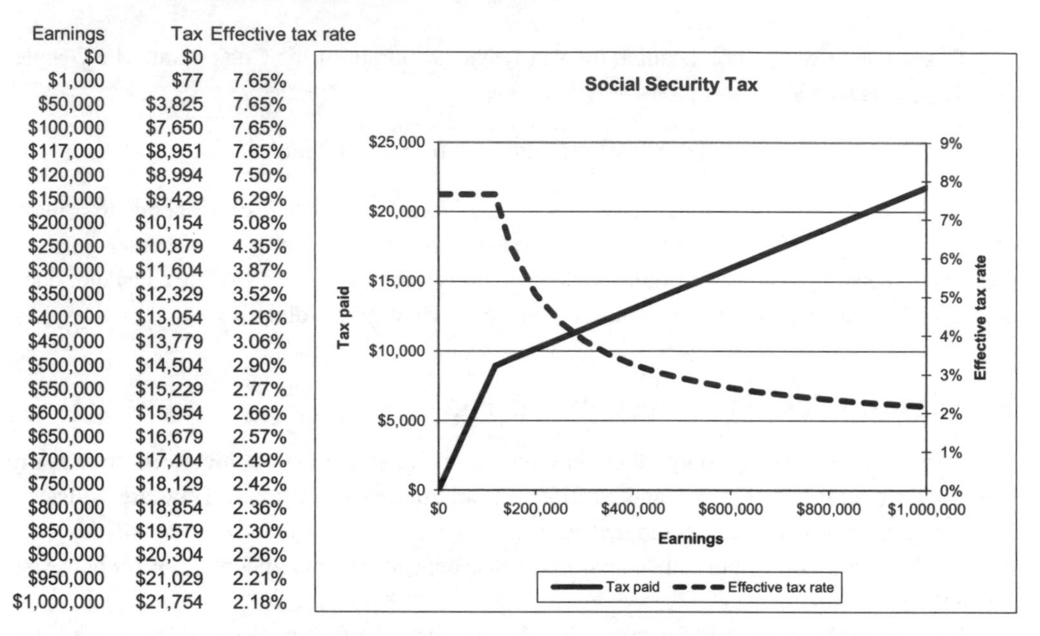

0.0620 Social security tax rate
0.0145 Medicare tax rate
$117,000 Maximum social security taxed earnings

Earnings	Tax	Effective tax rate
$0	$0	
$1,000	$77	7.65%
$50,000	$3,825	7.65%
$100,000	$7,650	7.65%
$117,000	$8,951	7.65%
$120,000	$8,994	7.50%
$150,000	$9,429	6.29%
$200,000	$10,154	5.08%
$250,000	$10,879	4.35%
$300,000	$11,604	3.87%
$350,000	$12,329	3.52%
$400,000	$13,054	3.26%
$450,000	$13,779	3.06%
$500,000	$14,504	2.90%
$550,000	$15,229	2.77%
$600,000	$15,954	2.66%
$650,000	$16,679	2.57%
$700,000	$17,404	2.49%
$750,000	$18,129	2.42%
$800,000	$18,854	2.36%
$850,000	$19,579	2.30%
$900,000	$20,304	2.26%
$950,000	$21,029	2.21%
$1,000,000	$21,754	2.18%

Figure 7.11. FICA (Social Security and Medicare) tax

Figure 7.11 shows part of the spreadsheet `SocialSecurityTax.xlsx`: a data table and a chart illustrating Social Security tax computations.

Income tax is a little more complicated. It's a *progressive graduated tax*. When you make more money you not only pay more tax, some of your income may be taxed at a higher rate. Table 7.12 shows the 2014 *tax brackets* for single taxpayers.

That table tells you that the first $9,075 of your income is taxed at 10%. If you make exactly that much, you pay $907.50 in tax. If you make more, the extra income is taxed at a higher rate—you have moved to a higher tax bracket. For example, if you make between $9,075 and $36,900 you will pay 15% of the amount you earn over $9,075. If you earn more than $36,900 you start paying at a 25% rate on the extra.

Table 7.12. 2014 single taxpayer brackets and rates

Bracket ($)	Marginal Tax Rate (%)
0–9,075	10
9,075–36,900	15
36,900–89,350	25
89,350–186,350	28
186,350–405,100	33
405,100–406,750	35
406,750+	39.6

Table 7.13. 2014 single taxpayer tax calculation [R180]

Taxable income		Tax		
more than	less than			
$0	$9,075			10% of taxable income
$9,075	$36,900	$908	+	15% of the excess over $9,075
$36,900	$89,350	$5,081	+	25% of the excess over $39,600
$89,350	$186,350	$18,194	+	28% of the excess over $89,350
$186,350	$405,100	$45,354	+	33% of the excess over $186,350
$405,100	$406,750	$117,541	+	35% of the excess over $405,100
$406,750		$118,189	+	39.6% of the excess over $406,750

Let's try an example. If you earn $50,000, then your total tax is

$$\begin{aligned} \text{total tax} &= 0.10 \times \$9{,}075 + 0.15 \times (\$36{,}900 - \$9{,}075) \\ &\quad + 0.25 \times (\$50{,}000 - \$36{,}900) \\ &= \$907.50 + \$4{,}173.75 + \$3{,}275.00 \\ &= \$8{,}356.25. \end{aligned} \tag{7.4}$$

Note carefully that when you are in a higher tax bracket the higher rate applies only to the extra income. The taxpayer in this example is in the 25% bracket, but that rate applies only to her earnings in that bracket. Table 7.13 explains this rule in another way.

The graphs in Figure 7.14 show that the dependence of tax on income is *piecewise linear*—built from pieces of straight lines that become steeper as income increases. You can find these graphs in the Excel spreadsheet `GraduatedTax.xlsx`.

Figure 7.15 shows the effective tax rate—the percentage of your income you pay in federal income tax. If you earn less than $9,075 you are in the lowest bracket and your effective tax rate

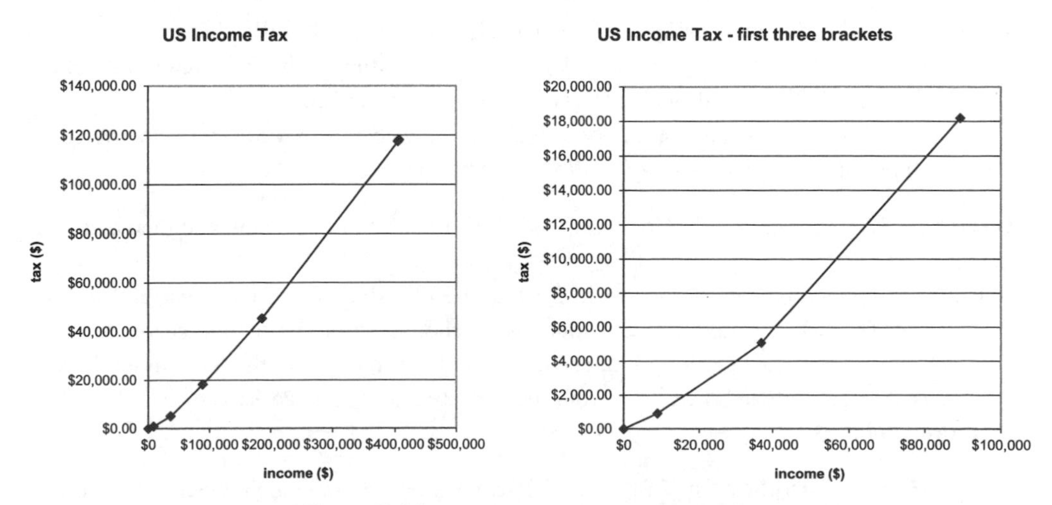

Figure 7.14. 2014 single taxpayer tax liability

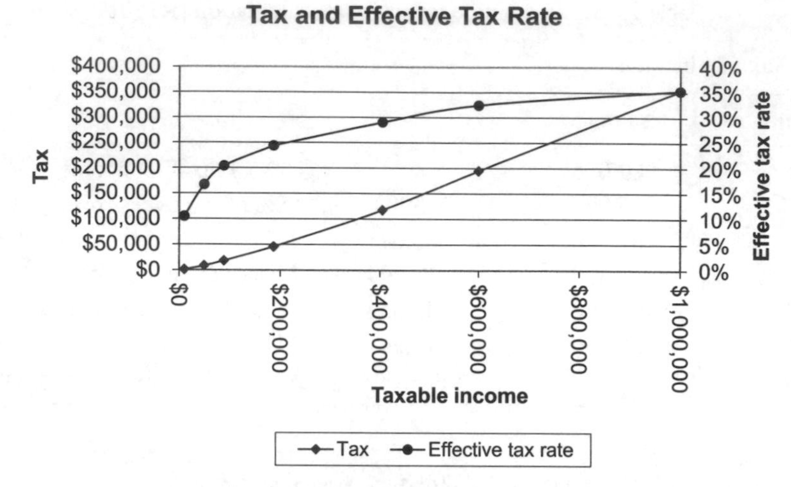

Figure 7.15. 2014 effective tax rates

is the rate in that bracket: 10%. If you earn $50,000 you are in the third bracket. You pay 10% of the first $9,075, 15% of the part in the second bracket, and 25% of the rest. The arithmetic in (7.4) shows that your total tax is $8,356.25. Your effective tax rate is $8,356.25/$50,000 = 16.71%. This is a weighted average of the three bracket rates 10%, 15% and 25%, with weights the amount of income taxed at each rate. Even if you make $500,000, which puts you in the top 39.6% bracket, your effective tax rate is still only about 31%, because you pay at a lower rate on the first part of your income.

In fact, the actual effective tax rate is lower than this for wealthier households because much of their income is taxed at a lower rate as capital gains rather than as ordinary income. Figure 7.16 shows the effective federal tax rate by household income for the year 2007.

7.8 Exercises

Exercise 7.8.1. [U][Section 7.2][Goal 7.2][Goal 7.3] Your electricity bill.

Verify the computations on your current electricity bill, either with a calculator or by modifying the Tamworth bill spreadsheet at `TamworthElectric.xlsx`.

If you don't have a current electricity bill, check the website for your local electric company, which probably provides a sample bill you can use instead.

Exercise 7.8.2. [U][Section 7.2][Goal 4.2][Goal 7.2] Electricity costs now and then and here and there.

Compare residential electricity cost in Boston in 2007 (the date of the NStar bill in Figure 7.2) to the cost where you live today. Your answer should take inflation into account.

If you have a current electricity bill, use it. If you don't, try to find the fixed monthly cost and the cost of electricity in $/kwh from your local electric company. Perhaps their website has that information.

Exercise 7.8.3. [R][S][Section 7.2][Goal 7.2][Goal 7.3] How much electricity does it use . . . ?

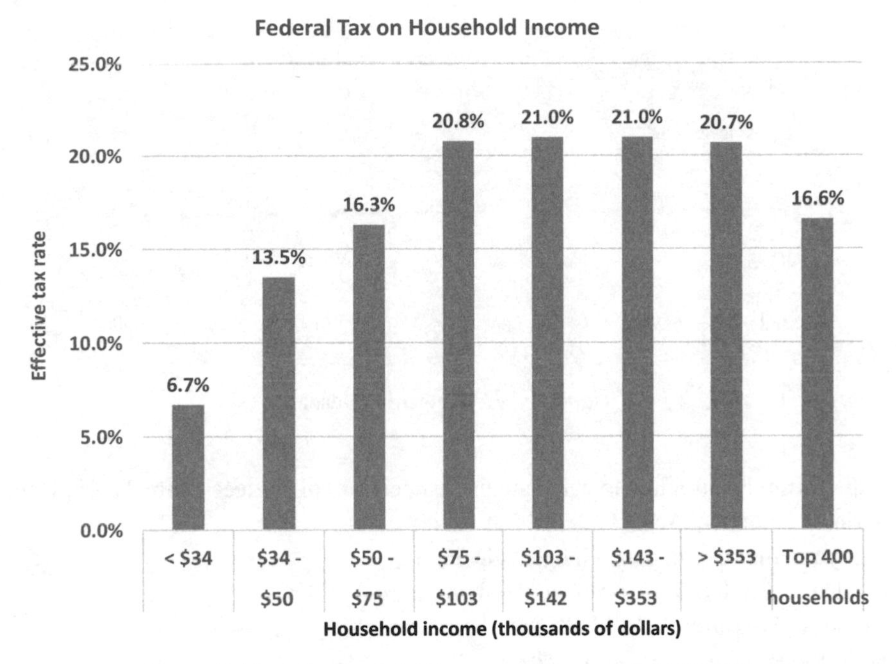

Figure 7.16. Effective federal tax rate by income, 2007 [R181]

The document `how_to_read_electricity_bills.doc` containing the Tamworth bill has several questions about how much electricity various appliances use. Answer them.

Exercise 7.8.4. [S][Section 7.2][Goal 7.1][Goal 7.7] Direct current.

In an article from *The New York Times* on November 17, 2011 headlined "From Edison's Trunk, Direct Current Gets Another Look" you can read that

> In a data center redesigned to use more direct current, monthly utility bills can be cut by 10 to 20 percent, according to Trent Waterhouse, vice president of marketing for power electronics at General Electric. Verizon Communications, a G.E. customer, expects to save 1 billion kilowatt-hours a year from a nationwide retrofit of its data centers, which translates to roughly enough to power 77,000 homes. [R182]

(a) How many watt-hours is a billion kilowatt-hours?
(b) About how much electricity is Verizon using now, given that they hope to save a billion kilowatt-hours per year?
(c) Verify the estimate that those billion kwh are "roughly enough to power 77,000 homes". (You won't find anything useful in the original article. How else will you search?)

Exercise 7.8.5. [S][Section 7.3][Goal 7.2] How hot was it?

Figure 7.17 shows a weather forecast for Hamilton, Ontario, Canada. The temperature there is displayed in degrees Celsius, marked °C.

(a) Look up the linear relationship for converting temperatures measured on the Celsius scale to temperatures on the Fahrenheit scale.

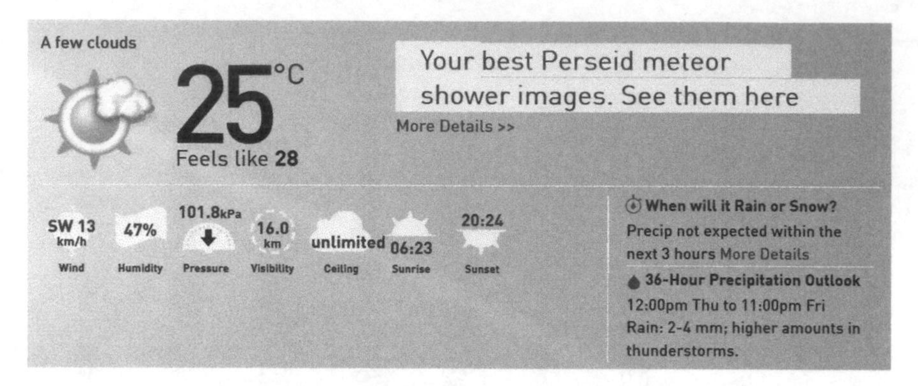

Figure 7.17. Weather in Canada

(b) Use the formula you found to calculate the temperature in degrees Fahrenheit in Hamilton, Ontario on Thursday August 13, 2015 at 12:45.
(c) Check your answer using the Google calculator.
(d) What does "Wind: SW 13 km/h" in the figure mean?
(e) What does "Pressure: 101.8 kPa" in the figure mean?
(f) What is the Perseid meteor shower?

Exercise 7.8.6. [S][Section 7.3][Goal 7.2][Goal 7.3][Goal 7.4] The Jollity Building.

In Exercise 4.7.11 there's an implicit linear model for Mr. Ormont's weekly income. Write the linear equation for that model. Clearly identify the independent and dependent variables and the units for the slope and the intercept.

Exercise 7.8.7. [S][Section 7.3][Goal 7.2][Goal 7.3] Newton trees.

In an article in the April 2012 issue of the Newton Conservators newsletter you can read that

> In the early 1970s there were approximately 40,000 trees lining the streets of Newton. Today, that number is about 26,000—a 35% loss. The current annual rate of decline is about 650 trees per year. At this rate, if unchecked, public street trees would diminish to approximately 10,000 within a generation (25 years), and in 40 years, public street trees would no longer be part of the Newton landscape. [R183]

(a) Check the arithmetic that leads to the claimed "35% loss".
(b) Check the arithmetic that leads to a "current annual rate of 650 trees per year".
(c) Check the predictions in the last sentence. Are they likely to come to pass?
(d) Write the equation for the linear model implicit in this quotation (use years since 2012 as the independent variable). Identify the slope and the intercept, with their units.

Exercise 7.8.8. [R][S][A][Section 7.4][Goal 7.2][Goal 7.3][Goal 7.5] Apple time.

Kuipers Family Farm, in Maple Park, IL, charged \$9 for admission to the apple orchard for the 2013 apple season. This includes a $\frac{1}{4}$ peck bag of apples and a hayride to the orchard. Visitors can pick additional peck bags of apples for \$15 each. (A bag of apples in the grocery store usually contains $\frac{1}{2}$ peck.)

(a) If you pick that additional peck of apples, how much do you pay?
(b) If you pick two additional pecks of apples, how much do you pay?
(c) If you just enjoy the free cider, your $\frac{1}{4}$ peck of apples and the sunshine, what do you pay?
(d) Write a linear function that shows how the total cost of the farm visit depends on the how many pecks of apples you pick. Identify the slope and intercept, with their units.
(e) Build an Excel spreadsheet to compute this function and use it to check the values you worked out by hand above. Include a chart in your spreadsheet.
(f) Use your spreadsheet to calculate your apple cost (in $/peck) when you pick no extra pecks, then when you pick 1, 2 or 10 extra pecks.

Exercise 7.8.9. [S] [A][Section 7.5][Goal 7.2][Goal 7.5] Comparing telephone calling plans.

A cell phone company has introduced a pay-as-you-go price structure, with three possibilities.

Plan 1	$10 a month	10 cents per minute
Plan 2	$15 a month	7.5 cents per minute
Plan 3	$30 a month	5 cents per minute

(a) For each plan, find a linear function that describes how the total cost for one month depends on the number of minutes used.
(b) Construct a table in Excel showing the total cost for one month for each of the three plans. Organize your data this way:
  - Create a sequence of cells in column A for the various possible numbers of minutes. Label that column. Start with 0 minutes. What's a good step to use? What's a reasonable place to stop?
  - Use columns `B`, `C` and `D` for each of the three plans. The fixed charge and charge per call should be in cells in those columns too, so you can use the same formula everywhere in the data table. (That will call for clever use of the `$` to keep Excel from changing row numbers and column letters when you don't want it to.)

(c) Use Excel to draw one chart showing how the monthly bill (y-axis) depends on the number of minutes you use the phone (x-axis) for all three plans.
(d) Write a paragraph explaining to your friend how she should go about choosing the plan that's best for her.

[See the back of the book for a hint.]

Exercise 7.8.10. [S][Section 7.5][Goal 7.2][Goal 7.5] Prepaid phones.

Until summer 2012, if you wanted an iPhone you needed to lock into a two-year contract. Then some mobile companies started selling the iPhone and letting you choose your own plan, with no contract. You can use the ideas from this chapter to make a quantitative comparison (we'll let you decide what other factors, such as paying a lot up front, matter in your decision).

Virgin Mobile began selling a 16GB iPhone 4S for $649.99, with no plan or contract. They offered a $55 per month "unlimited" data plan (in fact it was not unlimited, as once you go past 2.5GB of data they slowed the phone speed down considerably). If you purchased the phone through Sprint, it cost just $149.99. However, you had to sign up for a two-year contract. The least expensive option was $79.99 with what they called unlimited data.

(a) For each plan, find a linear function that shows how the total cost of the phone depends on the number of months you have it.
(b) Build a spreadsheet in Excel using your functions and fill in the cost of the two different options over several months.
(c) Graph the data in your spreadsheet.
(d) Write a short statement comparing the two plans. Clearly the Virgin Mobile plan is more expensive at first. When does it become the less expensive plan? If you had to choose one of the plans, which one would you choose and why? What other factors would you consider?

Exercise 7.8.11. [S][Section 7.5][Goal 7.3][Goal 7.5] Hybrid payback.

The "Best & Worst Cars 2011" issue of *Consumer Reports* provides the following data for new Toyota Camrys:

	conventional	hybrid
cost	$19,720	$26,575
fuel economy	26 MPG	34 MPG

Assume gasoline costs $3.50/gallon.

(a) Questions about the conventional Camry.
  (i) Once you own the car, how much does it cost to run, in dollars per mile? Does your answer make sense?
  (ii) Calculate the total cost (purchase plus gasoline) to drive the conventional Camry 10,000 miles.
  (iii) Write the linear equation that computes the total cost C of driving the conventional Camry M miles.
  (iv) Identify the slope and the intercept of this equation, with their units.
(b) Open the spreadsheet `ConventionalvsHybrid.xlsx`. Enter the numerical data from the table and the cost of gasoline in the appropriate cells. What formula should you enter in cell `C15` to check your answer to part (i)?
(c) Copy your formula to cells `B14:D29` to fill in the table. Where must you add $ signs to keep Excel from changing row and column references?

Create a properly formatted and labelled chart in Excel showing how the cost of driving each car depends on the number of miles driven. Use your graph along with the table to answer the following questions.
(d) If you drive 120,000 miles will you recover in gas savings the extra initial cost of the hybrid? Write a complete sentence or two and use appropriate precision for the numbers you use to make your argument.
(e) When will you recover the extra initial cost in gas savings if the government (re)instates a $3,000 tax rebate for hybrid purchases?
(f) With the original initial costs, how much would the price of gasoline have to be in order for the breakeven point to occur at 30,000 miles?
(g) Restore all the inputs to their original values. Arrange your spreadsheet so that it will print on one page, with the chart below the data table. (Use `Print Preview`.)

Table 7.18. Comparing cell phone plans [R184]

Plan	Phone cost	Monthly charge	Two year cost
Cricket Wireless	499.99	55	1,819.99
Virgin Mobile	649	30	1,369

(If you're thinking of buying a car, remember that there's a lot more that goes into the cost of driving one car or another (or any car at all) than just these simple computations using initial cost and miles driven.)

Exercise 7.8.12. [S][Section 7.5][Goal 7.3][Goal 7.5] Contract or not?

Table 7.18 provides data that appeared in a story in *The Boston Globe* on June 14, 2012 headlined "Pay full price for iPhone, avoid contract".

(a) How much would it cost (in total) to buy the Cricket phone and use it for two months?
(b) Write an equation for the total cost to buy and use the Virgin Mobile phone for M months.
(c) Identify the slope and the intercept of your equation, with proper units for each.
(d) Create an Excel spreadsheet and use Excel formulas to complete a table like this:

Months	Cricket	Virgin
0		
1		
...		
24		

(e) Check that your spreadsheet produces the answers in the table for 24 months.
(f) Create a properly labelled and formatted chart displaying the data in your table.
(g) When (in terms of months of use) would it be better to choose the Cricket phone?
(h) Suppose the monthly charge for the Cricket phone was just $45 while that for the Virgin phone increased to $35/month. Answer the previous question with this new data.

Exercise 7.8.13. [S][Section 7.6][Goal 7.2][Goal 7.1][Goal 7.7] Regenerative braking.

When you apply the brakes in a Toyota Prius the car uses some of the energy of the forward motion to recharge the battery. The dashboard displays a little car icon each time that recharging has collected 50 watt-hours.

(a) Estimate the energy equivalent of each icon in gallons of gasoline.
(b) Estimate the dollar value of that gasoline.
(c) Compare your estimate to the dollar value of 50 watt-hours of electricity in your house (what it would cost to keep a 100 watt bulb on for half an hour).
(d) Discuss the value of the display.

[See the back of the book for a hint.]

Exercise 7.8.14. [S][Section 7.6][Goal 7.1][Goal 7.7] Computers don't sleep soundly.

The website `michaelbluejay.com/electricity/computers.html` gives information about how much energy a computer uses while asleep, in standby mode, or in use. The iMac G5, for example, uses 97 watts while "doing nothing", compared to 3.5 watts while asleep.

(a) What do you think "doing nothing" means?
(b) If this type of computer is doing nothing all day (24 hours), how much electricity does it use? Express your answer in kilowatt-hours.
(c) Now suppose the computer goes to sleep after 15 minutes of doing nothing. How much electricity does it use in an idle day?
(d) If a kilowatt-hour of electricity costs 20 cents, how much money is saved in a day because the computer is smart enough to go to sleep?

Exercise 7.8.15. [S][C][Section 7.6][Goal 7.1][Goal 7.7] How Much Water Does Pasta Really Need?

On February 24, 2009 *The New York Times* published an article by Harold McGee addressing that question.

McGee's kitchen experiments convinced him that he could cook pasta in far less water than is customary. Since (he says) we consume about a billion pounds of pasta a year:

> My rough figuring indicates an energy savings at the stove top of several trillion B.T.U.s. At the power plant, that would mean saving 250,000 to 500,000 barrels of oil, or $10 million to $20 million at current prices. Significant numbers, though these days they sound like small drops in a very large pot. [R185]

(a) Verify the author's conversion of "several trillion B.T.U.s" to barrels of oil and then to dollars.
(b) Does McGee's estimate of a billion pounds of pasta per year make sense?
(c) How much water do Americans use cooking pasta? How much would they use if they followed McGee's advice?
(d) Does not boiling the extra water really save the amount of energy McGee claims?

Exercise 7.8.16. [S][Section 7.6][Goal 7.7] Solar energy.

On May 1, 2013, *The Arizona Republic* reported on the $500 million Arlington Valley Solar Energy II project near Phoenix. The article said

> [the project] will have 127 megawatts of capacity when finished. One megawatt is enough electricity to supply about 250 Arizona homes at once, when the sun is shining on the solar panels. [R186]

(a) How many watts of power does the average Arizona home need to run its appliances when the sun is shining?
(b) Research the power requirements of several typical home appliances: air conditioners, stoves, television sets, ... Then decide whether the article's claim about home power requirement on a sunny day in Arizona is reasonable.

Exercise 7.8.17. [S][C][Section 7.6][Goal 7.1][Goal 7.7] Solar power at Wellesley College.

The sign on a solar panel array at Wellesley College reads:

Solar Photovoltaic Array This 10-kilowatt Solar PV Array is composed of 48 panels, each 210 watts. It will generate approximately 13,000 kilowatt hours of electricity per year, enough to power 2 homes, 32 metal halide street lights or 85 LED street lights for an entire year. For real time electrical output please go to: `www.sunwatchmeter.com/home/day/wellesley-college` PLEASE KEEP OFF THE PANELS

You can see a picture at `www.theswellesleyreport.com/2010/09/wellesley-college-saves-the-planet/solar-panel/`

(a) Check the consistency of some of the numbers.
(b) How many hours of sunshine per day do the designers expect the installation to see?
(c) Visit the website on the sign and write about it. What do the graph and the meters represent? What is happening there now?

Exercise 7.8.18. [C][Section 7.6][Goal 7.7] Wind power.
From *The Los Angeles Times*, March 1, 2009:

> The U.S. last year surpassed Germany as the world's No. 1 wind-powered nation, with more than 25,000 megawatts in place. Wind could supply 20% of America's electricity needs by 2030, up from less than 1% now, according to a recent Energy Department report. [R187]

(a) What do these data say when you calculate wind power in megawatts per person, or as a percentage of the total power available?
(b) Explain why it might or might not be true to say that when this article appeared the U.S. now produced more wind energy than Germany?

Exercise 7.8.19. [S][Section 7.6][Goal 7.7][Goal 7.1] World solar power.
In a posting on their website on July 31, 2013, the Earth Policy Institute reported that

> The world installed 31,100 megawatts of solar photovoltaics (PV) in 2012—an all-time annual high that pushed global PV capacity above 100,000 megawatts. There is now enough PV operating to meet the household electricity needs of nearly 70 million people at the European level of use. [R188]

A graphic at the website `www.wec-indicators.enerdata.eu/household-electricity-use.html` shows that average household electricity consumption in Europe in 2013 was about 4,000 kwh/year. [R189]

(a) Use the data in this exercise to estimate the average number of hours per day that these solar panels are producing electricity.
(b) How does average household electricity consumption in the U.S. compare to that in Europe?

[See the back of the book for a hint.]

Exercise 7.8.20. [S][C][Section 7.6][Goal 7.1][Goal 7.7] Chilling out by the quarry.

On August 16, 2010 *The Boston Globe* described a local business's plan to cool its corporate facility with water from a nearby quarry rather than with conventional air conditioning.

> [Director of facilities] Dondero estimated that the cooling system, which eliminates the need for any type of refrigerant in the building, saves about $75,000 a year, reduces annual water use by one million gallons, and cuts yearly energy use by about 300,000 kilowatt hours—enough to power about 30 homes. [R190]

(a) What rate in dollars per kwh is Dondero using to support his assertion that this change will save $75,000 a year?
(b) Is the claim that 300,000 kilowatt-hours would power 30 homes for a year reasonable?

Exercise 7.8.21. [S][Section 7.6][Goal 7.1][Goal 7.7] Energy savings at MIT.

On March 26, 2011 Jon Coifman wrote in *The Boston Globe* that

> In just 36 months, [MIT and NStar] plan to cut the university's energy use 15 percent—enough to power 4,500 Massachusetts homes for a year. [R191]

(a) If all of MIT's energy use were devoted to powering Massachusetts homes, how many homes would that be?
(b) Compare your answer in part (a) to the number of homes in Cambridge.
(c) Estimate MIT's total annual energy use, in Btus.
(d) Convert your answer to the previous question from Btus to watt-hours.

Exercise 7.8.22. [S][Section 7.6][Goal 7.1][Goal 7.7] The Governor gets the units wrong.

On June 7, 2011 *The Norwich Bulletin* reported that

> Connecticut Governor Daniel Malloy recently signed off on a deal to tax electricity generators one quarter of one cent per kilowatt hour, or 25 cents per $100. [R192]

(a) What is wrong with the units in this quotation?
(b) Estimate the percentage change in the cost of electricity that would result from a one-quarter of one cent increase per kilowatt-hour.
(c) What do you think Governor Malloy intended to say?

[See the back of the book for a hint.]

Exercise 7.8.23. [U][Section 7.7][Goal 7.1][Goal 7.5][Goal 7.6] Your total federal tax bill.

Modify the graph in Figure 7.15 to show how total tax and the effective tax rate for (income tax + Social Security) depends on income.

Exercise 7.8.24. [S][C][Section 7.7][Goal 7.1][Goal 7.6] President Obama's income tax.

According to the White House website

> [The President] and the First Lady filed their [2013] income tax returns jointly and reported adjusted gross income of $481,098. The Obamas paid $98,169 in total tax.
>
> The President and First Lady also reported donating $59,251—or about 12.3 percent of their adjusted gross income—to 32 different charities. The largest reported gift to charity was $8,751 to the Fisher House Foundation. The President's effective federal income tax rate is 20.4 percent. ... The President and First Lady also

released their Illinois income tax return and reported paying $23,328 in state income tax. [R193]

The President itemized deductions, so he could deduct charitable contributions and state tax from his adjusted gross income:

> In the United States income tax system, adjusted gross income (AGI) is an individual's total gross income minus specific deductions. Taxable income is adjusted gross income minus allowances for personal exemptions and itemized deductions. [R194]

(a) With the information given, what is the largest possible value for the President's taxable income? What tax bracket would he be in?
(b) Use the Married Filing Jointly brackets and rates in the spreadsheet at `Federalindividualratehistory.xlsx` to compute the Obamas' 2013 federal income tax bill for your answer to part (a). If your result does not match the reported figure, what might explain the difference?
(c) If the Obamas had not made those charitable contributions the money would be part of their taxable income. Use your answers to the previous questions to answer these.
    (a) What would their taxable income have been? What bracket would that have put them in? What would their tax have been?
    (b) What fraction of the contribution was (essentially) made by the government?
    (c) What fraction of the Obamas' income did they contribute to charity?
    (d) Did they tithe?
    (e) How does their contribution compare to the national average?

Exercise 7.8.25. [S][Section 7.7][Goal 7.1][Goal 7.6] Using the tax table.
Use Table 7.13 to answer the following questions.

(a) Compute the tax due in 2014 on a net taxable income of $80K. Show your work.
(b) Compute the effective tax rate for that income.
(c) Check your answers with those in the spreadsheet `GraduatedTax.xlsx`.
(d) Compute the effective tax rate a second time, as a weighted average of the rates in the various brackets, using as weights the amount of income subject to tax at each rate. You should get the same answer.

Exercise 7.8.26. [S][Section 7.7][Goal 7.1][Goal 7.6] Taxes and inflation.
The spreadsheet `Federalindividualratehistory.xlsx` contains a complete history of income tax brackets and rates from the inception of the income tax in 1913 through 2013, in both dollars current in each year and adjusted for inflation (2012 dollars).

(a) Compute the tax due in 2003 for a single taxpayer with a net taxable income of $30K. What is her effective tax rate?
(b) Suppose that taxpayer received raises each year that kept up with inflation. Use an inflation calculator to calculate her net taxable income in 2013.
(c) Use the 2013 tax tables to compute her tax in 2013. What is her effective tax rate?
(d) Compare her 2003 tax and effective tax rate with her 2013 tax and effective tax rate, taking inflation into account. Has her tax gone up, or down, or stayed the same?

Exercise 7.8.27. [S][Section 7.3][Goal 7.2][Goal 7.3] Pandora growing fast!

On June 10, 2011, CNN Money reported that the internet music site Pandora is adding new users at the rate of one per second. Between February and April the number of users grew from 80 to 90 million. [R195]

Is the slope of one user per second correct based on the February and April numbers of users?

Exercise 7.8.28. [S][Goal 7.7] Bicycle power in Times Square.

On December 30, 2012 *The Boston Globe* reported on an Associated Press story about six bicycles that would help illuminate the famed falling ball in Times Square on New Year's Eve.

> Each bike will generate an average of 75 watts an hour. It takes 50,000 watts to light up the ball's LEDs. [R196]

Unfortunately, the Associated Press reporter is quite confused about the difference between energy and power. The "generate ... 75 watts an hour" in the quote makes no sense. We think what he or she is trying to say is that while someone is actually pedaling it each bike could power a 75 watt light bulb. All six bikes together could light up just 450 watts worth of LEDs.

(a) How many bikes would have to be pedaled simultaneously to light up all the ball's LEDs?
(b) Since there are only six bikes, people pedaling during the day will store the energy they generate in batteries, which will then be used to light the ball. Suppose the lights need to be on for two minutes while the ball drops at midnight.

How many hours of pedaling will it take to generate (and save) the electrical energy needed?

Exercise 7.8.29. [S] Flying twice as far.

A curious traveler asked this question on stack exchange:

> A flight from Los Angeles to Albuquerque is about 2 hours but is $\approx$ 670.2 miles.
>
> A flight from San Jose to Chicago is 4 hours but is $\approx$ 1859.0 miles.
>
> Can anyone explain why the travel time from San Jose to Chicago is not longer and closer to 5.75 hours?
>
> If the distance increase by 2, shouldn't the time increase by a factor of 2 as well? [R197]

(a) Write a linear model for this question. Takeoff and landing will take a fixed amount of time. Actual travel in the air will take time proportional to the distance traveled. Think about which of the variables (time and distance) is the independent variable, and identify the slope and intercept with their units.
(b) Use the data in the quotation to estimate the two constants in your linear model.
(c) Compare your answer to those at the link to the quotation.

Exercise 7.8.30. [S] Express lane?

In *The Boston Globe* on November 27, 2015 you could read that

> Amid the holiday grocery shopping madness, every line feels like the wrong one. And yet, some are wronger than others. Given equally capable cashiers, you are often better off bypassing the express lane. Research conducted at a large, unnamed,

> California grocery store found that while each item adds 3 seconds to the check-out time, it takes 41 seconds for a person to move through the line even before their items are added to the tally. Bottom line: The big time-consumers are not the items, but the small talk and the paying, says Dan Meyer, who has a doctorate in math education from Stanford University. [R198]

Suppose you have 10 items in your cart, so you are allowed to use the express lane. How much longer must the line there be (compared to the regular lane) to make the wait in the regular lane less?

You can answer that question with any strategy that makes sense to you, as long as you explain what you're thinking. If you need a starting place, one way is to use these steps:

(a) Write the linear equations showing how the time it takes a shopper to check out depends on the number of items in her cart. What are the slope and intercept, with their units?
(b) Suppose shoppers in the express lane buy 6 items (on average), while those in the regular lane buy about 20. Write the linear equations showing your waiting time in each line depends on the number of shoppers ahead of you.
(c) Now work on the main question—which line should you join when you have 10 items in your cart? How much longer must the express lane line be to make the wait on the regular lane line less?

[See the back of the book for a hint.]

Review exercises

Exercise 7.8.31. [A] If you drive at a rate of 50 miles per hour for 3 hours, how far have you driven? Identify each piece of this proportion: the quantities being measured and the proportionality constant, with the appropriate units.

Exercise 7.8.32. [A] You may remember from geometry that the circumference of a circle is directly proportional to the diameter of that circle. The relationship is

$$c = \pi d,$$

where c represents the circumference and d represents the diameter and $\pi \approx 3.14$ is the proportionality constant. If the diameter of a circle is doubled, how does the circumference change?

Exercise 7.8.33. [A] The cost of potatoes is proportional to the weight (in pounds) you buy. If potatoes cost $0.69 per pound, what is the cost for 3 pounds of potatoes?

Exercise 7.8.34. [A] The conversion from £ to U.S. $ is 1.53 $/£. How much is £200 worth in U.S.$?

Exercise 7.8.35. [A] Suppose that y is directly proportional to x. When $x = 16$, then $y = 4$.

(a) What is the proportionality constant?
(b) If $x = 32$, what is y?
(c) If $y = 32$, what is x?

Exercise 7.8.36. [A] Identify the slope and intercept in each of the following. When appropriate, state the units.

(a) $y = 2.5x + 6$.
(b) $y = -5x + 20$.
(c) $y = 300 + 40x$.
(d) $Q = 0.004E - 300$.
(e) He earns $9.25 per hour.
(f) To rent a car for one day, the cost is $25 plus $0.15 per mile.
(g) My new phone cost $25, plus a monthly charge of $15.
(h) The conversion from £ to U.S. $ is 1.53 $/£.
(i) The salesperson worked only on commission, earning 20% of the total amount sold.

Exercise 7.8.37. [A] Solve each problem.

(a) If $y = 2.5x + 6$ and $x = 4$, what is y?
(b) If $y = -5x + 20$ and $y = 0$, what is x?
(c) If $y = 300 + 40x$ and $x = -10$, what is y?
(d) If his salary is $9.25 per hour and he works 5 hours, how much does he earn?
(e) If the conversion from U.S. dollars to pounds sterling is 1.80 $/£, how much money would you get by changing $100 to £?
(f) If my new phone cost $25, and I pay a monthly charge of $15, what is my total cost after 10 months? When does my total cost reach $250?

8

Climate Change—Linear Models

Complicated physical and social phenomena rarely behave linearly, but sometimes data points lie close to a straight line. When that happens you can use a spreadsheet to construct a linear approximation. Sometimes that's useful and informative. Sometimes it's misleading. Common sense can help you understand which.

Chapter goals

Goal 8.1. Draw regression lines using Excel. Interpret regression lines.

Goal 8.2. Recognize when rounding too much distorts conclusions.

Goal 8.3. Think about causation vs correlation.

8.1 Climate change

Climate change (global warming) is a current hot topic. How rapidly is the Earth's average temperature increasing? What might the consequences be? What is the cause? What might we do about it? Should we try? The science is complex and the politics even more so. In a course like this we can't begin to unravel those complexities. But for just a taste of the analysis, we will briefly look at some data on the average temperature of the Earth and the concentration of carbon dioxide (CO_2) in the atmosphere in recent history. The spreadsheet `EarthData.xlsx` has data we downloaded from `www.earth-policy.org/data_center/C23`.

The chart on the left in Figure 8.1 shows a scatter plot of the average global temperature, in Celsius degrees, for the years 1960–2000. There is no formula for that relationship, but the points seem to trend upwards (on average). So we asked Excel to connect the dots to see the jagged rise and fall, and then we drew a line on the graph that looked like a reasonable approximation for the trend. The result is on the right. Then we used the line to predict a temperature of 14.58 degrees Celsius for 2010. In fact that average was 14.63 degrees Celsius. Given how up and down the data are (despite the long term average trend) we could hardly expect an accurate prediction. We added the textboxes and the arrows to the spreadsheet to explain how we drew the line.

The line we drew is a *model*—a mathematical construction that approximates something in the real world. This particular model is linear—the line that seems to match the data best. We could have used that model in 2000 to make a prediction for 2010—an estimate of what the temperature might be in a future year for which we didn't (at the moment) have data.

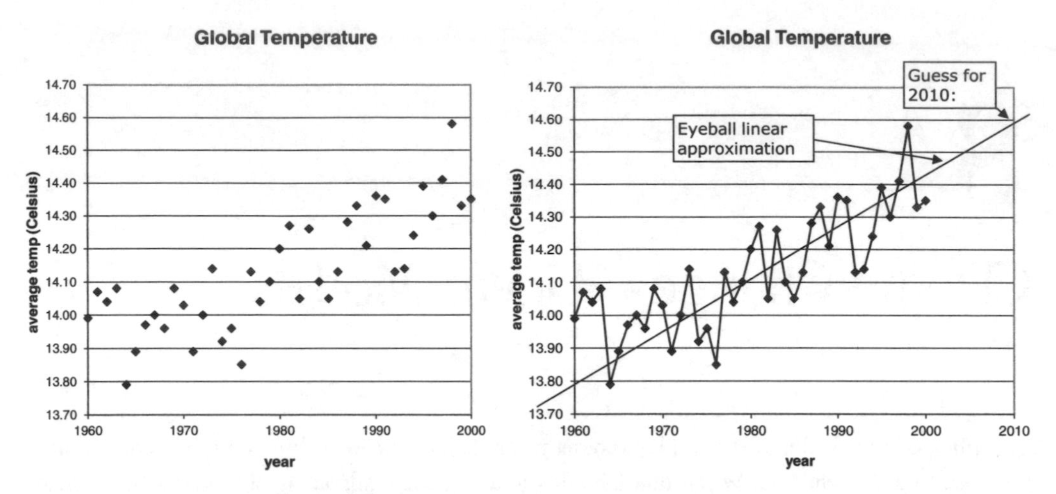

Figure 8.1. Global average temperature, 1960–2000

Excel knows the mathematics for finding the model line we guessed at "by eye". Figure 8.2 shows how invoke it: select the chart, select `Layout` from `Chart Tools`, select `Trendline` and then `Linear Trendline`. Excel draws the second line shown in Figure 8.3.

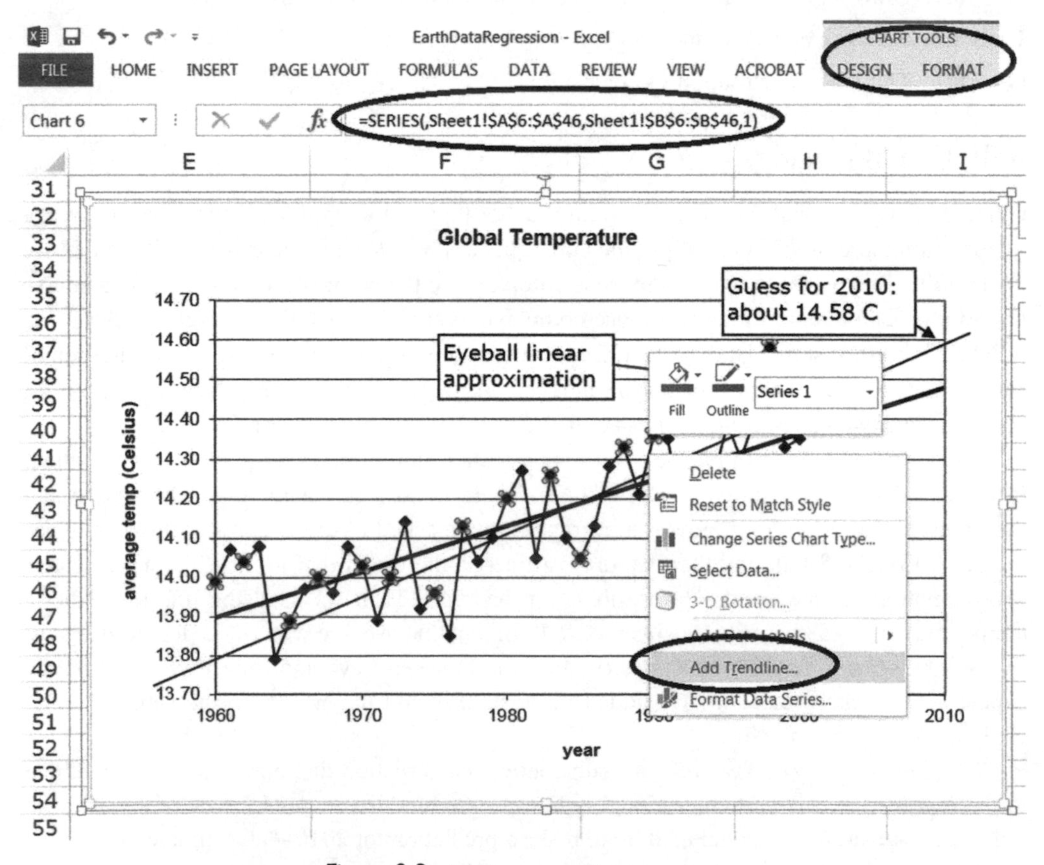

Figure 8.2. Adding a trendline to a chart

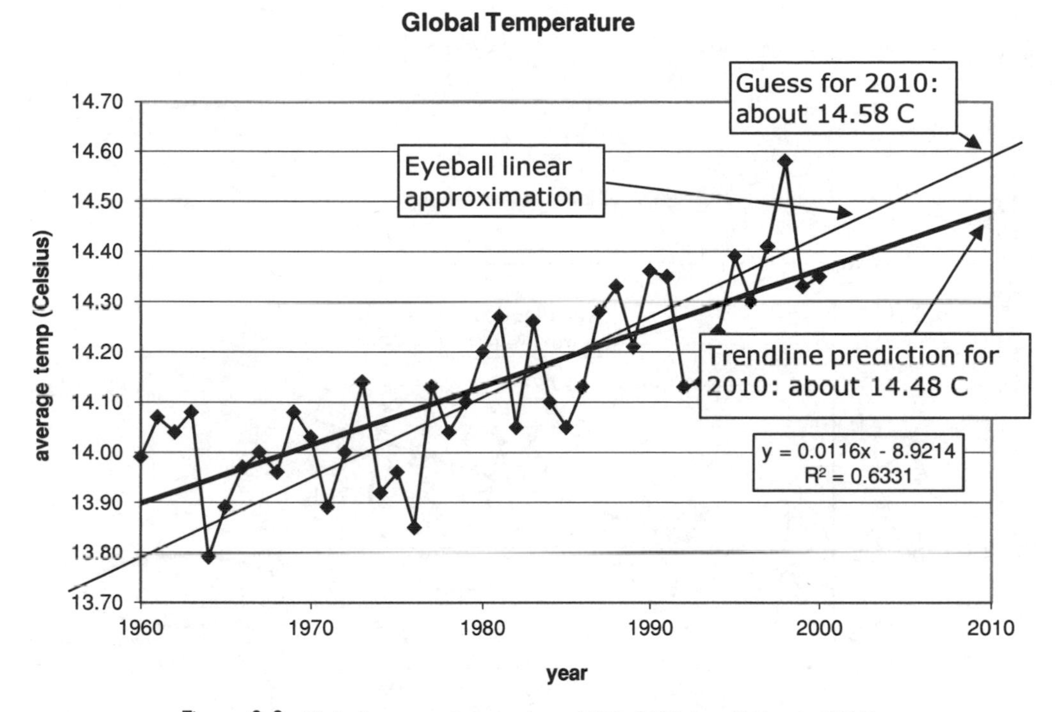

Figure 8.3. Global average temperature, 1960–2000 (prediction to 2010)

Figure 8.4 shows how to format the trendline: select it (by right clicking); select `Format Trendline ...`; `Forecast Forward` 10 periods (10 years). Check the boxes for `Display Equation` and `Display` *R*`-squared value`—we will need that data soon. You can change the `Trendline Name`, `Line Color` and `Line Style` if you wish.

Excel calls the line that best fits a scatterplot a trendline. Its official name is *regression line*. We learned (or remembered) in Chapter 7 that straight lines are described by linear equations. The one for the regression line in Figure 8.3 is

$$y = 0.0116x - 8.9214.$$

The slope of the regression line matters most. In this example it says that on average global temperature is increasing at a rate of

$$0.0116 \frac{\text{degrees (Celsius)}}{\text{year}}.$$

That's just over a hundredth of a degree (Celsius) per year, or a tenth of degree per decade. (Remember that the units of the slope are (units of y)/(units of x.)

The intercept for this linear equation, with its units, is

$$-8.9214\text{degrees (Celsius)}.$$

Supposedly, that is the temperature predicted (retroactively) by the regression line for year 0. That's nonsense, of course.

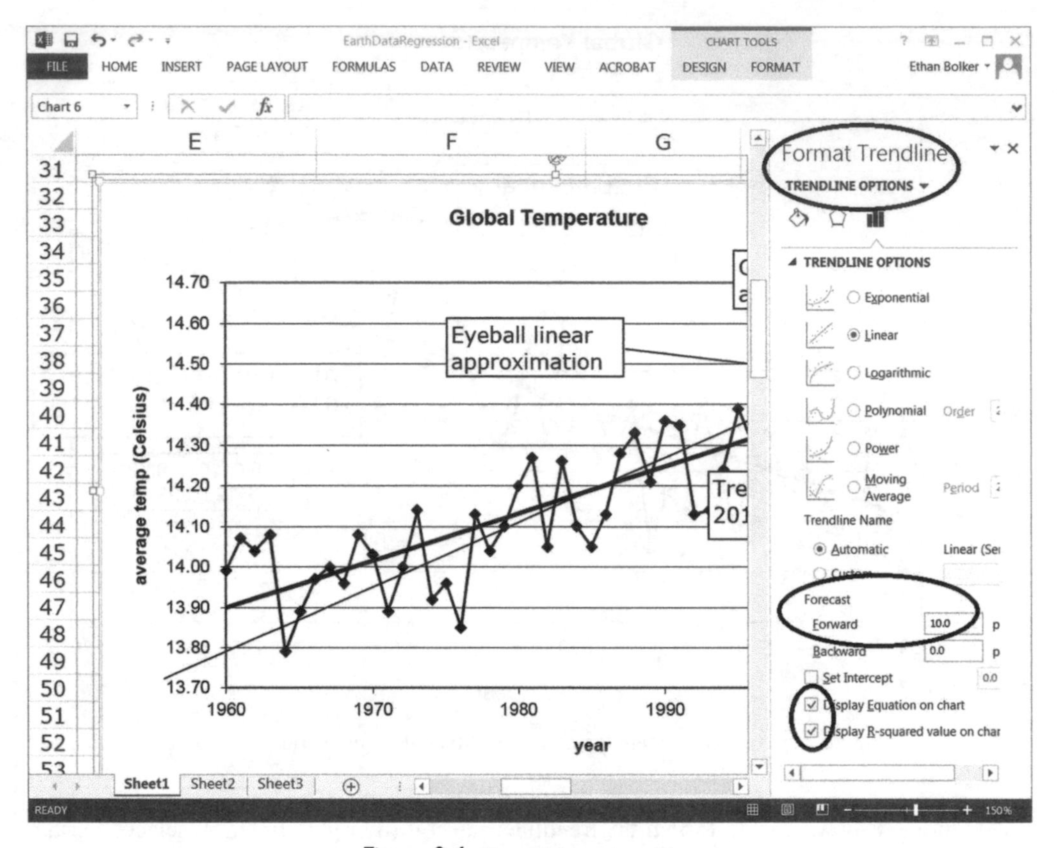

Figure 8.4. Formatting a trendline

In principle, we can use the equation of the line instead of our eyeball approximation to make our 2010 prediction. If we let $x = 2010$ we find the prediction

$$\begin{aligned}\text{average 2010 temperature} &= y \\ &= 0.0116 \times 2010 - 8.9214 \\ &= 14.2786 \\ &\approx 14.29 \text{ degrees Celsius.}\end{aligned}$$

Something is wrong! When we looked at the trendline that Excel drew, we had an estimate of 14.48 degrees. This calculation is not even close to that estimate. Stop and think: we estimated the 2010 temperature visually, using the Excel trendline, as 14.48 degrees Celsius. When we used the equation for that trendline to calculate the 2010 temperature, we got a number that didn't make sense with what we saw on the graph.

Be skeptical. Always ask whether the numbers from a newspaper or a website or a television commentator—or from a computer program—make sense. This one clearly doesn't. If we dig a little deeper we can see why.

It turns out that Excel rounded off the slope and intercept it showed on the chart. It knows the correct values, but thought all the digits were too ugly to display . To find them, enter the command

`=SLOPE(`

in a cell (we used cell `H27`, with a label in `G27`). Excel prompted for

`=SLOPE(known_y's, known_x's)`

so we selected the data

`=SLOPE(B6:B46,A6:A46)`

(the years 1960–2000) and Excel told us the correct value: 0.011642857. That's more precise than the rounded value 0.0116 shown on the chart. We found the intercept, −8.921393728, the same way, with the formula `=INTERCEPT(B6:B46,A6:A46)` (in cell `H28`). (In the `SLOPE` and `INTERCEPT` functions the y-values come first and the x-values second, even though in the data table the x-values are first and the y-values second.)

Then the correct equation for the model, before rounding, is

$$y = 0.011642857\,x - 8.921393728.$$

If we set $x = 2010$ in that equation Excel tells us $y = 14.48074913$ (cell `E30`). That rounds to our visual estimate of 14.48.

We are not the first to discover this problem. Microsoft's support page at `support.microsoft.com/kb/211967` outlines what they call a *workaround*, to show all the decimal places in the trendline equation displayed on the chart. Right click the formula for the trendline on the chart, then select `Format Trendline Label`.... In the `Number` selection there you can ask for the maximum number of decimal places: 30. Decide for yourself whether you like this method better than asking Excel directly for the `SLOPE` and `INTERCEPT`.

We've said repeatedly that it was wrong to show lots of decimal places when reporting approximate numbers, even when those decimal places appeared in your calculator or spreadsheet. But in this example we saw that too much rounding is wrong too. Using a slope rounded to four significant digits may give a ridiculous answer. The short answer to the question "when should you round?" is

> While you compute, use all the digits you have, even if it's more than you need. Round only when you're done.

Keep this in mind from now on—both when doing the problems in this text and when working with Excel (or another software program) in the future.

8.2 The greenhouse effect

Most climate scientists are convinced that the reason the Earth is warming is the increase in the concentration of greenhouse gases like carbon dioxide in the air.

A greenhouse is warm in the winter because sunlight enters through the glass roof, which prevents the inside air it heats up from escaping. Carbon dioxide (CO_2) behaves similarly in

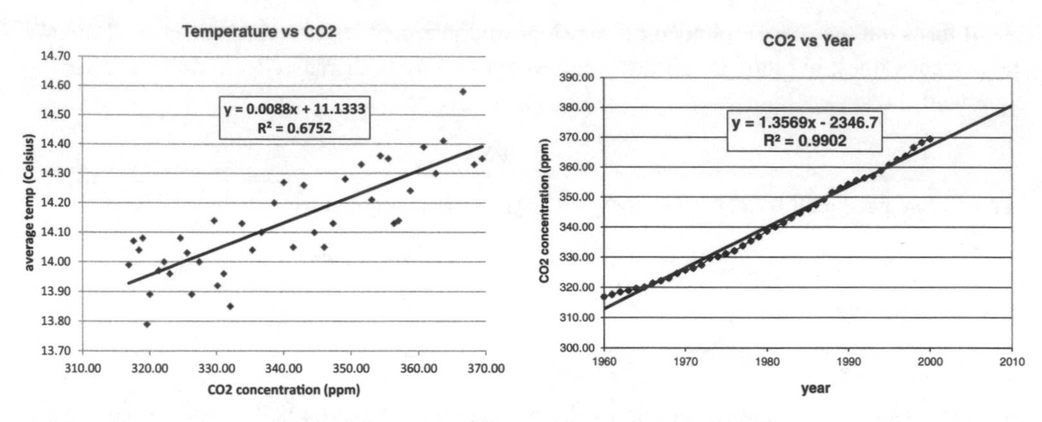

Figure 8.5. Temperature, CO_2 and time, 1960-2000

the atmosphere—it lets sunlight in but doesn't let heat out. The chart on the left in Figure 8.5 displays the data and the regression line showing how average temperature varies with the amount of CO_2 in the atmosphere. When we asked Excel to show the equation of the trendline this box appeared on the chart:

$$y = 0.0088x + 11.1333$$
$$R^2 = 0.6752$$

The slope of the trendline is

$$0.0088 \frac{\text{degrees Celsius}}{\text{part per million of } CO_2}.$$

An increase of one part per million of CO_2 corresponds to somewhat under one hundredth of a degree (Celsius) increase in temperature.

That's the trendline slope with four significant digits. If we ask Excel for more we see

$$0.00881808540214804 \frac{\text{degrees Celsius}}{\text{part per million of } CO_2}.$$

We would use that value in any computations we made.

The chart on the right in Figure 8.5 shows the increase in CO_2 concentration over the years (it does not mention temperatures at all). There the slope of the regression line is

$$1.3569 \frac{\text{parts per million of } CO_2}{\text{year}};$$

on average, every year the CO_2 concentration increases by about 1.36 parts per million.

The original data set contains three columns of information, listing the year, average global temperature and CO_2 concentration. In Section 8.1 we looked at the relationship between year and average global temperature, which documents the trend called "global warming" in the news. In this section we looked at the other two relationships, between temperature and CO_2 concentration and between CO_2 concentration and time, in hopes of understanding what might be behind the observed temperature trend. In the next two sections we'll think about what we may learn this way.

8.3 How good is the linear model?

How much a regression line helps understand the data and make predictions depends in part on how close the data points are to the line. Common sense tells you that the relationship between carbon dioxide concentration and time (on the right in Figure 8.5) is likely to be more reliable than that between carbon dioxide and temperature (on the left), which in turn looks better than that between temperature and time (Figure 8.3).

The official statistical measure of "close to the line" is a number between zero and one called "*R*-squared". The closer *R*-squared is to 1 the better the regression line fits the data. In Figure 8.3 R^2 is just 0.63321—not very good. That matches what we can see in the chart—the temperature seems to be increasing on the average, but can go up and down unpredictably from year to year. In the chart on the right in Figure 8.5 the R^2 value is 0.9902, which is very close to 1. In fact the measured 2010 concentration was 389.78 parts per million, so the relative error in the prediction is about −2.5%.

We are being deliberately vague about how close the R^2 should be to 1 to declare that the fit is "good". There are no rules for this. In the exercises below you will have a chance to develop your intuition.

We were careful to use the word "corresponds" when discussing the increase in CO_2 concentration and the increase in average temperature, not the word "causes". The data only say that the CO_2 concentration and the temperature are *correlated*—they trend together. They don't say one causes the other. Data can never tell you that. Climate scientists who work at understanding the physics and chemistry of carbon dioxide in the atmosphere have created scientific models that suggest causation. We will return to this distinction in Section 8.4.

There is much more to the climate change debate: some who accept the scientific models that say that greenhouse gases cause global average temperatures to increase are not convinced that the increase in greenhouse gases is due to human activity, and therefore see no need to change the way we use energy.

8.4 Regression nonsense

The graphic in Figure 8.6 resembles one that appeared in *The Boston Globe* on January 14, 2010 in a story headlined "Imaginary fiends", which began

> In 2009, crime went down. In fact it's been going down for a decade. But more and more Americans believe it's getting worse.[R199]

The data are from the FBI and the Gallup Poll. The FBI measures the crime rate in violent crimes per 100,000 people. The fear index is the percentage of people who say crime is going up.

The headline seems to announce a juicy story. The graph is drawn to accentuate the apparent contradiction, since the scales on both *y* axes don't start at 0. We will use these numbers to illustrate the kinds of nonsense arguments you can make with regression lines. There are three variables to play with: the year, the crime rate, and the fear index. We will focus on them two at a time and imagine different kinds of conclusions.

We started by entering the data in Excel, using the table in the online version of *Common Sense Mathematics* to save typing and prevent typing errors. To do that, select and copy the

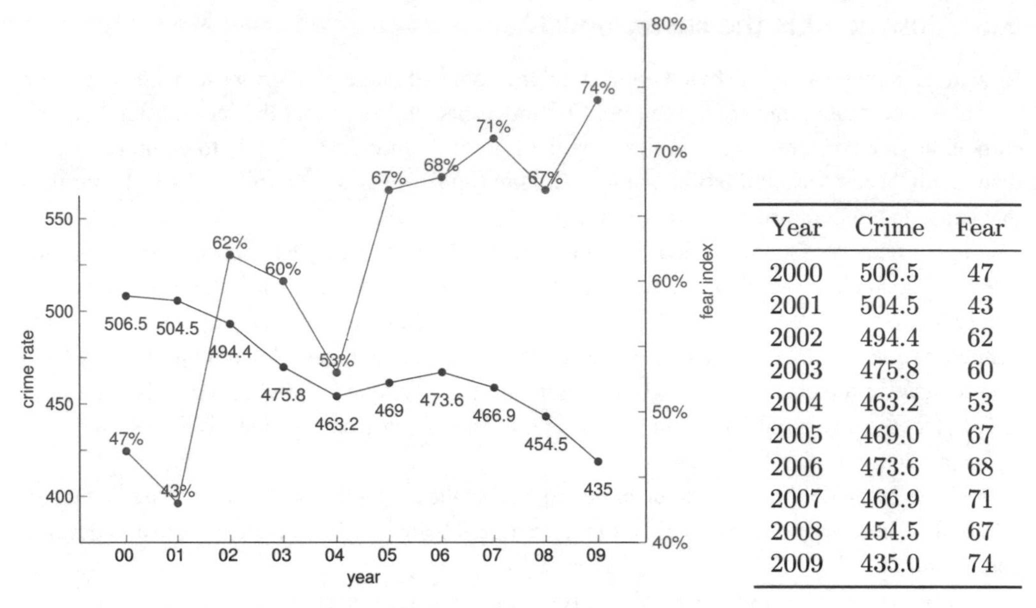

Year	Crime	Fear
2000	506.5	47
2001	504.5	43
2002	494.4	62
2003	475.8	60
2004	463.2	53
2005	469.0	67
2006	473.6	68
2007	466.9	71
2008	454.5	67
2009	435.0	74

Figure 8.6. Crime down, fear up [R200]

data from the table. Then paste it into Excel. The bad news is that it is then just text, all in one column. The good news is that Excel can separate the columns of data: open the `Data` tab and select `Text to Columns`. Then entering `Next` on all the dialog windows does the job.

Our work is in the spreadsheet `crimeDropsFearsRise.xlsx`.

The first graph in Figure 8.7 shows a scatterplot and trendline for the last two columns in the table. There we asked Excel to construct a graph with crime rate as the independent variable.

Since we chose crime rate as the independent variable it's easy to look at the graph—and the trend line—and conclude that the increase in crime rate is closely related to the decrease

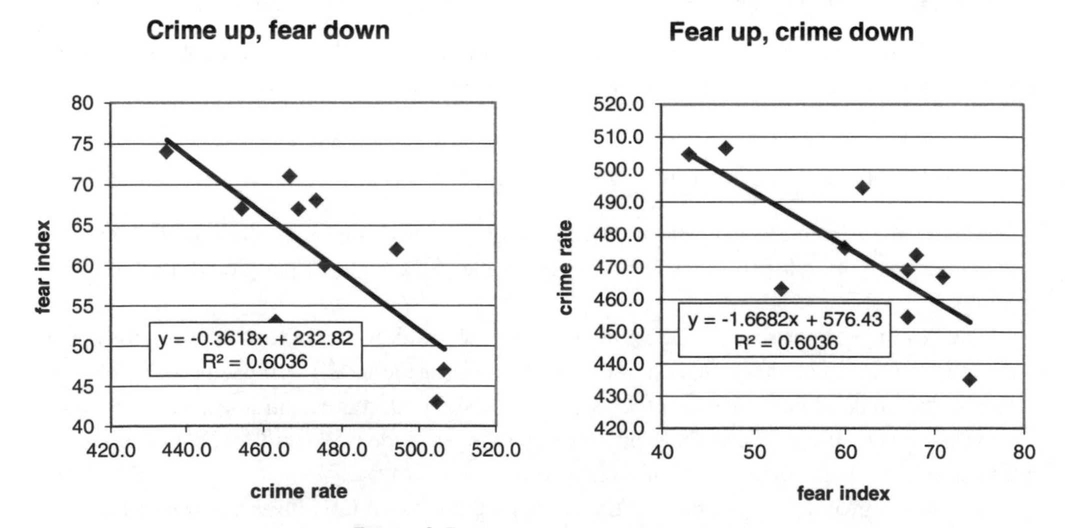

Figure 8.7. Crime vs fear regressions

in the fear index. The regression line slopes down—high crime rates seem to come along with decreased fear of crime. The R-squared value is 0.60—perhaps not compellingly high, but we won't let that stop us from thinking about the data. What might the correlation mean? Could an increase in crime (the independent variable on the x-axis) cause people to be less afraid? Here's an attempt at an explanation: Perhaps when crime is rare it's reported spectacularly in the news and people are frightened, while when it's common it gets less press and most people don't notice it as much because it isn't happening to them.

Does that make sense? Not to us, but it's the kind of argument you frequently see or hear—a simpleminded attempt to explain what seems to be a real "this is true because of that" connection, or perhaps what a politician would like you to believe is a real connection.

The second graph in Figure 8.7 shows the same data with the fear index as the independent variable. That changes our view of the data. Now we see crime dropping as fear increases. How might we explain that? Perhaps we'd argue that increasing fear of crime leads to more pressure on the police to arrest criminals, thus reducing the amount of crime. That's more plausible than the other way around, but still a shallow unconvincing analysis of complex social phenomena. Both the crime rate and the fear of crime are changing over time, one decreasing while the other increases, but just because we can find a trendline doesn't mean either change causes the other.

We can see the two trends separately if we plot each with time as the independent variable, as in Figure 8.8. With these charts we can create other nonsense arguments. The slope of the fear index regression line is about 3 percentage points per year. Since the index was at 74% in 2009, if the trend continues then in about 8 more years, in 2017, 98% of the population will believe that crime is getting worse every year. The second regression line says the crime rate is actually falling each year by about 7 violent crimes per 100,000 people, so in 2017 when everyone believes things are getting worse it will be down from 435 to about 380. Neither of these predictions carries much conviction.

The news story that prompted this discussion is misleading in another way. When we found the data on which it is based we discovered that in the previous decade, from 1990 to 2000, the crime rate and the fear index were both decreasing. The author of the article chose not to tell us

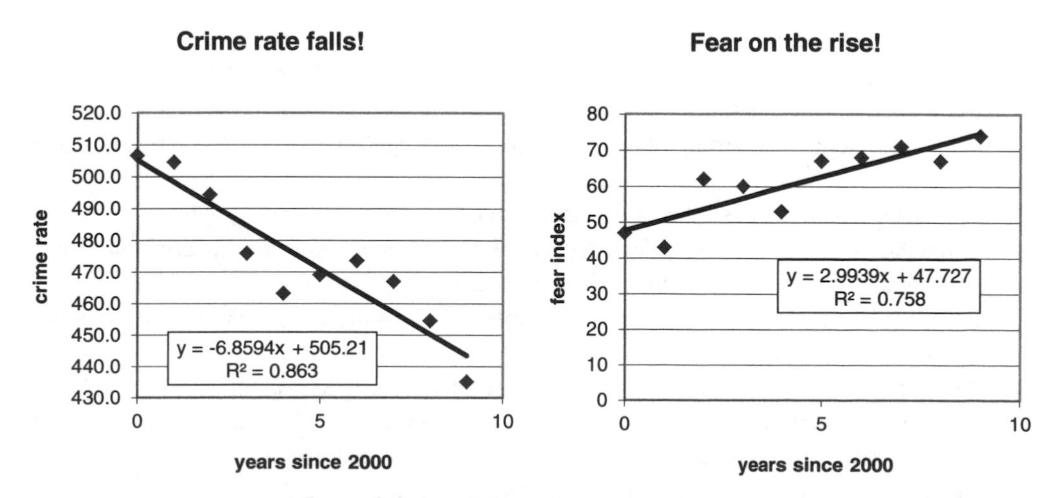

Figure 8.8. Fear index and crime rate over time

that. He *cherry-picked* the data to make his point (whatever it is) more dramatic. You can find all the numbers in our spreadsheet at `crimeDropsFearsRise.xlsx`.

The moral of this story:

Correlation is not causation.

It's very easy to use regression to link variables (crime rate and fear index, as in this example), to suggest trends and to make predictions or interpret correlation as explanation. Just because you can doesn't mean you should. It's often wrong. Watch out for people who do.

8.5 Exercises

Exercise 8.5.1. [S][W][Section 8.1][Goal 8.1] A trendline for linear data.

(a) What values would you expect to see for the slope, intercept and R-squared if you were to add a trendline to the Tamworth electricity bill in the spreadsheet `TamworthElectric.xlsx`?
(b) What would the trendline look like on the graph in Figure 7.5?
(c) Add the trendline and verify your predictions.

Exercise 8.5.2. [U][Section 8.3][Goal 8.1] Anscombe's quartet.

> Anscombe's quartet comprises four datasets that have nearly identical simple statistical properties, yet appear very different when graphed. Each dataset consists of eleven (x, y) points. They were constructed in 1973 by the statistician Francis Anscombe to demonstrate both the importance of graphing data before analyzing it and the effect of outliers on statistical properties. [R201]

Use the data in `AnscombesQuartet.xlsx` for the tasks that follow.

(a) For each data set, use Excel to find the mean of the x and y values. Label them in your spreadsheet.
(b) Do the mean values describe these four data sets very well? Explain.
(c) Graph each set of (x, y) values. Label each graph ("data set 1" etc.). Write a sentence or two describing the relationship between the x and y values, using what you see on the graph. Talk about how strong that relationship is (but don't calculate the R-squared value yet).
(d) Display the trendline, the trend line equation and the R^2 value on each graph.
(e) Round the slope and intercept to two decimal places. Write a sentence comparing the slope, intercept and R-squared value for each of the data sets.
(f) Explain in your own words how these examples demonstrate the importance of graphing data before analyzing it.
(g) The short description at the beginning of this problem also talked about the effect of "outliers" on statistical properties. In this context, an outlier is a number that lies outside most of the numbers in the data set. Does each of the data sets contain an outlier? If so, how does that outlier influence the basic statistics for each data set?

Exercise 8.5.3. [S][Section 8.1][Section 8.2][Goal 8.1] Faster than a speeding bullet.

The spreadsheet at `MarathonWinningTimes.xlsx` shows the history of the winning time in the Boston Marathon for men and women from 1966 (when women first ran) through 2013.

(a) Graph the men's and women's winning times depending on the year, properly label the axes and add a trendline for each data column.
(b) What is the average rate at which the men's finishing time changed from year to year?
(c) Use the trendline to predict when the men's winner will finish in two hours. How confident are you in that prediction?
(d) Use the trendline to predict when the men's winner will finish in one hour. How confident are you in that prediction?
(e) The trendlines suggest that in about six years the fastest woman will be as fast as the fastest man, and will be faster thereafter. Explain why the lines say that, and why it's nonsense.
(f) Make a better prediction about the long run relation between men's and women's winner finishing times.

[See the back of the book for a hint.]

Exercise 8.5.4. [U][Section 8.1][Goal 8.1] The leaning tower of Pisa.

The famous "Leaning Tower of Pisa" began to lean even while it was under construction in the 1170s. The table in Figure 8.9 shows the measured lean for the years 1975 through 1987: the distance in meters between where a particular point on the tower would be if the tower were straight and where it actually was.

Year	Lean (m)
1975	2.9642
1976	2.9644
1977	2.9656
1978	2.9667
1979	2.9673
1980	2.9688
1981	2.9696
1982	2.9698
1983	2.9713
1984	2.9717
1985	2.9725
1986	2.9742
1987	2.9757

Figure 8.9. The Tower of Pisa [R202]

(a) Construct the regression line for this data and estimate (visually) what the lean was in the year 2000.
(b) How good is that estimate likely to be?
(c) What is the slope of the regression line? What are its units? What does it mean?

(d) Check your estimate using the equation of the regression line. Can you use the formula as it appears in the chart, or do you need more decimal places?
(e) Explain why the actual numbers in the data table for the Tower of Pisa depend on the height of the "particular point" at which measurements were taken. What would the numbers be if the point were twice as high? Would the linear regression line be just as good?
(f) What has happened to the Tower of Pisa since 1987?

Exercise 8.5.5. [S][Section 8.1][Goal 8.1] Beverage consumption.

The spreadsheet at `BeverageConsumption.xlsx` contains data on the amounts of milk, bottled water and soft drinks consumed in the United States between 1980 and 2004.

(a) Use Excel to create a scatter plot of this data. Label the data series and the axes correctly.
(b) Explore correlations among the various categories (for example, between milk and water). Write about what you discover. In particular, which kinds of consumption are most closely correlated?
(c) Use the regression lines to make some predictions for years following 2004.
(d) Find the source of the data in `BeverageConsumption.xlsx`. If you find data for other years there, discuss the validity of your predictions.

[See the back of the book for a hint.]

Exercise 8.5.6. [S][Section 8.1][Goal 8.1] Energy consumption.

The Excel spreadsheet `EnergyConsumption.xlsx` contains a table showing the annual United States energy consumption, measured in terawatt-hours, between 1949 and 2005.

(a) Insert a new column labeled "years since 1949" in between the years column and the consumption column. Use Excel to fill in the cells for this column.
(b) Use Excel to find a linear trendline for this data. Include the equation and R^2-value for the trendline on the graph.
(c) Is this trendline a good fit for the data?
(d) What is the slope of this line? Include the units in your answer. Use your answer for the slope to complete the sentence: "For every additional year that passes, total energy consumption ..."
(e) Estimate total energy consumption in the years from 2006 to the present.
(f) Look for data with which to check the estimates from the previous part of the exercise.

Exercise 8.5.7. [S][Section 8.1][Goal 8.1] Supply and demand for office space.

The data in Table 8.10 appeared on page B5 in *The Boston Globe* on April 3, 2010.

Table 8.10. Lower rent, more vacancy

quarter	vacancy rate	rent ($/ft^2)
Q1 '06	11.8%	38.76
Q1 '07	7.5%	47.54
Q1 '08	6.0%	62.20
Q1 '09	9.0%	49.24
Q1 '10	11.1%	42.46

(a) Build and then discuss a linear regression line for the dependence of rent per square foot on vacancy rate.
(b) How do your conclusions change when you adjust rents to take inflation into account?

Exercise 8.5.8. [S][Section 8.1][Goal 8.1][Goal 8.3] Office rents.

On February 22, 2008 *The Boston Globe* ran a story under the headline "Office rents reach dizzying heights" that featured graphs like those in Figure 8.11.

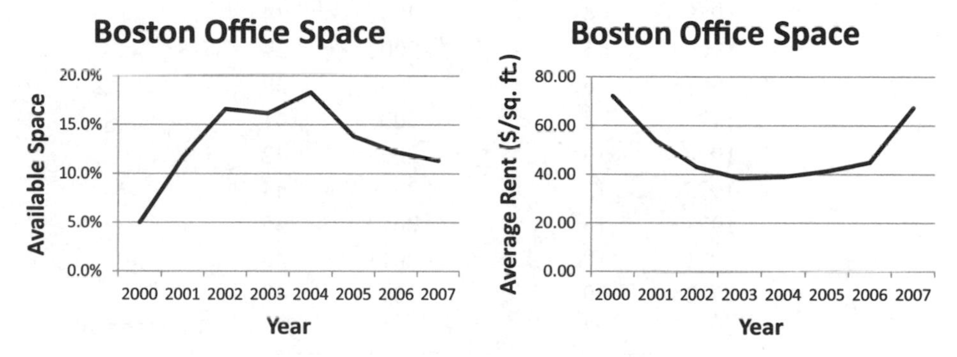

Figure 8.11. Boston office rental rates

The shapes of the curves illustrate the law of supply and demand—the more space is available the less you have to pay for it.

You can find the data in the spreadsheet `BostonOfficeRents.xlsx`.

(a) Show how rental cost depends on the percent of space available by creating a scatter plot using columns `D` and `F` and a regression line for that scatter plot. Identify the slope and its units. How good is the correlation?
(b) Use the graph and the formula to estimate office rent when the availability rate is 8%.
(c) The spreadsheet contains data on the vacancy rate as well as the availability rate. Create a scatterplot illustrating how the vacancy rate depends on the availability rate. Add a regression line and discuss what it tells you.

Exercise 8.5.9. [U][Section 8.3][Goal 8.1][Goal 8.1] First class mail.

Table 8.12 shows the cost of sending first class mail weighing up to one ounce.

(a) Copy and paste the data into Excel, then draw a graph of the data.
(b) Insert the trendline and display the trendline equation and the R-squared value on the graph.
(c) Write a sentence interpreting the slope of the trendline.
(d) Is this a strong correlation? Explain.

Exercise 8.5.10. [S][Section 8.1][Goal 8.1] College costs.

The spreadsheet `CollegeCosts2010.xlsx` shows the annual mean cost for tuition and fees at private and public four-year colleges in the U.S. between 1999 and 2010.

(a) Create a properly labeled graph showing how mean private and public education costs changed in the years 1999–2010.

Insert a linear trendline for each set of data. Use Excel to forecast the trendline out to 2015 (that is, 16 years past 1999).

Table 8.12. First class mail [R203]

Year	cents/oz	Year	cents/oz
1885	2	1988	25
1917	3	1991	29
1919	2	1995	32
1932	3	1999	22
1958	4	2001	34
1963	5	2002	37
1968	6	2006	39
1971	8	2007	41
1974	10	2008	42
1975	13	2009	44
1978	15	2012	45
1981	18	2013	46
1982	20	2014	49
1985	22		

(b) Write the equation for private education costs.
(c) Write the equation for public education costs.
(d) Interpret the numerical value of the slope in each trendline equation. That is, write a sentence explaining what the slope represents.
(e) Use your trendline equations to determine the projected mean tuition cost at both private and public four year colleges for 2015.
(f) Compare your answers from the previous questions with the graph. Are the answers consistent or do you need to use more digits in your calculation?

Exercise 8.5.11. [U][Section 8.1] Speed vs. MPG, revisited.

Exercise 3.10.52 looked at the relationship between speed and fuel consumption. You can do this problem even if you didn't do that one.

(a) Read data from the graph in Figure 3.7 and enter it in Excel.
(b) The information cited in Exercise 3.10.52 states that for each 5 mph you drive over 50 mph, your decrease in fuel economy means that you pay an additional $0.25 for gas. Use Excel to graph the data corresponding to speeds above 50 mph. Construct a regression line for this data. What does the slope of the regression line tell you about how fuel economy changes as speed increases? If your speed increases by 5 mph, how does your fuel economy change, on average?
(c) Use Excel to convert the data in your table from mpg to gallons per 100 miles. Graph the data again and insert the regression line. What does the slope of the regression line tell you about how fuel economy changes as speed increases? Is it easier to explain how fuel economy changes when your speed increases by 5 mpg?

Exercise 8.5.12. [S][Section 8.4][Goal 8.1] Playing with regression lines.

Use the spreadsheet `PlayWithRegression.xlsx` to explore the following questions.

(a) What happens when all the y-values are the same?
(b) What if all but one of the y-values are the same and you vary that one?
(c) What if y decreases as x increases?
(d) What if the x and y values match?

Exercise 8.5.13. [S][Goal 8.1][Section 8.4] Should businesses use private jets?

On May 26, 2012 *The Boston Globe* published a letter to the editor from David V. Dineen, Executive director of the Massachusetts Airport Management Association. He observed that companies using their own private jets had earnings 434 percent higher than those using commercial airlines. [R204]

Explain how and why Dineen is using the statistic he quotes to encourage readers to confuse correlation with causation.

Exercise 8.5.14. [S][Section 8.4][Goal 8.3] Cherry-picking.

In Section 8.4, we discovered that the author had "cherry-picked" the data. Find out what "cherry-picking" means, and where the phrase comes from. Find and discuss some examples.

Exercise 8.5.15. [S][Section 8.4][Goal 8.1][Goal 8.3] Watch TV! Live Longer!

The data in the spreadsheet `TVData.xlsx` show the life expectancy in years for several countries, along with the number of people per television set in those countries. (The idea (and the data) for this problem come from the article `www.amstat.org/publications/jse/v2n2/datasets.rossman.html`.)

(a) Which countries have the highest and lowest life expectancy at birth? Which have the highest and lowest number of people per television set?
(b) Use Excel to create a properly labelled scatter plot of the life expectancy and people per television data. Find the trendline and display the equation and the R-squared value on your graph.
(c) What is the slope of the trendline (with its units)? Explain its meaning in a sentence.
(d) Does a small number of people per television set improve health? Would people in countries with low life expectancy live longer if we sent them shiploads of television sets?
(e) Does living longer increase the number of television sets? If we improved the life expectancy in a country by providing better medical care would that cause there to be fewer people per television set?
(f) What else could be going on here? Why might high life expectancy be strongly correlated with a low ratio of people per tv set?

Exercise 8.5.16. [S] [W] Crime rates revisited.

(a) Use the data in `crimeDropsFearsRise.xlsx` to redo the analysis for the entire period from 1990 to 2009.
(b) Are the crime rates in this exercise consistent with those in the example we studied in Chapter 2?

[See the back of the book for a hint.]

Exercise 8.5.17. [S][Section 8.4][Goal 8.1][Goal 8.3] The Mississippi River.

> In the space of one hundred and seventy-six years the Lower Mississippi has shortened itself two hundred and forty-two miles. That is an average of a trifle over one mile and a third per year. Therefore, any calm person, who is not blind or idiotic, can see that in the Old Oolitic Silurian Period, just a million years ago next November, the Lower Mississippi River was upwards of one million three hundred thousand miles long, and stuck out over the Gulf of Mexico like a fishing-rod. And by the same token any person can see that seven hundred and forty-two years from now the Lower Mississippi will be only a mile and three-quarters long, and Cairo and New Orleans will have joined their streets together, and be plodding comfortably along under a single mayor and a mutual board of aldermen. There is something fascinating about science. One gets such wholesale returns of conjecture out of such a trifling investment of fact.
>
> Mark Twain
>
> Life on the Mississippi [R205]

Discuss this linear model for the length of the Mississippi river. What's the slope? Can you verify Twain's arithmetic?

Exercise 8.5.18. [Section 8.4][Goal 8.3] Well, maybe.

Explain the joke in the cartoon in Figure 8.13 reproduced from `xkcd.com/`.

Figure 8.13. Well, maybe. [R206]

9

Compound Interest—Exponential Growth

In this chapter we explore how investments and populations grow and radioactivity decays—exponentially.

Chapter goals

Goal 9.1. Understand that exponential growth (or decay) is constant relative change.

Goal 9.2. Understand how compound interest is calculated.

Goal 9.3. Work with exponential decay.

Goal 9.4. Reason using doubling times, half-lives, rule of 70.

Goal 9.5. Fit exponential models to data.

9.1 Money earns money

Imagine that you have $1,000 to invest. One of the points of investing money is to earn interest on your investment. Would you rather earn $100 per year in interest, or 8% per year in interest? In each case the interest is added into your principal (the balance in your account) each year, and you never make any withdrawals.

The first scenario is called simple interest. You find the new balance by adding $100 each year to the previous balance. After one year the balance would be $1,100, after two years $1,200, and so on.

The second scenario is a little more complicated. The interest each year is a fixed percentage of the balance. The one-plus trick does the job in one step: after one year, the new balance would be $1.08 \times \$1,000 = \$1,080$. After two, it would be $1.08 \times \$1,080 = \$1,166.40$. This pattern is called compound interest.

Table 9.1 shows your balance in each case for the first four years.

So far simple interest offers a better return on your investment. What would the numbers be in 10 years? We could continue building the table a year at a time by hand (which is tedious), we could have Excel calculate for us (we'll do that in a minute) or we could look for a pattern and find a formula for each scenario, so that we can compute for any year we like without having to do the work for all the years in between. We'll do that first.

Table 9.1. Simple and compound interest

Year	Balance	
	Simple interest	Compound interest
now	\$1,000	\$1,000.00
1	\$1,100	\$1,080.00
2	\$1,200	\$1,166.40
3	\$1,300	\$1,259.71
4	\$1,400	\$1,360.49

Simple interest leads to a linear equation. Each year the balance increases by a fixed amount, \$100, so the slope is \$100/year. The intercept is the starting value, \$1,000. The linear function is

$$B = 1000 + 100 \times T$$

where B represents the balance, in dollars, and T the number of years. If you leave your money growing until $T = 10$ years, your balance will be $1000 + 100 \times 10 = 2000$ dollars.

The function describing compound interest isn't linear. The percentage increase is constant but the amount of interest changes from year to year. In the first year you earn \$80, while in the second you earn \$86.40. To see what kind of function to use, we unwind the arithmetic in the compound interest column of Table 9.1.

Year 1: $1080.00 = 1000.00 \times 1.08$

Year 2: $1166.40 = 1080.00 \times 1.08 = (1000 \times 1.08) \times 1.08 = 1000 \times 1.08^2$

Year 3: $1259.71 = 1166.40 \times 1.08 = (1080 \times 1.08^2) \times 1.08 = 1000 \times 1.08^3$

It's clear that the function describing this growth is

$$B = 1000 \times (1.08)^T \tag{9.1}$$

where, as before, B represents the balance, in dollars, and T the time, in years. It's an exponential function, because the independent variable T is the exponent of 1.08. The 0.08 in $1.08 = 1 + 0.08$ is the constant relative change for each additional year. The 1000 is where we start: the value of B when $T = 0$. (You may remember but not have enjoyed the fact that $1.08^0 = 1$. If so, perhaps it makes a little more sense in this context. After 0 years you've received no interest, so your balance should be multiplied just by 1.)

Suppose you want to compare the balances at simple and compound interest after year 10. With simple interest you will have $1000 + 10 \times 100 = 2000$ dollars. But how can you compute 1000×1.08^{10} without boringly multiplying by 1.08 ten times? The calculator in Figure 2.2 isn't powerful enough. For that job you need a *scientific calculator*, one with a key labeled $\boxed{y^x}$ or $\boxed{x^y}$.

There are many online. Here are two: `www.math.com/students/calculators/source/scientific.html`, `web2.0calc.com/`. Each will tell you that at the end of year 10 the balance will be about \$2159, so the exponential growth has caught up with the linear.

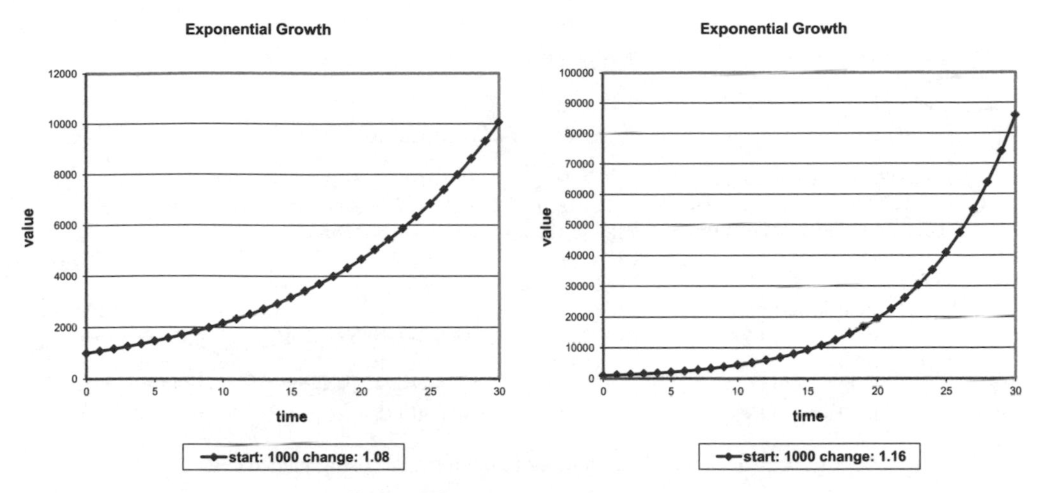

Figure 9.2. Two exponential graphs

You can do the computation with the Google calculator's buttons but to use the search bar or Excel you need to know how to enter the exponent from the keyboard without a y^x key. Both use the caret character "^" to raise a number to a power. That's meant to suggest literally "raising" the exponent. You just type

```
1000 * 1.08 ^ 10
```

into the Google search bar, or as a formula (preceded by an equal sign) in a cell in Excel to check the arithmetic in the previous paragraphs.

In *Common Sense Mathematics* we rarely put things to remember in boxes, but the moral of this discussion deserves that treatment:

> In linear growth, the absolute change is constant.
> In exponential growth, the relative change is constant.

Interest isn't the only place exponential growth happens. In Exercise 9.6.2 we ask you to think about others.

9.2 Using Excel to explore exponential growth

The spreadsheet `exponentialGrowth.xlsx` answers "what if" questions about exponential growth by allowing you to change the values 1,000 and 1.08 in (9.1). Figure 9.2 shows two examples, for the equations

$$B = 1000 \times (1.08)^T$$

and

$$B = 1000 \times (1.16)^T.$$

Each swoops upward at an increasing rate. That shape is the signature for exponential growth.

Let's take some time to see how Excel updates the calculations when we change the constants in the equation. Click on one of the cells in the exponential column, say `B17`. The

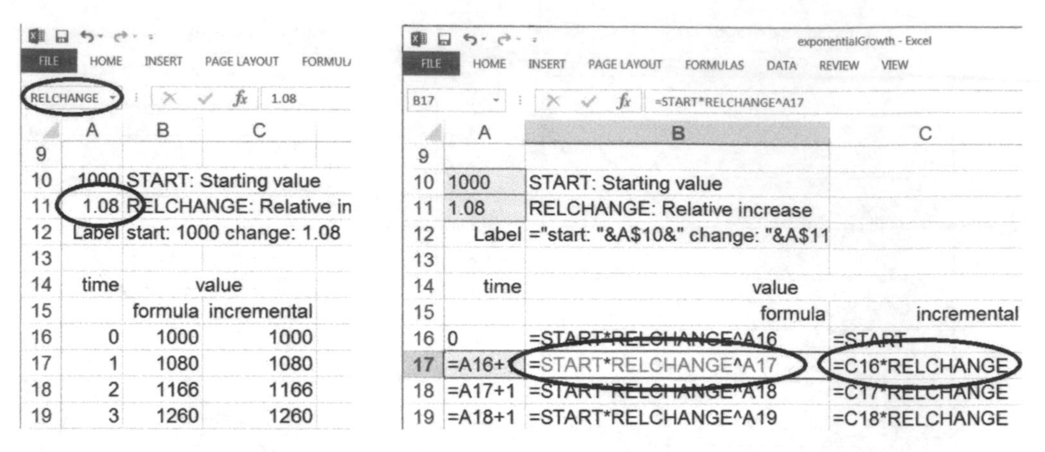

Figure 9.3. Screenshots showing Excel formulas using named cells

formula bar reads

$$= \texttt{START*RELCHANGE}^\wedge\texttt{A17}, \tag{9.2}$$

which is Excel's version of (9.1). We labeled cells `A10` and `A11` as `START` and `RELCHANGE` so that we could use (9.2) instead of

$$= \texttt{\$A\$10*\$A\$11}^\wedge\texttt{A19}.$$

The version using cell labels is much easier to understand than the one with cell references, and it doesn't need the dollar signs to tell Excel not to change those references when we copy from one row to another.

To label a cell, click on it. The Name Box at the left of the Formatting Toolbar will contain the address of the cell, so if you click on cell `H4` you will see `H4` there. You can highlight the contents of the box and type in your own name.

Figure 9.3 shows two screen shots of our spreadsheet, the first with cell values, the second with cell formulas.

The numbers in columns `B` and `C` are the same. Excel computes them in different ways. We've seen how `B17` uses the algebra in (9.2). The value in cell `C17` comes from the previous value in `C18` instead:

$$= \texttt{C16*RELCHANGE}.$$

The spreadsheet has a second tab (labeled "Compare two growth trajectories") that shows two exponential curves on the same set of axes. Figure 9.4 provides an example. There it's really clear how much faster growth is at 16% than at 8%.

In the hypothetical investment comparison at the start of this chapter linear growth starts out better but by year 10 exponential growth leads to a higher balance. To explore what happens in more detail, use the spreadsheet `linearExponential.xlsx`. It extends Table 9.1 to cover 15 years. The graph in Figure 9.5 shows that starting at year 7, the value of the exponential function is larger than the linear.

Now we can answer "what-if" questions. Suppose, for example, our money earned 7% interest instead of 8% interest. To redo the calculations we need to change just one number:

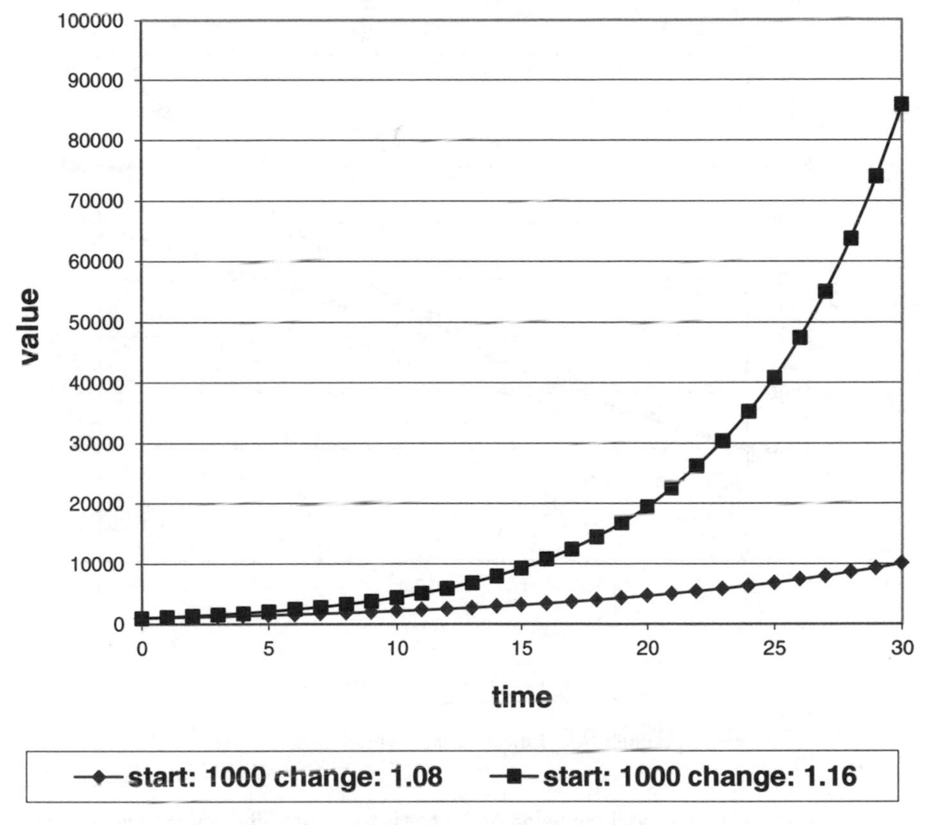

Figure 9.4. Two exponential graphs on the same set of axes

replace the 1.08 in cell `A9` with 1.07. Excel recomputes the values of the exponential function in column `C` and redraws the graph. Then you can see that with this lower interest rate, we have to wait 11 years before the exponential growth of compound interest gives us a better return.

9.3 Depreciation

It's always easier to think about increases (adding and multiplying) than decreases (subtracting and dividing) but sometimes things do decrease.

Suppose you buy a new car for \$20,000. As soon as you drive it out of the dealer's lot it's worth less. In fact it's worth less each year: it *depreciates*. Its value depends on its age.

If the car is a business expense you might choose linear depreciation for tax purposes—suppose the value decreases by \$1,800 each year. The equation that determines the value V as a function of the age A is

$$V = 20{,}000 - 1{,}800A.$$

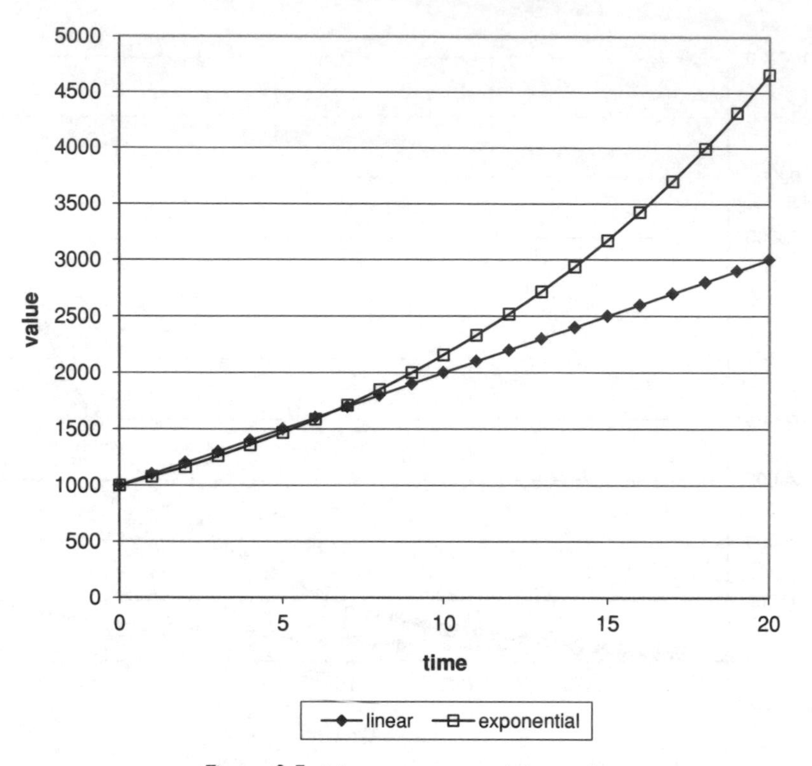

Figure 9.5. Linear vs exponential growth

But a more realistic way to model the value of the car is to assume that the percentage decrease is the same each year. Suppose it's 13%. Then each year its value is 87% of what it was the year before. The corresponding equation is

$$V = 20{,}000 \times 0.87^{A}.$$

We can use our old friend `linearExponential.xlsx` to draw Figure 9.6 showing what the car is worth over time in each case. Set `START` to 20,000, `ABSCHANGE` to −1,800, and `RELCHANGE` to $1 - 0.13 = 0.87$. (The relative change is still positive. It's a decrease rather than an increase because it's less than 1.)

When the depreciation is linear the car is worthless (at least on paper) after about 11 years. Excel doesn't know that, so it continues the graph on into negative values. If we wanted to use this graph in a more formal presentation we'd have to prevent that, and change the labels. Leaving it this way exhibits the power of thinking abstractly in Excel—the original spreadsheet can manage shrinking just as easily as growth.

9.4 Doubling times and half-lives

How long will it take to double your money? (We'll assume you're clever enough to insist on compound interest.) The answer depends on the interest rate and the initial balance. The

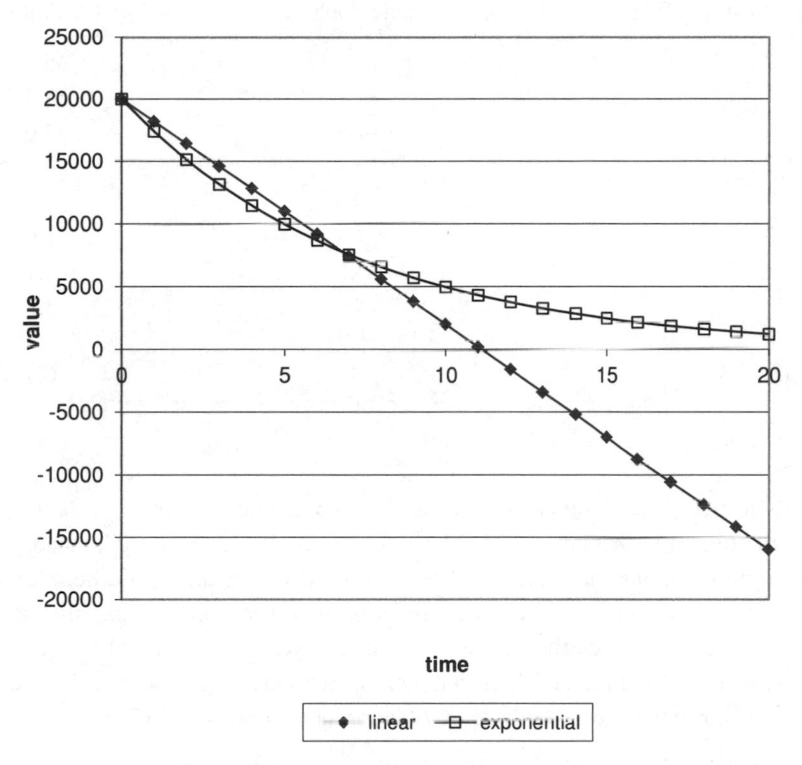

Figure 9.6. Linear vs exponential depreciation

`exponentialGrowth.xlsx` spreadsheet shows that at 8% interest with an initial investment of \$1,000 the balance is \$2,000 after 9 years (the table shows \$1999.004627, which is quite close enough to double).

If you change the initial investment to \$100 then Excel shows a balance of \$200 after the same 9 years. Experimenting with many different initial investments always shows the same doubling time. So the time it takes to double your money does not depend on the amount you start with.

What about a different interest rate? If you use 5% interest in the spreadsheet the doubling time seems to be between 14 and 15 years. We can do a calculation: $1.05^{14.5} = 2.028826162$, so 14.5 is a good guess.

At 2% interest it takes more than 30 years to double your money, so the spreadsheet doesn't give us the answer. We could find it by adding some rows, but we'll use another method instead. We'll try to guess the value of T in the equation $1.02^T = 2$ and adjust our guess until we're close enough. Perhaps the answer is $T = 40$ years:

$$1.02^{40} = 2.208039664.$$

Too big, so we need less time. Try 35:

$$1.02^{35} = 1.999889553.$$

Bingo!

Table 9.7. Double your money

interest rate (%)	approximate doubling time	70/rate
2	35	35.0
3	24	23.3
5	14.5	14.0
7	10.5	10.0
8	9	8.8
10	7.5	7.0
15	5	4.7
20	4	3.5
50	1.7	1.4
100	1	0.7

We've collected these results and a few more in the second column of Table 9.7.

The third column in that table shows the results from the "Rule of 70", that says that you can estimate the compound interest doubling time by dividing the magic number 70 by the annual interest rate as a percent. The approximation is better when the interest rate isn't too large; those are just the cases that matter most in everyday investing. The most commonly quoted consequence of the Rule of 70 is that money invested at 7% will double in 10 years.

Figure 9.8 shows how good the Rule of 70 is for interest rates up to 20%.

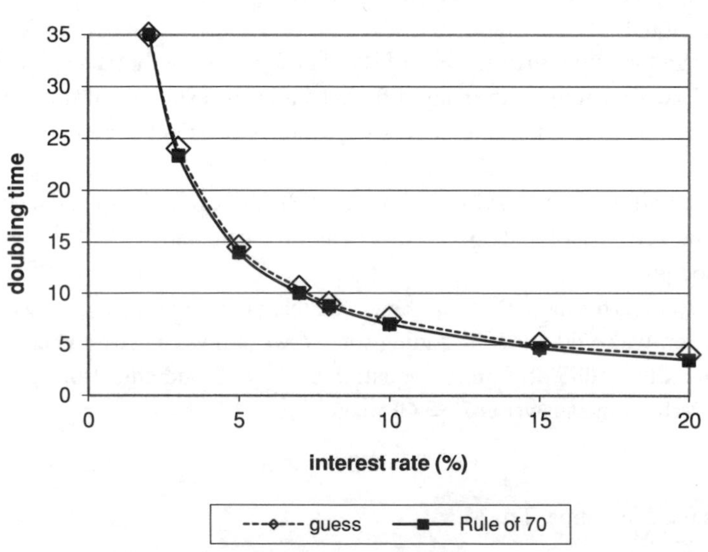

Figure 9.8. Doubling times

When a relative increase occurs repeatedly the doubling time is independent of the initial value. So if inflation is 5% per year, all prices will double in 14 years.

Knowing the doubling time helps you make quick calculations. Since 5% inflation doubles prices in 14 years it will quadruple them in 28 years. In 42 years they will be eight times as large. The Bureau of Labor Statistics inflation calculator says that inflation in the 42 years from 1968 to 2010 increased the cost of a $100 item to $626. That's not quite eight times as much, so the average inflation rate for those years was not quite 5% per year.

The Rule of 70 applies to depreciation as well—it tells you the *half-life*. That's the equivalent for depreciation of the doubling time—the time until half the original value has disappeared. Like doubling time, the half-life depends on the depreciation rate, but not on the original value. For 13% annual depreciation it's approximately $70/13 = 5.38461538 \approx 5.4$ years.

The term half-life comes from atomic physics, where it describes the way the quantity of a radioactive element decreases over time. The following quotation from `www.nirs.org/factsheets/hlwfcst.htm` provides food for quantitative thought.

> After ten half-lives, one-thousandth of the original concentration [of a radioactive substance] is left; after 20 half-lives, one millionth. Generally 10-20 half-lives is called the hazardous life of the waste. Example: plutonium-239, which is in irradiated fuel [from a nuclear power plant], has a half-life of 24,400 years. It is dangerous for a quarter million years, or 12,000 human generations. [R207]

This is the kind of quotation that begs to have its numbers checked.

First let's look at "ten half-lives". After one half-life the concentration is half what it was at the start. After two half-lives it's half of a half, or 1/4. After three half-lives it's 1/8. So after 10 half-lives it's

$$\frac{1}{2} \times \frac{1}{2} \times \cdots \times \frac{1}{2} = \frac{1}{2^{10}}$$

of what it was.

We saw when we studied the metric prefixes that $2^{10} = 1{,}024 \approx 1{,}000$. That's why "kilo" means "1,000" most of the time but "1,024" when computers are involved. Now we use the same fact to see why $1/2^{10}$, which is exactly 1/1,024, is approximately 1/1,000—one one thousandth.

What about twenty half-lives? In that time the original concentration will be reduced to 1/1,000 of 1/1,000 of what it was at the start. Since a thousand thousand is a million, that's one one-millionth.

According the quotation, plutonium-239 will be dangerous for at least ten 24,000 year half-lives. That's 240,000 years, which is indeed about a quarter of a million years. Is it 12,000 generations? Yes, if you calculate with $240{,}000/12{,}000 = 20$ years per generation. That's perhaps a little low for the developed world, but good enough to highlight the danger of nuclear waste.

9.5 Exponential models

For compound interest and radioactive decay the equation for exponential change gives exact answers, just as the linear equation gives exact answers for simple interest and electricity bills.

Table 9.9. Bacteria growth, Time: hours, Population: millions of organisms per milliliter [R208]

	population		
time	W	R	S
4	7.9	8.5	17.5
8	17.3	13.5	42.1
12	44.7	48.9	225.2
17	119.3	268.7	407.3
20	41.3	98.0	149.3
24	41.3	64.0	160.0

When change is approximately linear a regression line may be useful. When it's approximately exponential, we can construct an *exponential trendline*. Most elementary texts discuss the reproduction of bacteria as a toy example of exponential growth. Bacteria reproduce by dividing, so each individual gives rise to 2, then 4, then 8 descendants, and so on. The number of bacteria grows exponentially. But that can't go on forever. Since exponential growth curves quickly become steep, it can't go on for very long. Eventually, crowding or diminishing resources cause growth to slow, perhaps even to reverse as organisms die faster than new ones are born.

Table 9.9 records the population of three different strains of the *E. coli* bacterium in a one day experiment conducted by Professor Vaughn Cooper at the University of New Hampshire.

The raw data points in the graph on the left in Figure 9.10 (built in the spreadsheet `bacteriaGrowth.xlsx`) suggest that the growth of each strain was exponential until about hour 17. To construct the graph on the right we plotted the data for each strain for that period. In the resulting chart we right-clicked on a data point for each strain, selected `Add Trendline ...`, and chose `Exponential` on the `Type` tab. As usual, we asked for the equations and the *R*-squared values. Those are all pretty close to 1; best for strain W, worst for strain S.

Let's look at the equation for the exponential trendline for strain W, which you can see in the figure:

$$y = 3.3663\, e^{0.2109x}. \tag{9.3}$$

In Chapter 8 we discovered that the linear trendline equation reported on the graph might not provide enough significant digits. The same is sometimes true for the equations for exponential trendlines, but recovering the more accurate values is trickier, and not worth trying to master. In this example the displayed numbers are good enough.

What is the "*e*" in this equation? The complete answer to that question calls for much more mathematics than you need to know to apply common sense to quantitative arguments. But since Excel uses it you may encounter it somewhere so we'll discuss it briefly.

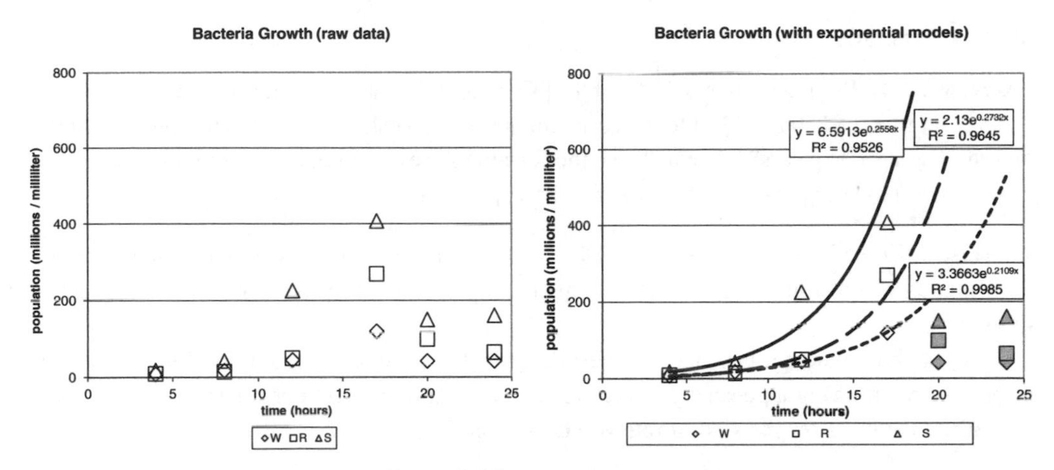

Figure 9.10. Bacteria growth

The simplest explanation is that e is just a particular number—approximately 2.7183. Like $\pi \approx 3.1416$ it's one of those numbers whose decimal expansion "goes on forever", so the first few decimal places give just an approximation. The number e appears naturally in discussions of exponential growth just as π appears in discussions of circles. It was named by the prolific mathematician Leonhard Euler (1707–1783) who was the first to recognize its importance.

Back to our exponential growth function for the strain W. We see that

$$e^{0.2109} \approx (2.7183)^{0.2109} \approx 1.2348.$$

You don't have to remember the value of e to compute with it, since both Excel and the Google calculator provide the built-in function EXP to do the job. Entering `=EXP(0.2109)` in a cell in Excel or the same thing without the equal sign in the Google search bar will give you the same answer.

Now we can substitute 1.2348 for $e^{0.2109}$ in (9.3) and rewrite it as

$$y = 3.3663 \times 1.2348^x.$$

In that form we recognize this behavior as exponential growth at a constant rate of about 23% per hour. The Rule of 70 tells us to expect a doubling time of about $70/23 \approx 3$ hours. That matches the data, which does indeed show the strain W population doubling about every three hours for most of the day.

If you experiment with `EXP` you will find that $e^{0.7} = 2.0137527\ldots \approx 2$. It's the 0.7 in the exponent that leads to the Rule of 70. To learn just how, go on to take a course in calculus.

Using the exponential trendline for constant growth rate to predict the future populations would fail in this experiment. About two thirds of the way through the 24 hour day a different reality appears. The gray bullets in Figure 9.10 show that all three populations drop dramatically. (A newspaper reporter wanting to emphasize the drama might say, incorrectly, that the populations dropped "exponentially.")

It's the biologist's job to understand why. The analysis of the exponential growth at the start has told him only how the bacteria grow when there's lots of food, lots of room and no competition.

9.6 Exercises

Exercise 9.6.1. [S][R][Section 9.1][Goal 9.1][Goal 9.2] Compound interest computations.

If you invest $1500 at 7% interest compounded every year, how much will you have at the end of 10 years? 15 years? 20 years? Use the formula for exponential growth; then check your answers with the `exponentialGrowth.xlsx` spreadsheet.

Exercise 9.6.2. [S][Goal 3.1][Section 9.1][Goal 9.1] When do you expect exponential growth?

In each of the following situations, explain why you would expect linear or exponential growth.

Think about whether the change is best described as an absolute rate (like dollars per hour or gallons per mile) or a percentage (like percent per year or percent per washing).

Write the units for the kind of rate you decide on.

(a) Price increases from year to year due to inflation.
(b) How the amount of gas you use depends on how far you drive.
(c) The amount of money left on your public transportation debit card as the days go by and you commute to school or work.
(d) The amount of sales tax you pay, depending on how much you buy.
(e) The amount of dirt left in your kid's filthy jeans when you wash them over and over again.
(f) The population of the world as the years go by.
(g) Your credit card balance if you stop making payments. (We will study credit cards in the next chapter.)
(h) The height of the snow as it accumulates in a big storm.
(i) The number of people sick in the first weeks of the flu season.
(j) The number of subscribers to a hot new social network in its first days.

Think about your answers before you look at the hints.

[See the back of the book for a hint.]

Exercise 9.6.3. [S][Section 9.1][Goal 9.1] Is it really exponential?

In everyday usage the phrase "growing exponentially" is just a vibrant synonym for "growing rapidly". It's rare that it really means a constant relative change.

Find instances of "exponential" growth in the media where what's meant is just very rapid growth.

Exercise 9.6.4. [S][Section 9.1][Goal 9.1] Health care spending.

In Chapter 3, Exercise 3.10.16, we used data from the 2010 National Health Expenditures report to compute the absolute and relative changes in health care spending per person from 2007 to 2008.

(a) Use the results of those calculations to build linear and exponential models for the growth of health care spending per person.
(b) Use each model to predict when health care spending will reach $10,000 per person per year.

Exercise 9.6.5. [S][Section 9.1][Goal 9.1] Malthus.

In 1798 Thomas Malthus, an English economist and clergyman, wrote "An Essay on the Principle of Population". He said there:

> I think I may fairly make two postulata.
>
> First, That food is necessary to the existence of man.
>
> Secondly, That the passion between the sexes is necessary and will remain nearly in its present state.
>
> These two laws, ever since we have had any knowledge of mankind, appear to have been fixed laws of our nature, and, as we have not hitherto seen any alteration in them, we have no right to conclude that they will ever cease to be what they now are, without an immediate act of power in that Being who first arranged the system of the universe, and for the advantage of his creatures, still executes, according to fixed laws, all its various operations.
>
> ...
>
> Assuming then my postulata as granted, I say, that the power of population is indefinitely greater than the power in the earth to produce subsistence for man.
>
> Population, when unchecked, increases in a geometrical ratio. Subsistence increases only in an arithmetical ratio. A slight acquaintance with numbers will shew the immensity of the first power in comparison of the second.
>
> ...
>
> The power of population is so superior to the power in the earth to produce subsistence for man, that premature death must in some shape or other visit the human race. The vices of mankind are active and able ministers of depopulation. They are the precursors in the great army of destruction; and often finish the dreadful work themselves. But should they fail in this war of extermination, sickly seasons, epidemics, pestilence, and plague, advance in terrific array, and sweep off their thousands and ten thousands. Should success be still incomplete, gigantic inevitable famine stalks in the rear, and with one mighty blow levels the population with the food of the world. [R209]

Malthus claimed that the food supply grows in a linear fashion. As a unit of food supply he used the amount of food needed for one person for one year. He estimated food production in Britain in 1798 as 7,000,000 food units and that food production might increase by a constant 280,000 units each year.

Malthus also believed that the population of Britain was growing at a rate of 2.8% each year. In 1798, the population was about 7,000,000.

(a) Write a linear function that models food production.
(b) Write an exponential function that models population growth.
(c) Was there enough food for each individual in Britain in 1798?
(d) Using Malthus's models, determine whether there would be enough food for each individual in Britain in 1800.
(e) Malthus claimed that the population in Britain would eventually outstrip the food supply—a prediction we now call "the Malthusian dilemma." He didn't have Excel to do the arithmetic for him, but we do. Use it to estimate when Malthus's predicted disaster would occur. Was Malthus right?

Exercise 9.6.6. [S][Section 9.1][Goal 9.1][Goal 9.2] The pawn shop business model.

On April 9, 2011 *The New York Times* reported on a pawn shop that opened in an ex-Blockbuster store:

> The borrowers are given 60 days to pay back the loan, and La Familia charges a 20 percent interest rate per month. (So for a $100 loan, the borrower would need to pay back $140 after 60 days.) [R210]

(a) Explain why 20% interest per month on a $100 loan for two months would actually require repayment of a little more than $140.
(b) What is the annual interest rate when this business lends money?

Exercise 9.6.7. [S][W][Section 9.2][Goal 9.1][Goal 9.2] Playing with exponential growth.

Open the spreadsheet `exponentialGrowth.xlsx` and describe what happens to the graph when you make each of the following experiments. If you can see easily what happens to the numbers, describe that too.

(a) Change the value of `START` from 1,000 to 10, then 100, then 10,000. Change it to some other random positive values that aren't as nice.
(b) Change the value of `START` from 1,000 to $-1{,}000$.
(c) Change the value of `RELCHANGE` to 1.
(d) Change the value of `RELCHANGE` to 1.01 (1% growth). Fit a linear trendline to the data. What is the R-squared value? What does it tell you?
(e) Change the value of `RELCHANGE` to 2. Why does the graph look flat at 0 as far as $T = 20$? Is it really flat?
(f) Change the value of `RELCHANGE` to 10.
(g) Change the value of `RELCHANGE` to 0.9 (a 10% decrease).
(h) (Optional) Can you figure out how we got the label on the graph to incorporate the values of `START` and `RELCHANGE`?

Exercise 9.6.8. [S][Section 9.2][Goal 9.1] Five percent.

If you try to use `linearExponential.xlsx` to see when exponential growth at 5% catches linear you see that it's still behind at 20 years, which is as far as the table goes.

Modify the spreadsheet to determine when it catches up.

Exercise 9.6.9. [U][Section 9.2][Goal 9.1][Goal 9.2] Playing with linear vs exponential growth.

Use the spreadsheet `linearExponential.xlsx` to answer these questions.

(a) How does changing the initial investment change the time it takes for the exponential function to catch up with the linear function?
(b) Does doubling or tripling both the initial investment and the absolute change affect the time it takes for the exponential function to catch up to the linear function?

Exercise 9.6.10. [S][Section 9.2][Goal 9.1][Goal 9.2] Deals you can't believe.

The data in this problem aren't real. But the problem is interesting and instructive, so it's worth spending time on.

Table 9.11. Three car deals

Month	Payment		
	Deal 1	Deal 2	Deal 3
(down) 0	10,000	5,000	1.00
1	100	50	0.01
2			
⋮			
24			
Total			

Suppose you are shopping for a car and find three deals advertised:

- Make a $10,000 down payment and pay only $100 per month for two years.
- Just $5000 down, monthly payments start at a low $50 and increase by $50 each month for two years.
- Give me $1.00 today and take the car home! Pay 1 penny for the first month. Then double your payment each month. After two years, the car is yours.

(a) Before you do any calculating, which deal do you think is best? Why?
(b) What would your monthly payments be in the second and tenth months if you take the second dealer's offer?
(c) What would your monthly payments be in the second and tenth months if you take the third dealer's offer?
(d) For each deal, write an algebraic expression that gives the monthly payment.
(e) Use Excel to calculate your total payments for the 24 months. Set up four columns as in Table 9.11. Then tell Excel how to fill in the columns to 24 months. Finally, use the `SUM` function to add up the payments.
(f) Now use what your calculations tell you to compare the three deals. Which is best? Which is worst?

Exercise 9.6.11. [S][C][Section 9.2][Goal 9.1] Green energy in China.

In the December 21 & 28 2009 issue of *The New Yorker* Evan Osnos wrote in his essay "Green Giant: Beijing's crash program for clean energy" that China's spending on R & D, now seventy billion dollars a year, has been growing at an annual rate of about twenty percent for two decades. [R211]

(a) What does "R&D" stand for?
(b) Use Excel to build a chart of annual Chinese R&D expenditures for the years 1989–2008.
(c) Add a data column showing the annual expenditures adjusted for inflation (use the United States cost of living index) and display that data on your chart.

Exercise 9.6.12. [S][Section 9.3][Goal 9.1] Car excise tax.

In Massachusetts you pay excise tax each year on the current value of your automobile. Assume for the sake of this problem that the rate is 3%, so you would pay $600 in excise tax in the first year you owned a new $20,000 car.

Use Excel to answer the following questions.

(a) Suppose the car depreciates linearly at a rate of $1,800 per year. What formula calculates the amount of excise tax you pay as a function of the age of the car?
(b) If you own the car for ten years, what will the car be worth then and how much total excise tax will you have paid?
(c) Answer the same questions if it depreciates at the rate of 13% per year.
(d) Find real data on the way a new car depreciates in value. Is an exponential model a good approximation?

Exercise 9.6.13. [S][Section 9.4][Goal 9.4] Iodine 131.

An article in *The New York Times* on April 6, 2011 soon after the Fukushima disaster discussed levels of radioactive iodine (iodine 131) in fish caught near Japan. The article noted that Japan recently revised the safety limit for iodine 131 in fish to 2,000 becquerels per kilogram. (A becquerel is a measure of radiation.)

Radioactive iodine has a half-life of about 8 days.

If a fish contained 10,000 becquerels of iodine 131 per kilogram, how long would it take for the iodine to decay to a "safe" level?

Exercise 9.6.14. [U][Section 9.4][Goal 9.1][Goal 9.4] Quadrupling time.

(a) Explain why the quadrupling time in exponential growth is just twice the doubling time.
(b) Show that quadrupling time is given by a "Rule of 140" analogous to the rule of 70.

Exercise 9.6.15. [S][Section 9.4][Goal 9.1][Goal 9.4][Goal 9.2] Tripling time.

Suppose you invest $1000 at 10% interest compounded every year. (That's a pretty good rate of return if you can get it—don't trust a Madoff promise!)

(a) How long will it be until your balance is $3000?

[See the back of the book for a hint.]

(b) Find the tripling time for some other interest rates.
(c) Check that the tripling time in exponential growth is given (approximately) by a "Rule of 110."
(d) Check that $e^{1.1} \approx 3$.

Exercise 9.6.16. [U][Section 9.4][Goal 9.4][Goal 9.2] Compounding very frequently.

(a) Calculate the effective rate for 8% annual interest when it's compounded weekly, daily, hourly and once every second.
(b) Estimate the effective rate if the interest is compounded every instant.

[See the back of the book for a hint.]

(c) Redo the calculations starting with a 25% annual increase. (Not realistic for interest on a bank account!) Show that the Rule of 70 for doubling times is more accurate the more frequently you compound the interest.

Exercise 9.6.17. [S][C][W][Section 9.4][Goal 9.1][Goal 9.4] How fast does information double?

In the Preface to the Carnegie Corporation report *Writing to Read* Vartan Gregorian wrote that he's been told that the amount of available information doubles every two to three years. [R212]

(a) What growth rate in percent per year would lead to a doubling time of two to three years?
(b) Who is Vartan Gregorian?
(c) Can you verify his assertion?

[See the back of the book for a hint.]

Exercise 9.6.18. [S][Section 9.5][Goal 9.5][Goal 9.4] Bacteria doubling time.

Find the approximate doubling times for strains R and S in the bacteria growth example in Section 9.5.

Exercise 9.6.19. [S][Section 9.5][Goal 9.1][Goal 9.5] When will R catch S?

The population of strain S outnumbers that of strain R for the entire first 17 hours of the experiment discussed in Section 9.5. But the exponential trendline equation shows that strain R is growing faster. If the exponential growth were to continue (which it didn't) when would strain W catch up?

[See the back of the book for a hint.]

Exercise 9.6.20. [U][Section 9.5][Goal 9.5] The magic number *e*.

(a) Find the value of *e* in Excel using the formula `=EXP(1)`.
(b) Find the value from Google with the same formula (without the equal sign). Check that the answers agree as far as they go together.
(c) Which provides more digits?
(d) Can you get more precision from Excel by formatting the cell in which the number appears?
(e) Find even more digits with an internet search.

Exercise 9.6.21. [S][Section 9.5][Goal 9.1] Educating mothers saves lives, study says.

On September 17, 2010 *The Boston Globe* carried an Associated Press report on a study that found that

> for every extra year of education women had, the death rate for children under 5 dropped by almost 10 percent. In 2009, they estimated that 4.2 million fewer children died because women of childbearing age in developing countries were more educated.
>
> In 1970, women aged 18 to 44 in developing countries went to school for about two years. That rose to seven years in 2009. [R213]

(a) How much did the death rate for children under 5 decline from 1970 to 2009?
(b) Build as much as you can of the exponential model implicit in this quotation. What are the independent and dependent variables? What is the annual relative change?

Exercise 9.6.22. [S][Section 9.5][Goal 9.1][Goal 9.5] Email.

On May 30, 2011 Virginia Heffernan blogged in *The New York Times* that the number of email accounts grew from about 15 million in the early 1990s to 569 million by December 1999, and that today [when she was writing] there are [were] more than 3 billion. [R214]

(a) Is this exponential growth?

(b) Can you use these numbers to make predictions?

[See the back of the book for a hint.]

Exercise 9.6.23. [S][Section 9.5][Goal 9.5] When will India pass China?

In an article dated April 1, 2011 on the website About.com you could read that India, the world's second largest country, had a population of 1.21 billion. India was expected to pass China by 2030, when it would have more than 1.53 billion people. China's population then would be just 1.46 billion.

The article noted that India's growth rate of 1.6% per year doubles its population in less than 44 years. [R215]

(a) Is the article correct in stating that an annual growth rate of 1.6% means India's population will double in 44 years?
(b) Assuming that India's growth rate remains 1.6% annually, what will its population be in 2030 when it surpasses China's?
(c) Assuming that India's growth rate remains 1.6% annually from 2011 on, what will its population be in the year 2100? Compare that figure to the current population of the world. Do you think India's growth rate can in fact continue at 1.6% for the 89 years from 2011 to 2100?

Exercise 9.6.24. [S][Section 9.5][Goal 9.5] Health care expenditures grow.

The National Health Expenditures report, released in January 2009, stated that overall health care spending in the United States rose from $7062 per person in 2006 to $7421 per person in 2007.

(a) Calculate both the absolute change and percentage change in health care spending per person from 2006 to 2007.
(b) Using 2006 as your starting year (2006 = year 0), determine an exponential equation that calculates the amount of health care spending over time assuming the annual percentage change stays the same. Clearly identify the variable names and symbols in your equation.
(c) Using 2006 as your starting year (2006 = year 0), determine a linear equation that calculates the amount of health care spending over time assuming the annual absolute change stays the same. Clearly identify the variable names and symbols in your equation.
(d) Create an Excel spreadsheet to compare the two growth models' predictions for health care spending through the year 2021. Include a chart showing both models.
(e) Which model first predicts that U.S. health care spending will reach a level of $10,000 per person? In what year will that occur?

Exercise 9.6.25. [S] Joe Seeley, *in memoriam*.

Joe Seeley died at age 50 in the fall of 2012.

He was a brave and witty blogger at `joes-blasts.blogspot.com/` throughout his hospitalization, creating virtual lemonade from the sourest of lemons. We think his words helped him; I know they helped those who cared for him to cheer him on. They will help the hospital staff care better for patients who come after him. And they will help you learn a little mathematics.

Figure 9.12 appeared in the blog at a hopeful moment in his odyssey. It shows Joe's white blood cell counts on days following a stem cell transplant. He chose white for the bars, to

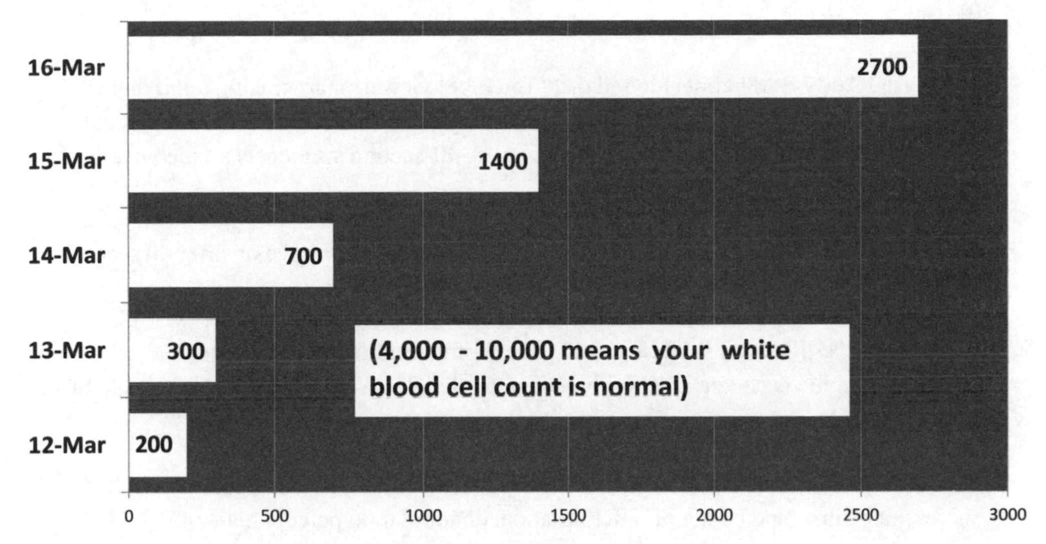

Figure 9.12. Proliferating white blood cells

symbolize white blood cells, and red for the background, for blood in general. I wrote him about it:

> March 18, 2011 6:58 AM
> Ethan Bolker said . . .
>
> Exponential growth is good! . . . Will you still have a daily double after the predicted short dip? May I use your data for my quantitative reasoning class at UMass Boston?
>
> March 18, 2011 9:48 AM
> Joseph Seeley said . . .
>
> I will not see doubling again, unless something is wrong. Over the next few months, the counts will rise and fall, sometimes for no reason that the doctors can determine.
>
> I hereby authorize the use of my blood count data for any and all educational purposes.

(a) Enter the data in Excel. Reproduce Joe's chart. Match the formatting (labels, colors, sizes, fonts) as well as you can.
(b) Create an exponential trendline for the data.
(c) Use common sense or your trendline to predict when Joe's white blood cell count will be in the normal range.
(d) His white blood count on March 17 was 5100. Does that match your prediction?
(e) Modify your chart to include this new information. Mark it with a suitable exclamation!

Exercise 9.6.26. [S][Section 9.3][Goal 9.3] How impatient are you?

The Boston Globe reported on September 24, 2012 that:

> [MIT grad Andy Berkheimer] found that [YouTube] viewers start closing out if there's even a two-second delay. Every one-second delay after that results in a 5.8 percent increase in the number of people who give up. A 40-second wait costs a video nearly a third of its audience. [R216]

Show that at this rate more than ninety percent of the viewers would give up after 40 seconds—not the "nearly a third" in the quote.

Exercise 9.6.27. [U][Section 9.1][Goal 9.1] Does compounding always matter?

On page 32 of his excellent and highly recommended *The Signal and the Noise* Nate Silver wrote

> [Over the] 100-year-period from 1896 through 1996 ... sale prices of houses had increased by just 6 percent *total* after inflation, or about 0.06 percent annually. [R217]

Silver clearly divided six percent by 100 to reach his 0.06 percent annually conclusion. But that's not how percent increases work. The 0.06 percent annual increase must be compounded.

Would taking compounding into account change Silver's fundamental point?

Exercise 9.6.28. [S][Section 9.1][Goal 9.2] Cuba, you owe us $7 billion.

On April 18, 2014 Leon Neyfakh wrote in *The Boston Globe* that property confiscated by the Cuban government in the 1959 revolution was

> ... originally valued at $1.8 billion, which at 6 percent simple interest translates to nearly $7 billion today. [R218]

(a) Is the simple interest calculation in the quotation correct?
(b) What would the value be today at 6 percent compound interest?
(c) What would the value be today simply taking inflation into account?
(d) Discuss which of the three valuations makes the most sense.

Exercise 9.6.29. [S][Section 9.1][Goal 9.2] "As Time Goes By".

On December 13, 2012 you could read in *The New York Times* that the piano from Rick's place in the 1942 movie *Casablanca* is up for auction.

> Sotheby's expects [it] to sell from $800,000 to $1.2 million in the auction on Friday. That is between 34 to 48 times what [Ingrid] Bergman was paid for sharing top billing with Humphrey Bogart. [R219]

(a) How much was Ingrid Bergman paid for her role in the film? Calculate this two ways using the data in the quotation and comment on what you discover.
(b) Would adjusting her pay to take inflation into account allow her to bid on the piano in 2012?
(c) What compound interest rate would she have to have earned on her pay to bid on the piano in 2012?
(d) Find out what happened at the auction.

Review exercises

Exercise 9.6.30. [A] You invest \$500 in an account that earns \$10 in interest each year.

(a) At the end of 24 months, what is the balance?
(b) At the end of 30 months, what is the balance?
(c) At the end of 5 years, what is the balance?
(d) Find the linear equation that gives the balance after t years.

Exercise 9.6.31. [A] You buy a car for \$15,000 and for tax purposes you depreciate it at a rate of 11% per year.

(a) At the end of 24 months, what is the value of the car?
(b) At the end of 5 years, what is the value of the car?
(c) Find the exponential equation that gives the value of the car after t years.
(d) Does the value of the car ever reach \$0?

Exercise 9.6.32. [A] Calculate the percentage.

(a) What is 8% of \$2000?
(b) What is 108% of \$2000?
(c) What is 3.25% of \$800?
(d) What is 103.25% of \$800?

Exercise 9.6.33. [A] Use a calculator to evaluate these expressions using exponents. (You may find typing into the Google or Bing calculator much faster than using one that requires you to press keys, either with your fingers or with a mouse.)

(a) 1.03^4
(b) 0.89^5
(c) 140×1.03^4
(d) 80×0.89^5
(e) $\frac{1}{3^8}$
(f) $(\frac{1}{3})^8$
(g) 1.25^0
(h) 1.25^1
(i) e^2
(j) e^{15}

Exercise 9.6.34. [A] In the exponential functions below, identify the relative change and the initial amount.

(a) $P = 100 \times (1.05)^T$.
(b) $y = 400 \times (0.88)^x$.
(c) $S = 550 \times (1.22)^Q$.
(d) $P = 96 \times (0.50)^T$.

Exercise 9.6.35. [A] Excel gives the following best-fit exponential function for a set of data: $y = 2.099 \times e^{1.344x}$. Find the constant growth rate and rewrite the function without using e.

10

Borrowing and Saving

When you borrow money—on your credit card, for tuition, for a mortgage—you pay it back in installments. Otherwise what you owe would grow exponentially. In this chapter we explore the mathematics that describes paying off your debt.

Chapter goals

Goal 10.1. Examine how credit cards work.

Goal 10.2. Study balance and interest when paying off a loan periodically.

Goal 10.3. Calculate monthly mortgage payments and examine the costs and benefits of home ownership.

Goal 10.4. Understand periodic compounding, APR and other interest terms.

Goal 10.5. Understand the basics of saving money with a long-term goal like retirement.

10.1 Credit card interest

We've learned how to read an electricity bill—now we'll tackle a credit card statement. Figure 10.1 shows a sample from `www.practicalmoneyskills.com/`.

If you have a credit card you get a statement like this once a month. John Doe (the owner of this card) charged \$125.24 in merchandise during January. (The billing cycle seems to end on the 13th of the month, but we'll assume it's the first.) He's decided not to use this card any longer, and will settle his debt by paying \$20 each month. When will he be debt free, and how much interest will he have paid?

He makes a minimum payment of \$20 in February for his January purchases, so his balance is $\$125.24 - \$20 = \$105.24$. He's paid no interest so far. But now that changes. The credit card company charges interest on the balance he carries in February. The FINANCE CHARGE SUMMARY shows a periodic (that is, monthly) rate of 1.65% so he will pay $\$105.24 \times 0.0165 = \$1.74$ in interest. The 1+ trick tell us that at the beginning of March he owes

$$\$105.24 \times 1.0165 = \$106.98.$$

After his \$20 payment on March 1 his balance is

$$\$105.24 \times 1.0165 - \$20 = \$86.98.$$

	CREDIT CARD STATEMENT		SEND PAYMENT TO Box 1244 Anytown, USA
ACCOUNT NUMBER 4125-239-412	NAME John Doe	STATEMENT DATE 2/13/09	PAYMENT DUE DATE 3/09/09
CREDIT LINE $1200.00	CREDIT AVAILABLE $1074.76	NEW BALANCE $125.24	MINIMUM PAYMENT DUE $20.00

REFERENCE	SOLD	POSTED	ACTIVITY SINCE LAST STATEMENT	AMOUNT
483GE7382		1/25	PAYMENT THANK YOU	-168.80
32F349ER3	1/12	1/15	RECORD RECYCLER ANYTOWN, USA	14.83
89102DIS2	1/13	1/15	BEEFORAMA REST ANYTOWN, USA	30.55
NX34FJD32	1/18	1/18	GREAT ESCAPES BIG CITY, USA	27.50
84RT3293A	1/20	1/21	DINO-GEL GASOLINE ANYTOWN, USA	12.26
973DWS321	2/09	2/09	SHIRTS 'N SUCH TINYVILLE, USA	40.10

Previous Balance	(+)	168.80	Current Amount Due	125.24
Purchases	(+)	125.24	Amount Past Due	
Cash Advances	(+)		Amount Over Credit Line	
Payments	(-)	168.80	Minimum Payment Due	20.00
Credits	(-)			
FINANCE CHARGES	(+)			
Late Charges	(+)			
NEW BALANCE	(=)	125.24		

FINANCE CHARGE SUMMARY	PURCHASES	ADVANCES	For Customer Service Call: 1-800-xxx-xxxx
Periodic Rate	1.65%	0.54%	For Lost or Stolen Card, Call: 1-800-xxx-xxxx
Annual Percentage Rate	19.80%	6.48%	24-Hour Telephone Numbers

Figure 10.1. A credit card statement [R220]

The credit card company used $1.74 of the payment for the February interest. The rest they subtracted from his balance.

Table 10.2 tells the rest of the story. Figure 10.3 shows the Excel formulas we used in `PayOffDebt.xlsx` to calculate the values in that table, along with a graph showing how the balance decreases each month at a slightly faster rate, until it reaches 0.

John's last payment is for the $10.86 balance.

In seven months he's paid $5.80 in interest. That doesn't seem too terrible. It's $\$5.80/\$125.24 = 0.046311 \approx 4.6\%$. But don't be fooled. This is a monthly statement, so that 4.6% isn't the annual interest rate. John didn't borrow that money for a year. Some of it he had for seven months, some for just one.

The law requires the credit card company to tell you the *APR* (*annual percentage rate*) somewhere on the monthly statement. You can find it on this one in the **FINANCE CHARGE SUMMARY** section: it's 19.80%. You can check their arithmetic: $1.65\% \times 12 = 19.80\%$. We will have more to say about the APR in Section 10.4.

That large interest rate is why the credit card company wants you to pay just the small minimum each month. The full balance appears at the top of the statement labelled NEW

Table 10.2. Seven months to pay it off

Month	Balance	Interest
Jan	125.24	0.00
Feb	105.24	1.74
Mar	86.98	1.44
Apr	68.41	1.13
May	49.54	0.82
Jun	30.36	0.50
Jul	10.86	0.00
total		5.80

BALANCE—but the only payment shown is the MINIMUM PAYMENT DUE. You have to know that it's best for you to pay the full balance at once.

There are other ways credit card companies make you pay for the convenience of borrowing their money. One kicks in if you miss a payment by even a day or so. Then they charge a substantial late payment fee and may also increase the already large interest rate. The law requires credit card companies to print a warning on your statement. Here's what one says:

> **Late Payment Warning:** If we do not receive your minimum payment by the date listed above you may have to pay up to a $39.00 late fee and your APRs will be subject to increase to a maximum Penalty APR of 29.99%.

Moreover, the late payment will show up on your credit report, so when you go to a bank later to take out a mortgage on the condo you want to buy, they may charge you a higher interest rate too.

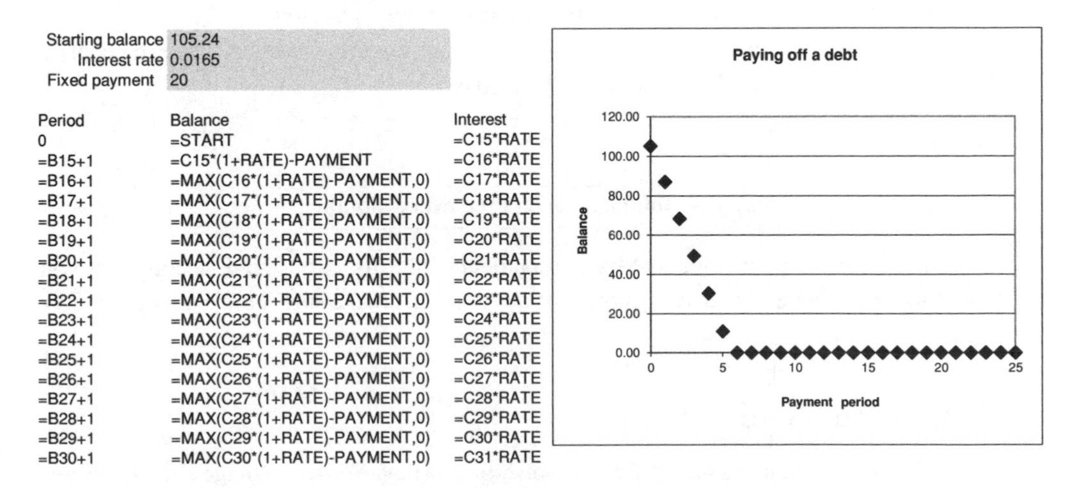

Starting balance	105.24
Interest rate	0.0165
Fixed payment	20

Period	Balance	Interest
0	=START	=C15*RATE
=B15+1	=C15*(1+RATE)-PAYMENT	=C16*RATE
=B16+1	=MAX(C16*(1+RATE)-PAYMENT,0)	=C17*RATE
=B17+1	=MAX(C17*(1+RATE)-PAYMENT,0)	=C18*RATE
=B18+1	=MAX(C18*(1+RATE)-PAYMENT,0)	=C19*RATE
=B19+1	=MAX(C19*(1+RATE)-PAYMENT,0)	=C20*RATE
=B20+1	=MAX(C20*(1+RATE)-PAYMENT,0)	=C21*RATE
=B21+1	=MAX(C21*(1+RATE)-PAYMENT,0)	=C22*RATE
=B22+1	=MAX(C22*(1+RATE)-PAYMENT,0)	=C23*RATE
=B23+1	=MAX(C23*(1+RATE)-PAYMENT,0)	=C24*RATE
=B24+1	=MAX(C24*(1+RATE)-PAYMENT,0)	=C25*RATE
=B25+1	=MAX(C25*(1+RATE)-PAYMENT,0)	=C26*RATE
=B26+1	=MAX(C26*(1+RATE)-PAYMENT,0)	=C27*RATE
=B27+1	=MAX(C27*(1+RATE)-PAYMENT,0)	=C28*RATE
=B28+1	=MAX(C28*(1+RATE)-PAYMENT,0)	=C29*RATE
=B29+1	=MAX(C29*(1+RATE)-PAYMENT,0)	=C30*RATE
=B30+1	=MAX(C30*(1+RATE)-PAYMENT,0)	=C31*RATE

Figure 10.3. Paying off credit card debt

Does all this mean that using a credit card is a bad idea? No, as long as you're careful. Then you can take advantage of some of the good things credit can do for you:

- The statement above shows that John Doe paid his last balance of $168.80 on time in full, avoiding all finance charges. So, in fact, he borrowed that money for a month from the credit card company at no cost. If he kept it in a savings account until it was time to pay his bill he'd even have made a few pennies in the meanwhile.
- Some credit cards give you back a reward at the end of the year—perhaps 1% of your total purchase dollars or some frequent flyer airline miles.
- The credit card companies don't rely only on interest charges and penalties to make money. They also collect a fee from the businesses where you use your card—perhaps 2%. So when making a large purchase you may be able to negotiate a discount for paying by cash or check since there will be no fee for cash while the fee for a check might be smaller than that for credit. Even if you can't get a better bargain you can help keep a small local merchant in business by paying with cash.
- We've seen that if you miss a payment your credit rating may suffer. But just avoiding credit errors won't get you a good credit rating. For that you have to prove you can manage credit—by having a credit card and paying the balance in full when due. Then when it's time to borrow money for a car or a condo your good credit rating may get you a lower interest rate.
- If you have a balance on an existing card that you can't afford to pay off immediately, consider opening a second card and transferring the balance. The new card company may offer you 0% interest for a while to encourage you to switch. If you do that and then don't use the new card you can pay off the old balance over time without any further interest charges. Be sure to read the small print before you do this—the transferred debt may be interest free, but often there's a charge (perhaps three or four percent) to make the transfer.

  Federal legislation passed in response to the 2009 financial crisis forced credit card companies to change their policies so that "payments above the Minimum Payment due will be applied first to higher interest rate balances." That notification appeared on one of the authors' statements, along with the kind thought that "This may help you to pay off your highest interest rate balances more quickly and reduce your interest charges." They did not reveal how much money they spent lobbying in Washington against the regulation.
- Finally, you may find a credit card issued by one of your favorite charities. Then the charity collects a small fraction of the fees the merchants pay.

Do remember:

Pay your full balance on time every month.

You can even arrange to have that happen automatically from your bank account, so you don't have to remember and you save the cost of a stamp. Just make sure there's enough money in the bank.

10.2 Can you afford a mortgage?

There's a $250,000 condominium in Denver you want to buy. You've managed to scrape together $50,000 for the down payment (savings, your parents, ...) but will have to take out a mortgage for the \$200,000 balance. Can you afford it? There are many websites that provide a place to start. We visited `www.wellsfargo.com/mortgage`, filled out the Home Lending Rate & Payment Calculator and discovered that on August 14, 2015, in Denver, Colorado you could get a 30 year fixed rate mortgage at 4.250% annual interest with a monthly payment of \$984.00 or a 15 year fixed rate mortgage at 3.375% with a monthly payment of \$1,417.

In this section we'll look at what those numbers mean, see how they are calculated and discuss a few important issues (some quantitative, some not) that you should think about when making a decision like this one.

Paying off a mortgage is like paying off a credit card balance when you make no new purchases. There's an annual rate. Your balance at the end of a month includes interest computed at one twelfth of the annual rate. Each month you pay all the current interest and some of the principal. Since the principal is decreasing, there's less interest each month so more of the payment goes toward the principal. One difference is that the credit card company sets the minimum payment; then it takes as long as it takes to pay off the balance, while the mortgage payment is figured out in advance so that everything is paid off at a particular time—usually 15 or 30 years.

The mortgage company uses this formula to calculate the monthly payment:

$$P \times \frac{r/12}{1 - (1 + \frac{r}{12})^{-12y}} \tag{10.1}$$

where P is the principal (the amount of your mortgage), r is the annual interest rate, and y is the length of the mortgage, in years.

It is probably the most complicated formula in *Common Sense Mathematics*. We won't explain where it comes from, and you need not memorize it. But you can understand some parts of it. It has the form

$$P \times (\text{complex expression involving } r \text{ and } y)$$

which tells you that your monthly payment is proportional to P. The complex part is the expression in parentheses—the *effective monthly rate*. That's the number of dollars in your payment for each dollar you borrow. There the $r/12$ finds the monthly rate from the annual rate. The product $12y$ is the number of months in y years.

You can use that formula to check that the Wells Fargo calculator finds the right monthly payment of \$984 for a 30 year \$200,000 loan at 4.250% interest. We did the arithmetic in Excel, with the formula

```
=(STARTBALANCE*INTERESTRATE/12)/(1-(1+INTERESTRATE/12)^(-12*YEARS))
```

in cell `C11` on the `mortgage` worksheet in `PayOffDebt.xlsx`. There you can see the principal balance at the end of each year and the total interest paid. On that 30 year mortgage it's \$154,196.72.

When you borrow you always pay back more than the amount you borrowed. In this case, lots more. Should that frighten you? Maybe or maybe not. Is it worth it? Perhaps, for several reasons.

- It would take a long time to save up the full purchase price (to avoid borrowing). Saving would be difficult because you would be paying rent the whole time. So you can think of the mortgage payments as money spent instead of rent.
- The condo may well be worth more after 15 or 30 years than the total you paid for it—even including the interest on the mortgage.
- Inflation is pretty nearly inevitable over the years. These computations are all made in dollars computed in the year you make the purchase, but the actual value of that money when you pay it to the bank in later years will be less, in then current dollars. Think of it this way: your salary is likely to increase at least as fast as inflation, so the fixed monthly mortgage payments will be a smaller and smaller percentage of your take home pay.
- That said, you do want to minimize the amount of interest you pay, by paying attention to the significant difference between a 15 and a 30 year mortgage. The short one has a lower interest rate (2.5% instead of 3.5%) and a much lower total interest cost: about $40,000 instead of $123,000. So you should choose it if you can manage the extra $430 per month in payments.
- You will also get a lower rate if you have established a good credit rating in the years before you apply for your mortgage. So start now to use a credit card wisely.

Words of warning. This discussion shows how, in principle, you pay off a loan by paying some interest and some principal periodically. That's just one of the financial things you'll need to understand when you think about buying a house or condo. Just asking the bank or shopping online for an interest rate isn't sufficient. As with most other topics in this book our hope is to provide a quantitative foundation for further questions. Some of those will address these issues:

- There are other up front costs: legal fees, title searches, inspections, points.
- The cost of owning is more than just the cost of the mortgage. You must be prepared for expenses that your landlord would cover if you were renting—things like real estate tax, insurance, repairs.
- Variable rate mortgages generally start out with lower rates than fixed rate mortgages—but payments can balloon when the initial rate expires.

There are many books and web pages that may help—here's just one we found with a simple search: `www.ourfamilyplace.com/homebuyer/checklist.html`.

10.3 Saving for college or retirement

In Chapter 9 we studied how money accumulates when you invest a big chunk and let the interest compound. But you rarely save a big chunk of money all at once. A more realistic way to save, for college (for your children) or for retirement, is to put away a fixed amount on a regular basis.

The kind of calculations we made above to study paying off a debt will help now to study how money saved regularly accumulates. Suppose you can invest $1200 a year and make your payment at the end of the year. We'll see soon why it's better to deposit $100 every month instead.

You think you can get a 6% return on your investment, since you're willing to take some short term risk for the sake of long term return. At the end of the second year you will have

$$\$1200 \times 1.06 + \$1200 = \$2472$$

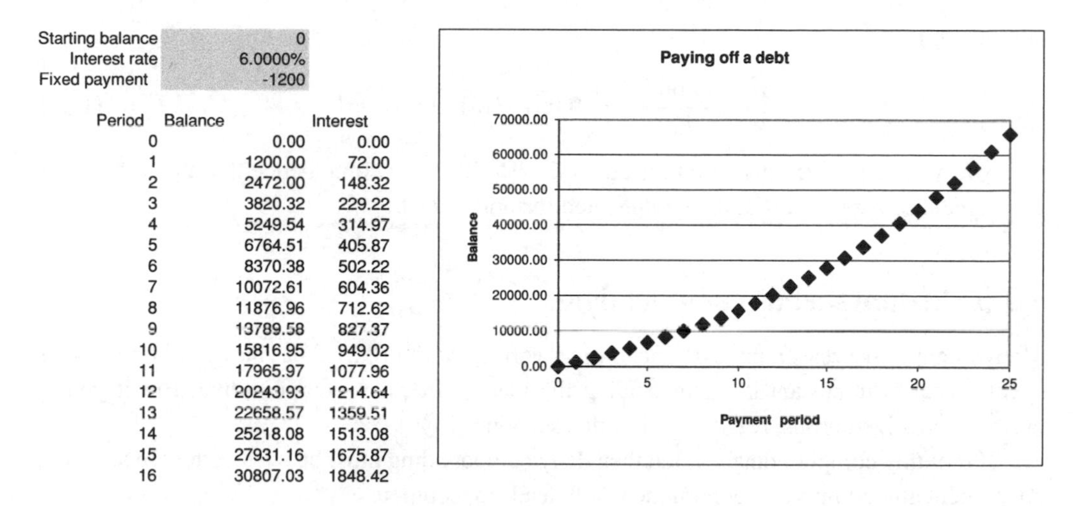

Starting balance	0
Interest rate	6.0000%
Fixed payment	-1200

Period	Balance	Interest
0	0.00	0.00
1	1200.00	72.00
2	2472.00	148.32
3	3820.32	229.22
4	5249.54	314.97
5	6764.51	405.87
6	8370.38	502.22
7	10072.61	604.36
8	11876.96	712.62
9	13789.58	827.37
10	15816.95	949.02
11	17965.97	1077.96
12	20243.93	1214.64
13	22658.57	1359.51
14	25218.08	1513.08
15	27931.16	1675.87
16	30807.03	1848.42

Figure 10.4. Saving for retirement

and after the third

$$\$2472 \times 1.06 + \$1200 = \$3820.32.$$

These calculations look just like the ones we made for paying down credit card debt, except that in this case we add the periodic payment to the balance rather than subtracting it. That means we can use the debt payment spreadsheet to see how our money accumulates by entering a negative "payment" to be added to the growing balance. Figure 10.4 shows the result (we haven't changed the labels).

In 25 years you will have accumulated nearly \$66,000. Of that amount, you contributed just $\$1,200 \times 25 = \$30,000$. The rest, more than \$35,000, is interest. Making precise sense of the total accumulation in cell `C40` and the total interest in cell `D41` is a little tricky. You have to think carefully about whether these values are computed at the beginning or the end of the year, before or after interest is credited. You don't need to do this detailed analysis to understand the principle.

10.4 Annual and effective percentage rates

We saw in Chapter 9 that compounding is a good thing for your investments. When you borrow money it's a good thing for the credit card company. Suppose in the example in Section 10.1 John Doe carried a balance of \$100 for a year, making no payments. Recall that the monthly interest rate is 1.65%, one twelfth of the 19.80% APR. At the end of the year interest will have been compounded twelve times and he would owe

$$\$100 \times 1.0165^{12} = \$121.70,$$

which corresponds to an *effective annual rate* (EAR) of 21.7%—almost 2 full percentage points higher than the already outrageous advertised APR of 19.8%.

The effective rate works for you rather than against you when you invest rather than borrow. If you make monthly deposits to your retirement account earning 6% annual interest the computation

$$\left(1 + \frac{0.06}{12}\right)^{12} = 0.0616778119 \approx 1.0617. \tag{10.2}$$

tells you your effective interest rate is about 6.17%. It's not a coincidence that the left side of this equation matches part of the complicated formula in (10.1).

10.5 Instantaneous compounding

Many quantitative reasoning texts spend time working with (10.2). That may not be time well spent. What is important is understanding the idea of frequent compounding. But if you're curious and adventurous, read the rest of this section.

If monthly compounding is good then daily compounding must be even better. To see what six percent annual interest compounded daily leads to, compute

$$\left(1 + \frac{0.06}{365}\right)^{365} = 1.06183131.$$

That corresponds to an effective annual rate of 6.183%. Hourly compounding gives

$$\left(1 + \frac{0.06}{8765}\right)^{8765} = 1.06183633,$$

which is just a tiny bit better. Compounding every minute results in 1.06183654, which differs only in the seventh decimal place. These computations suggest that as you compound more and more often you do better and better, but by less and less. There seems to be a limit. In fact there is. You can find it with the magic number e and the Excel function `EXP` we discussed in Section 9.5. If you compound 6% annual interest *every instant* the computation

$$e^{0.06} = \text{EXP}(0.06) = 1.061836547$$

tells you the effective interest rate to nine decimal places.

To compare monthly and instantaneous compounding in terms that are easier to understand, suppose you invested a thousand dollars. Then the Google calculator tells you

```
1 000 * ((e^0.06) − ((1 + 0.06/12)^12)) = 0.158734681
```

so the difference after one year is about 16 cents. That's chump change for a thousand dollar investment.

Finally, suppose you could find someone to pay you 100% interest annually. Then without compounding, one dollar would double, and become two. If you compounded instantaneously your dollar would turn into $e =$ `EXP(1)` $= 2.72$ dollars in a year.

10.6 Exercises

Exercise 10.6.1. [W][Section 10.1][Goal 10.1] Your credit report.

> The Fair Credit Reporting Act (FCRA) requires each of the nationwide credit reporting companies—Equifax, Experian, and TransUnion—to provide you with a free

> copy of your credit report, at your request, once every 12 months. The FCRA promotes the accuracy and privacy of information in the files of the nation's credit reporting companies. The Federal Trade Commission (FTC), the nation's consumer protection agency, enforces the FCRA with respect to credit reporting companies. [R221]

To get your reports, visit `www.annualcreditreport.com/index.action`.

That's not a clickable link. You have to type it into your browser. Here's why, from `www.annualcreditreport.com/aboutThisSite.action`.

> AnnualCreditReport.com is the official site to get your free annual credit reports. This right is guaranteed by Federal law. To verify that this is the official site, visit www.consumerfinance.gov/askcfpb/311/how-do-i-get-a-copy-of-my-credit-report .html.
>
> Don't be fooled by look-alike sites. You can be sure that you are on the right site if you type `www.annualcreditreport.com` in your browser address line. Don't come to this site by clicking on a link in another site or in an email.

Now you have your credit *report*. That's not the same as credit *score*.

> Based on the information in your credit report, lenders calculate your credit score so they can assess the risk you pose to them before they decide whether they will give you credit. The higher your score, the less risk you pose to creditors.
>
> The information in your credit report is used to calculate your FICO (the acronym stands for Fair, Isaac and Company) score. Your score can range anywhere from 300–850. Aiming for a score in the 700s will put you in good standing. A high score, for example, makes it easier for you to obtain a loan, rent an apartment or lower your insurance rate. [R222]

Look for a place on the web that will give you an estimate of your credit score. Wherever you go, be sure to read the fine print, and don't pay for anything.

Write about what you discovered.

Exercise 10.6.2. [S][Section 10.1][Goal 10.1] How long to pay it off?

Starting in 2010 credit card companies were required to provide the information in Table 10.5 each month. The numbers there are for a bill with a balance of $2,020.37, a minimum payment amount of $40.00 and an annual percentage rate of 12.24%.

If you use the `PayOffDebt.xlsx` spreadsheet to work on this exercise you will need the `mortgage` worksheet, since the `plain` worksheet only covers 25 payment periods.

(a) Verify the three year time to pay off the balance at a rate of $67 per month.

(b) Show that a constant monthly payment of $40 is much more than is needed to pay off the balance in 18 years. How can the 18 year claim be correct?

(c) The 2010 Consumer Credit Law allows banks to raise the minimum payment on an account to a constant amount sufficient to pay off the balance in five years. What would that minimum payment be for this bill?

Exercise 10.6.3. [Section 10.2][Goal 10.3] Build your own mortgage.

Table 10.5. Paying off a credit card balance

If you make no additional charges using this card and each month you pay . . .	You will pay off the balance shown on this statement in about . . .	And you will end up paying an estimated total of . . .
Only the minimum payment	18 years	$3,843
$67	3 years	$2,426 (Savings = $1,417)

Redo the computations in Section 10.2 for a house or condo of your choice in your town. Start with a reasonable cost and down payment. Find rates from at least two separate on line sites; check them with the formula and the `PayOffDebt.xlsx` spreadsheet.

Exercise 10.6.4. [U][Section 10.2][Goal 10.3][Goal 10.2] Using the debt payoff spreadsheet.

The debt payoff spreadsheet can reproduce some of the computations from the exponential growth spreadsheet we introduced in Chapter 9. Test that by setting the monthly payment to 0 and the annual interest rate to 12 times the growth rate you want to study.

In particular, what happens if the annual growth rate is 1200%, the starting balance is 1, and the monthly payment is 0?

Exercise 10.6.5. [R][S][Section 10.2][Goal 10.2][Goal 10.3] Jumbo loans.

On November 20, 2010 a story in *The Boston Globe* headlined "Rates for big loans tumble" said that

> Over the past year, the average interest rate for so-called jumbo loans—$523,750 and up in the Boston area—has fallen from 6 percent to about 5 percent for a 30-year, fixed-rate mortgage. That translates into a monthly savings of about $375 on a $600,000 loan. [R223]

(a) What monthly payment will retire the loan when the interest rate is 6%?
(b) What monthly payment will retire the loan when the interest rate is 5%?
(c) Is the newspaper's claim of a $375 monthly saving correct?

Exercise 10.6.6. [R][S][Section 10.2][Goal 10.2][Goal 10.3][Goal 10.4] Mortgages in the news.

A March 4, 2011 article in *The New York Times* headlined "Without Loan Giants, 30-Year Mortgage May Fade Away" claimed that the monthly payment on a 30 year mortgage at six percent interest would be $600 but just $716 for a 20 year mortgage. [R224]

On the same day in an article in *The Boston Globe* headlined "The end of 30-year fixed-rate mortgage?" you could read that

> The difference between a 15- and 30-year mortgage amounts to well over $600 per month on a $300,000 loan, a substantial amount that may prevent wide swaths of the middle class from buying homes. [R225]

Verify the calculations in each of these quotations.

Exercise 10.6.7. [U][W][Section 10.3][Goal 10.5][Goal 10.4] Retirement planning.

Find an online retirement income calculator. Use it with data you imagine for yourself. Write down what you do as you proceed. (Screenshots would be nice.) Record what it tells you at the end.

Their calculator is much more sophisticated than the simple one in Excel we introduced in this chapter. See if you can use ours to get answers that match what it told you.

Exercise 10.6.8. [S][Section 10.4][Goal 10.4] Payday loans.

The Boston Globe on New Year's Day 2009 reported that

> New Hampshire is giving payday lenders the gong as it rings in the new year. A law that takes effect today caps the interest rate on small loans at 36 percent a year, which the industry has said will put it out of business.
>
> Payday lenders typically charge $20 per $100 for two-week loans backed by the borrower's car title or next paycheck. That amounts to 1.43 percent interest per day, an annual rate of 521 percent.
>
> The cap translates to a daily interest rate of about 0.1 percent, or total interest charges of $1.38—a dime a day—on a $100, two-week loan. [R226]

(a) What is a "payday loan"?
(b) Verify the computation that 1.43% interest per day is 521% interest annually.
(c) If the 1.43% daily interest is compounded daily then the true annual rate of interest is in fact much more than 521%. How much is it?

[See the back of the book for a hint.]

(d) Verify that paying interest of $1.38 on a two week loan of $100 is just about a "dime a day" and corresponds to a daily interest rate of about a tenth of a percent. What annual rate does that represent?
(e) Visit a payday loan website and report on what you discover there about interest rates.

Exercise 10.6.9. [S][C][Section 10.4][Goal 10.5][Goal 4.2] Supporting a hospital bed.

The headline "Charity sues R.I. hospital over donation in 1912" accompanied an article in the City and Region section of *The Boston Globe* on February 23, 2008. The article described a gift intended to provide a free bed in perpetuity for needy patients.

There you could read that

> Mark E. Swirbalus, a Boston lawyer representing Children's Friend, said that "as far as we know, the hospital never set aside a bed and never set aside the money."

> The $4,000, if conservatively invested by the hospital in 1912, would be worth about $1.5 million today, he said. [R227]

(a) Is Swirbalus's claim about a "conservative investment" correct?

[See the back of the book for a hint.]

(b) What the hospital should have done was invest the money and use just the interest each year to fund the bed. That would work—if only there were no inflation that made the cost of the bed increase.

Suppose the hospital got a 6.5% percent yearly return on investment and annual inflation was 3.5%. Explain why it would be able to spend about $120 on the bed in 1912, and could keep spending at that rate as the years went on.

(c) If the cost of providing a hospital bed in 1912 was $120, what would it be in 2008 if all you had to do was adjust for inflation?

Exercise 10.6.10. [S][Section 10.2][Goal 10.3] Half the time, more than double the benefit.

Show that taking out a 15 year mortgage instead of a 30 year mortgage (for the same loan amount at the same annual rate) doesn't double your monthly payment and more than halves the total interest you pay on your loan.

(The advantages are usually even greater since you can usually negotiate a lower interest rate for a shorter mortgage.)

Exercise 10.6.11. [S][Section 10.3][Goal 10.5] Saving $50,000.

(a) Use the spreadsheet `PayOffDebt.xlsx` to figure out how much you'd have to save per year at 3% interest (compounded annually) to have a balance of $50,000 in your account at the end of 25 years.

(b) Then use the mortgage tab in that spreadsheet to answer the same question if you save a fixed amount each month rather than each year.

Exercise 10.6.12. [S] What merchants pay for credit card services.

From *The Nilson Report*:

> Spending for goods and services in 2013 by general purpose-type credit, debit, and prepaid cards issued in the United States reached $4.530 trillion. Market share is shown for credit, debit, and prepaid.
>
> 1. Credit Purchase Volume $2.399 Tril.
> 2. Debit Purchase Volume $1.949 Tril.
> 3. Prepaid Purchase Volume $0.182 Tril. [R228]

The Boston Globe reported that "merchants in the United States spent $71.7 billion on fees [for these transactions] last year."

(a) What is the average merchant fee, as a percentage?

(b) Make sense of the $4.530 trillion total: think about it in units like dollars per person per day, dollars per transaction, ...

Exercise 10.6.13. [U] Excel templates from the internet.

At `www.excely.com/template/loan-calculator.shtml` you can download a Loan Calculator Excel Template.

(a) Do the calculations there match those in `PayOffDebt.xlsx`?
(b) Find out where that template uses Excel's built-in `PMT` function. Compare how it works to the formula in the spreadsheet `PayOffDebt.xlsx`.

11

Probability—Counting, Betting, Insurance

Pierre de Fermat and Blaise Pascal invented the mathematics of probability to answer gambling questions posed by a French nobleman in the seventeenth century. We follow history by starting this chapter with simple examples involving cards and dice. Then we discuss raffles and lotteries, fair payoffs and the house advantage, insurance and risks where quantitative reasoning doesn't help at all.

Chapter goals

Goal 11.1. Compute probabilities for games of chance by counting outcomes.

Goal 11.2. Calculate fair price of a bet as a weighted average.

Goal 11.3. Calculate house advantage as (payout)/(income).

Goal 11.4. Understand insurance as a lottery.

11.1 Equally likely

In its everyday qualitative meaning "probably" is just a synonym for "likely" or "I think so but I'm not sure." In this chapter we start with simple examples where we can make "probably" quantitative by counting the possibilities.

To think about the chance of some particular event involving coins, dice, cards or raffles happening, count the possible equally likely outcomes, then count how many match what you're looking for and write down the appropriate fraction.

- The probability of heads when tossing a fair coin is $\frac{1}{2}$.
- The probability of rolling a 6 with a fair die is $\frac{1}{6}$.
- The probability of drawing an ace from a well shuffled deck is $\frac{4}{52}$.

In words:

$$\text{probability of an event} = \frac{\text{number of outcomes that match the event}}{\text{number of possible outcomes}}.$$

Writing probabilities as fractions helps you remember what they mean. But since they're just numbers, we can write them as decimals if we wish. Since they are numbers between 0 and 1, we often express them as percentages.

- The probability of heads tossing a fair coin is $\frac{1}{2} = 0.5 = 50\%$.
- The probability of rolling a 6 with a fair die is $\frac{1}{6} \approx 0.167 \approx 17\%$.
- The probability of drawing an ace from a well shuffled deck is $\frac{4}{52} \approx 0.077 = 7.7\%$.

Events that can never happen have probability 0. Events with probability 1 are certain to happen.

- The probability of rolling a 7 with a die is $\frac{0}{6} = 0 = 0\%$. It doesn't matter whether the die is fair or not.
- The probability of drawing a heart, a club, a spade or a diamond from a deck of cards is $\frac{52}{52} = 1 = 100\%$. It doesn't matter whether the deck is well shuffled or arranged in some nice order.

There are other probability problems you can solve by counting, as long as you're careful to count the right things.

Many state lotteries offer a prize if you pick the right six numbers in the some range. The numbers must be different, with no repetitions, but the order in which you pick them doesn't matter. To find the probability that your pick will win you have to count how many ways there are to pick six numbers. That's a problem for a math course more advanced than this one: the answer is 20,358,520 when the range is numbers from 1 to 52. So the probability of winning pick-six is about one twenty-millionth. If twenty million people play, expect about one winner.

If you pick a state at random, what's the probability that it has a state lottery? Wikipedia (probably reliable for these data) says that (as of 2015) all but five states have lotteries. Then the probability is

$$\frac{\text{number of states with lotteries}}{\text{number of states}} = \frac{45}{50} = 0.9 = 90\%.$$

That's just the percentage of states with a lottery.

You have to be careful when you interpret this number. It's not the same as the probability that a random person in the United States lives in a state with a lottery, because the states without lotteries tend to be smaller than average. To find that probability you would have to count

$$\frac{\text{number of people living in states with lotteries}}{\text{total population}}.$$

This is a weighted average of the states, where the weights are the state populations, with a zero for states with no lottery.

We will have much more to say about lotteries in Section 11.4.

11.2 Odds

Another way to describe a coin toss is to say "the odds are fifty-fifty." Heads and tails are equally likely—the odds are even.

Here are the odds for some gambling events; we usually write odds with a colon (:) and read the colon out loud as "to".

- The odds for rolling a 6 with a fair die are $1:5$ or one to five. The odds against are $5:1$, or five to one.
- The odds for drawing an ace from a well shuffled deck are $4:48$, or $1:12$. The odds against are twelve to one.
- The odds for heads tossing a fair coin are $1:1$.

These examples illustrate how to find the odds for an event when you can count the equally likely possibilities and decide which ones are favorable. You compute

$$\text{(number of favorable cases)} : \text{(number of unfavorable cases)}\,.$$

The odds against the event are

$$\text{(number of unfavorable cases)} : \text{(number of favorable cases)}.$$

Odds are fractions in disguise, so the odds against drawing a spade from a deck of cards may be expressed as $39:13$ (counting all the possibilities) or simply as $3:1$ (three to one).

The odds against a winning pick-six ticket are about 20 million to 1.

You can convert back and forth between odds and probabilities. Since the odds against drawing a spade are $39:13$, the probability that you won't draw a spade is $\frac{39}{52} = \frac{3}{4}$. In general, if the odds for an event are $a:b$ then its probability is $a/(a+b)$.

If you start out knowing that you will draw a spade with probability 25% you know too that the probability that you'll draw a heart, a diamond or a club is 75%. With both those probabilities it's easy to find the odds: they are $0.25:0.75$ for drawing a spade. That's just our old friend $1:3$ in disguise.

In general, if the probability of an event is p then the odds for that event are $p:(1-p)$. The odds against are $(1-p):p$.

The few formulas in this section are just common sense. If you understand them you won't have to memorize them. If you try to memorize them without understanding them you may end up using them in the wrong places.

11.3 Raffles

Simple raffles are gambles with computable probabilities. Tickets are sold, some are chosen at random and the people who hold those tickets get prizes. You may be familiar with fundraising raffles run by school parent teacher organizations.

Suppose the PTO sells 500 tickets for a raffle with a single prize.

Since each of the 500 tickets has an equal chance of being selected, the odds of a ticket winning are $1:499$, or $499:1$ against. The probability that any particular ticket wins is $\frac{1}{500} = 0.002 = 0.2\%$, or two tenths of a percent.

The probability that a particular person wins may be different. If you buy 10 tickets then you win with probability $\frac{10}{500} = 0.02 = 2\%$. If you don't play, the probability is 0. If you buy all of the tickets then you win with probability $\frac{500}{500} = 1 = 100\%$.

Now let's connect probability with money, as the inventors of the mathematics of probability did centuries ago. Suppose the PTO wants to offer a \$1000 prize to the winner. Then the *fair price* of a ticket is

$$\text{fair price} = \frac{\text{total prize money}}{\text{number of tickets}} = \frac{\$1{,}000}{500\text{ tickets}} = 2\frac{\$}{\text{ticket}}. \tag{11.1}$$

Using what we learned in Chapter 5 we can rewrite this computation as a weighted average. One of the tickets is worth \$1000; the others are worthless, so

$$\begin{aligned}
\text{fair price} &= \frac{\text{total value of tickets}}{\text{number of tickets}} \\
&= \frac{499 \times \$0 + 1 \times \$1000}{500} \\
&= \frac{499}{500} \times \$0 + \frac{1}{500} \times \$1000 \\
&= \text{probability of losing} \times \text{value of losing ticket} \\
&\quad + \text{probability of winning} \times \text{value of winning ticket} \\
&= 0.998 \times \$0 + 0.002 \times \$1000 \\
&= \$2.
\end{aligned}$$

In the fourth line of the computation the ticket counts disappear. The fair price is the weighted average value of a ticket, weighted by the probabilities for each kind of ticket.

That average is the fair price of a ticket because all the money collected is returned in prizes. That may make for an exciting evening at the PTO meeting, but it won't raise any money. So the PTO decides to charge \$3.00 for each ticket, keep the prize at \$1,000, and use the other \$500 to buy classroom supplies for the kids.

Since the total prize money and the number of tickets have not changed, the fair price is still \$2. So on average each ticket loses

$$\text{cost of ticket} - \text{fair price of ticket} = \$3 - \$2 = \$1.$$

Of course you never lose exactly one dollar with one ticket. You either collect \$1000 for a net gain of \$997 or get nothing and lose your \$3 bet.

Yet another way to calculate the average loss is to see that the prize is just 2/3 of what the PTO collects, so the fair price is 2/3 of the \$3 cost, or \$2. Then on average each ticket loses the other 1/3, or \$1.

Would you buy a \$3 ticket when the fair price is just \$2, knowing that on average you will lose \$1.00? Perhaps. Even though you're very likely to lose your three dollars you can feel good about supporting the school. Perhaps the thrill you get anticipating what you will do with the prize if you win despite long odds makes the probable loss more bearable.

11.4 State lotteries

We found this quote on the web:

> Lotteries rank first among the various forms of gambling in terms of gross revenues: total lottery sales in 1996 totaled \$42.9 billion. 1982 gross revenues were \$4 billion, representing an increase of 950% over the preceding 15 years, 1982–1996.

> Lotteries have the highest profit rates in gambling in the U.S.: in 1996, net revenues (sales minus payouts, but not including costs) totaled $16.2 billion, or almost 38% of sales. They are also the largest source of government revenue from gambling, in 1996 netting $13.8 billion, or 32% of money wagered, for governments at all levels. [R229]

The numbers in this 1996 report are stale, although still useful for explaining important ideas. In Exercise 11.9.8 we've asked you to update them.

The payoff rules for state lotteries are very complex, and vary widely from game to game. It's hard to think about the fair price of any particular ticket. But with a little careful reading of the numbers in the quote you can compute the expected average return on each dollar you bet. That number, which will be less than a dollar, is the fair price.

The bookkeeping[1] for analyzing these numbers is

$$\text{total from ticket sales} = \text{prizes awarded} + \text{overhead} + \text{net revenue to state.}$$

In 1996 gross revenues—that is, ticket sales, dollars bet—were $42.9 billion.

The $16.2 billion in the second paragraph is "sales − payouts", so the payouts must be $42.9 − $16.2 = $26.7 billion. Then

$$\frac{\text{payouts}}{\text{sales}} = \frac{\$26.7 \text{ billion}}{\$42.9 \text{ billion}} = 0.622377622 \approx 62\%$$

so for each lottery dollar in 1996, players got back (on average) a little more than 62 cents in prize money. That is the fair price of a one dollar ticket. The other 38 cents is the 38% of sales that count as total revenue for the government—the $16.2 billion not returned to bettors as prizes. Some of that money was overhead. After subtracting that, the net revenue available for other use was $13.2 billion.

The table at `www.census.gov/govs/state/10lottery.html` lists the following data for the 2010 Texas lotteries:

Income	Apportionment of funds		
Ticket sales (excluding commissions)	Prizes	Administration	Proceeds available
3,542,210	2,300,182	184,980	1,057,048

Since

$$\frac{2,300,182}{3,542,210} = 0.649363533 \approx 65\%$$

Texas returned 65 cents on the dollar in lottery payouts in 2010—three cents more than the national average.

[1] One of our favorite words. We don't know another with three double letters in a row.

11.5 The house advantage

Raffles and lotteries are designed to make money. So is casino gambling—for the casinos. They make a profit, and states tax the proceeds to raise revenue.

Before you lay down your bet at a casino, you should think about how much you will pay to play—the difference between a dollar bet and the fair price of that bet (the average amount returned to you for your dollar). That difference is called the *house advantage*.

In Section 11.4 we discovered that the house advantage on state lotteries is about 35%. At gambling casinos it's much smaller. As in the state lotteries, the house advantage varies from game to game. It's the highest (about 10%) for slot machines—and there is no way to know that when you decide to play. But for roulette we can actually calculate the house advantage.

A fair roulette wheel is a circle divided into 36 equal wedges numbered from 1 to 36, colored alternately red and black. A ball runs around the rim of the wheel, slowing down until it falls into a random wedge. Before the wheel spins you place your bet, perhaps:

- on the number 17 ("straight-up"), with a winning probability of $\frac{1}{36}$. The odds are 35 to 1 against.
- on red, with a winning probability of $\frac{18}{36} = \frac{1}{2}$. Even odds.
- on odd, at even odds.
- on one of the numbers 1 through 12 (a "dozen bet"), with a winning probability of $\frac{12}{36} = \frac{1}{3}$. Two to one against.

What would be a fair return on a $1 bet?

- If you bet straight-up the payoff should be $36.
- If you bet on red, the payoff should be $2.
- If you bet on odd, the payoff should be $2.
- For a dozen bet the payoff should be $3.

There are several ways to see that these are fair. We'll work them out with the $36 payoff for the straight-up dollar bet on a single number.

- Imagine the spin of the wheel as a raffle with 36 tickets. A dollar bet on 17 is like buying one of the tickets. Imagine that others have bought the other 35 for $1 each. Then the casino has collected $36. The fair thing to do would be to pay that to the winner—then all the money collected is awarded as prizes.
- Using the technique we learned in Section 11.3, we can check the numbers in this equation:

  $$\text{winning probability} \times \text{winning payoff} + \text{losing probability} \times \text{losing payoff}.$$

  For the straight up bet that equation says

  $$\frac{1}{36} \times \$36 + \frac{35}{36} \times \$0 = \$1$$

  which is a perfectly fair average return on a $1 bet!
- What happens when you play for a long time? Since you pay $1 for each spin of the wheel and win about $\frac{1}{36}$ of the time you should collect $36 for each win in order to break even in the long run.

- The odds for winning are 1 : 35. Your $1 bet on 17 is a bet against the casino. They put up $35 to match your $1. The winner takes all $36.

The other computations (to check the fair payoffs for bets on red, or odd, or 1-12) work the same way.

In real life the casino must cover expenses, pay the state its share of the take and still turn a profit, so the average value of a bet must be less than the fair price. The difference is the house advantage.

In the PTO raffle the organization assumes it sells all the tickets and decides how much of what it collects to return as prizes. But the casino can't count on people betting on all the numbers, and can't know how many people will bet.

Figure 11.1 shows how they collect the house advantage in roulette. The picture shows a wheel with an extra green wedge numbered 0. An American roulette wheel will have another green wedge numbered 00.

Figure 11.1. The house advantage in roulette [R230]

The casino uses the old fair price payoffs: $36 for a winning $1 straight-up bet on 17. But the extra wedges change the probabilities. Here is the calculation for an American wheel with 38 wedges:

$$\frac{1}{38} \times \$36 + \frac{37}{38} \times \$0 = \$0.94736842105$$

which means that on average you lose more than a nickel of every dollar you bet. The house advantage is just over 5.25%.

Does this mean you shouldn't play? Not necessarily. You may be willing to pay the house advantage in return for the thrill of the gamble. But before you do, you should understand the odds for the game you choose.

There are casino games in which a skilled player can win—slowly, and with great effort. At the poker table you are competing with other gamblers, not with the house, which pays its expenses and profits by taking a fraction of the ante or pot on each deal. So the house always wins, but a skilled poker player can win too by beating the other players.

In principle, you can also win at blackjack. We'll think about why in the next chapter.

11.6 One-time events

In our discussions so far we've assumed each example is "fair" (even if payoffs weren't)—coins and dice and roulette wheels are properly balanced, decks of cards are properly shuffled, no one peeks when drawing the winning raffle ticket. In each case all possible outcomes are equally likely so we could compute probabilities just by counting cases.

To test for whether a particular coin or die is really "fair" you could imagine repeating an experiment many times. A fair coin should come up heads about half the time (but not exactly half the time, which would be very unlikely). A fair die should show a 5 about 1/6 of the time. We'll return to this topic in Section 12.2.

There are many situations in real life where probabilities and odds appear but can't be computed by simple counting or checked by repeated experiments. Will the Chicago Cubs win the World Series? Which horse will win the Kentucky Derby? Who will be elected? Will it rain tomorrow?

Suppose you bet your Chicago friend that the odds against the Cubs winning the World Series are 99 : 1. You put up $99, she puts up $1 and the winner takes home $100 when the season is over. That means that (in principle) you believe that probability of that Cubs World Series win is just $1/100$.(Those might be the right odds, since as of October 2015 the Cubs haven't won a World Series for more than a century. But they are in the playoffs as we go to press.)

When lots of people have an opinion they are willing to bet on, they can decide the probability collectively.

There's a way in which many state lottery payoffs depend on what the bettors think: the total prize money for a winning pick-six combination is divided among the people who bet on that combination. The odds for any particular number combination don't change, but the payoff does. Exercise 11.9.11 pursues this idea.

In horse racing the odds at the track are determined by the bets placed, in what's called *parimutuel* betting. Most readers of this book won't be playing the horses, and those who do will (or should) know all about this kind of betting. We discuss it here anyway since it provides an example where we can actually see how the bets determine the odds.

Before the race the punters place their bets at the tote. ("Punter" and "tote" are racing terms. You can look them up if you don't know what they mean.) After the race the track skims its *take* or *commission*— a percentage of the total amount bet. The winning bettors share the remainder in proportion to the amount each bet.

Table 11.2 shows the amount bet on each of six horses in an imaginary race. The horses are real—all winners of the Kentucky Derby—but we made up the numbers.

Table 11.2. Race of champions

Horse	Bets (K$)
Barbaro	239.0
Spend a Buck	333.2
Donerail	18.6
Twenty Grand	904.4
Apollo	155.0
Dark Star	66.3
total	1,716.5

The favorite horse is Twenty Grand precisely because more people have bet on him to win.

Since the total amount bet is \$1,716,500, the collective wisdom at the track says that Apollo will win with probability $155{,}000/1{,}716{,}500 = 0.0903000291 \approx 9\%$. The fair payoff is $1{,}716{,}500/155{,}000 = 11.0741935 \approx 11$ dollars per dollar bet.

That corresponds to odds against of about 10 to 1. If Apollo wins, each dollar bet will collect \$11: the original dollar plus the ten the other bettors put up in vain.

We had fun choosing the amounts bet on each of these six horses so that the odds of each are close to their odds in the Derby they won. We've included the longest shot of all, Donerail, and a favorite, Twenty Grand, who ran at less than even odds.

The $10 : 1$ odds for Apollo do not take into account the race track's commission. We don't know how much that was, or even whether it was the same in all six races. Suppose that in this fantasy race it's 10%.

Suppose Apollo wins. The track pays 90% of the take to those who bet on Apollo—$0.9 \times \$1{,}716{,}500/\$155{,}000 = 9.96677\ldots \approx 10$ dollars per dollar, instead of 11 dollars per dollar. The odds are effectively just $9 : 1$ against. The track makes money by lowering the odds, which no longer reflect the probabilities determined by the bets.

The moral of the story: you can win at the race track if you really know better than most people which horse is likely to win. There's that word "likely" again—you need to know a lot about the horses and play a lot for your knowledge to pay off. Perhaps the best way to win is to sell suckers a system...

11.7 Insurance

When you buy insurance you're gambling. In this case the gamble is one you hope to lose—you don't want to get sick, or have your house burn down, or total your car. In each of those situations you've made a small advance payment you hope and expect to lose in order to cover your losses when a catastrophic event with small probability happens.

Insurance companies estimate probabilities in order to determine the fair price for their policies, then add what they need to cover their administrative expenses and make a profit—their "house advantage." In the long run, on average, their customers never get all their money back. Therefore you want to think things through when you're deciding whether to buy insurance for more than the fair price. Sometimes you may be better off accepting the risk yourself.

Here's a sample of the kind of advice you can find on the web. It's from Liz Pulliam Weston, writing for *MSN Money*.

> Say you have a 10-year-old Honda that's worth \$4,000 in a private-party sale and have a \$500 deductible. Your risk is \$3,500. If your premiums for collision and comprehensive are more than \$350 a year, it may be wiser to bank that money toward a newer car. [R231]

If we make a simple assumption we can think about this using probabilities. Suppose that the only kind of accident to worry about is one that totals the car. Then Weston's advice is reasonable if you think that the probability that you'll have such an accident is less than 10%. Here's why. Imagine that the insurance policy is a lottery ticket, which "wins" if you have an accident. A winning ticket is worth \$3,500. If you think you have a 10% chance of winning, then the fair price (for you) is \$350. If you think your chance of totaling your car is less than 10% then the fair price is more than \$350, so perhaps you shouldn't buy the insurance.

Of course the real decision isn't this easy. You should take into account the fact that your accident might not total the car. You have to think about making this decision every year—sometimes your car will be worth more than $4,000, sometimes less. But the principle is clear. If the premiums are very high compared to your estimate of your risk, you should consider not buying collision and comprehensive insurance. Over the course of a driving lifetime you will probably save money.

However, there are often good reasons to pay more than the fair price for insurance. If you don't have the money to replace a totaled car and you must have one, then you need that insurance. Even if you have the money, the cost to you of a large loss may be more than you can afford, or may feel like more than the dollar amount.

For a discussion of answers to the question "Why buy insurance?" visit `money.stackexchange.com/questions/54561/why-buy-insurance`.

Sometimes you may be required to buy insurance. In order to drive, you must carry liability insurance to cover the cost of injuries to others in an accident you caused. If you have a mortgage on a house the bank will insist on fire insurance to protect their interest in the money they've lent you. The taxes you pay to support the police and fire departments can be considered a kind of insurance. You will probably never need their services, but you want them to be there when you do. Healthy people buy health insurance (and may even be required to do so) to spread the cost of catastrophic medical bills.

George Bernard Shaw wrote about this in *The Vice of Gambling and the Virtue of Insurance*. His essay is well worth reading. You may be able to find it in your library as Chapter 3 of T. W. Korner's *Naive Decision Making: Mathematics Applied to the Social World* or in Volume 3 of James R. Newman's *The World of Mathematics*. There's a section on health insurance that's the clearest argument we've seen for a "public option". Too bad it was written a century ago by a socialist.

11.8 Sometimes the numbers don't help at all

About thirty five years ago Joan Bolker had to decide whether to invest three years of hard work in hopes of earning a clinical psychology license.

Only after more than two thousand hours of clinical internships (which she would have to arrange) could she petition to have her doctorate in education count as appropriate postgraduate preparation for her new career. Only if that petition were granted would she be allowed to take the psychology licensing examination, much of which covered material she had not studied in any course.

Clearly the odds were long. She faced a significant investment of time, energy and lost income, with an unknown and hard to estimate probability of success at the end. She took the risk. She won her gamble, with a combination of talent, persistence and luck.

The moral of the story: sometimes numbers don't help. "Not everything that can be counted counts, and not everything that counts can be counted." (A quote often (wrongly) attributed to Albert Einstein). [R232]

In this case there was no way to quantify the costs, the benefits and the probabilities in order to make what might look like a rational choice. The kind of back-of-an-envelope probability calculations we've studied about playing the lottery or buying insurance are often of little help when making life-changing one-of-a-kind choices.

11.9 Exercises

Exercise 11.9.1. [R][S][Section 11.1][Goal 11.1] What's in a name?

(a) What is the probability that the name of a state (of the United States) chosen at random begins with the letter "A"?
(b) What is the probability that the name of a state (of the United States) chosen at random begins with the letter "Z"?
(c) How much more likely is it that a state name begins with "M" than with "A"?

Exercise 11.9.2. [U][Section 11.1][Goal 11.1] "Probably" in everyday English.

(a) Use the index to this book to find places where we used the words "probably" or "likely" other than in the chapters devoted to studying probability. Discuss the meaning of the word there. When it makes sense, provide a numerical estimate of the probability.
(b) Do the same for two or three occurrences of "probably" or "likely" in the media.

Exercise 11.9.3. [S][Section 11.1][Goal 11.1] Is it safe to swim?

In an article in *The Boston Globe* reprinted from the *Washington Post* on March 4, 2012 you could read a story headlined "Possible cut to beach testing a health threat, critics say". The story reports on Environmental Protection Agency estimates that say that the average person goes to a beach, lake or river about 10 days a year, and that about 3.5 million people get sick from splashing in bacterial contaminated water. [R233]

What is the probability that a visit to the beach will make you sick?

Exercise 11.9.4. [S][Section 11.2][Goal 11.1] It's a horse race.

Use the data in Table 11.2 to compute

(a) The odds and payoff for Donerail, the long shot.
(b) The odds and payoff for Twenty Grand, the favorite.
(c) The payoff for these two horses if the track takes a 10% commission before paying off any bets.

[See the back of the book for a hint.]

Exercise 11.9.5. [S][Section 11.2][Goal 11.1] Extended warranties.

What is missing from this list from `tv.about.com/od/warranties/a/buyexwarranty.htm` about points to think about when deciding whether to buy an extended warranty for your new TV? [R234]

1. Value of item being purchased
2. Price of extended warranty
3. Length of manufacturer's warranty
4. Length of extended warranty and date coverage begins

Exercise 11.9.6. [S][R][Section 11.3][Goal 11.2] Which average?

In the raffle discussed in Section 11.3 there are 500 tickets and a $1000 prize. We found that the average value of a ticket was $2.

(a) Which average is that—mean, median or mode?
(b) Compute the other two "average" ticket values.

Exercise 11.9.7. [S][R][Section 11.3][Goal 11.2] Multiple prizes.

Suppose a lottery with 1,000,000 tickets has a first prize of $200,000, three second prizes of $60,000 each and 100 third prizes of $200 each.

(a) What is the probability that a ticket wins the first prize?
(b) What is the probability that a ticket wins some prize?
(c) What is the fair price of a ticket?
(d) How much should the state charge for a ticket if it needs 10% of the revenue for overhead and wants to make $500,000 profit?

Exercise 11.9.8. [U][Section 11.4][Goal 11.2] 1996 was a long time ago.

The quotation that starts Section 11.4 comes to us courtesy of the University of North Texas CyberCemetery:

> The University of North Texas Libraries and the U.S. Government Printing Office, as part of the Federal Depository Library Program, created a partnership to provide permanent public access to the Web sites and publications of defunct U.S. government agencies and commissions. This collection was named the "CyberCemetery" by early users of the site. [R235]

Update the numbers from that quote (go back to Section 11.4) so that you can rewrite the paragraph referring to a much more recent year than 1996.

[See the back of the book for a hint.]

Exercise 11.9.9. [S][W][Section 11.4][Goal 11.2] Massachusetts Lottery statistics.

- The Massachusetts Lottery Commission reported that in 2006 they distributed over $761 million in Direct Local Aid to the Cities and Towns of the Commonwealth.
- From `www.masslottery.com/winners/faqs.html`

> What happens to the revenue which the Lottery generates from sales?
>
> 1. A minimum of 45% of revenues stays in the State Lottery Fund to be paid out in prizes. The Lottery's current prize percentage is over 69.
>
> 2. A portion of revenues is transferred to the commonwealth's General Fund for the expenses incurred in administering and operating the Lottery. The administrative and operating expenses of the Lottery are appropriated by the legislature as part of the annual state budget. Operating expenses cannot exceed 15%. Currently, operating expenses are under 8%. These operating expenses include 5.8% in commissions and bonuses paid to the sales agents who sell the tickets and under 2% in administrative expenses due to Lottery operation.
>
> 3. After prizes and expenses, the remaining Lottery revenues (approximately 23%) is transferred to the Local Aid Fund and returned to the cities and towns of the Commonwealth in the form of local aid. [R236]

- Several years later, on January 5, 2011 *The Boston Globe* reported that

> Massachusetts lottery agents have sold about $26 million in tickets over the course of this 16-draw series, which has failed to produce a jackpot winner. The tickets have raised $11 million for cities and towns. [R237]

(a) Sketch a pie chart showing how the money collected by the Lottery Commission was distributed among the three categories prizes, overhead and aid to Cities and Towns. Label each slice with its percentage, and one of the slices with an amount of money.
(b) What was the total dollar amount collected by the Lottery Commission in 2006?
(c) What was the fair price of a $5 ticket?
(d) How much on average did people in Massachusetts spend on lottery tickets in 2006? On average, how much did they get back in prizes? Is this "average" the mean, the median or the mode?
(e) Does the 2011 payout for the 16-draw series match the prize percentage reported in 2006?

Exercise 11.9.10. [S][Section 11.4][Goal 11.2] Megabucks changes the odds.

The first two paragraphs of an article in *The Boston Globe* on March 21, 2009 said that

> Like anyone who plays the lottery, Dean Thornblad was hoping to get rich quickly. He studied the odds of winning the various games before shelling out $150 for three season tickets that automatically enter him in twice-weekly drawings of Megabucks. At 1 in 5.2 million, the odds of hitting the jackpot, long by any standard, seemed to him at least "somewhat imaginable."
>
> But even his boundless optimism is being stretched by the lottery's latest proposal. The agency, under mounting pressure to return more money to cash-strapped cities and towns, is planning to make the odds of winning even slimmer, reducing them to 1 in 13.9 million beginning May 2, by making players match six numbers between 1 and 49, instead of six between 1 and 42. [R238]

What has happened to the expected value of Thornblad's ticket?

Exercise 11.9.11. [S][C][Section 11.4][Goal 11.2] Uncommon numbers.

In many state lotteries the customer picks the numbers she thinks will win. The prize is then divided among all the people who happened to pick the winning numbers. Much as we try to analyze only real situations, the real Massachusetts lottery is too complicated for this class. (Many people find it too complicated to choose the numbers they want to bet on, and elect "quick picks" instead.) So this question is about an imaginary lottery.

Here is how our lottery works. Tickets cost $1. Each person buying a ticket chooses the number between 1 and 100 that she thinks will win. When all the tickets have been sold, the state picks a number at random between 1 and 100. All the people who have chosen that number divide 70% of the total collected among themselves. (The other 30% the state uses for overhead and local aid.) So the fair price for a $1 ticket is $0.70 or 70 cents.

Of course the winners collect much more than the fair price (since the losers collect nothing). For example, if 1000 people bought tickets, 39 was the winning number, and 8 people chose 39, each would get $(\$1000 \times 0.7)/8 = \$87.50$.

If everyone buying tickets used "quick pick" then the 1000 tickets would (more or less) consist of 10 for each of the 100 numbers, ten people would have the winning number and the typical payoff would be $\$1000 \times 0.7/10 = \$70$.

Now that you've read this far and understood the game, we can ask an interesting question. Suppose you know that people are so afraid of the number 13 that no one ever picks it. You think (correctly) "If I buy a ticket and choose 13, I'm probably not going to win. But if I do win, I will win big because I won't have to share the prize." So every day you buy one of the 1000 tickets, and choose 13, knowing that no one else will. You lose with probability $99/100 = 0.99 = 99\%$ and win with probability $1/100 = 0.01 = 1\%$.

In the long run, how much money do you win (on the average) each day?

[See the back of the book for a hint.]

At `blogs.wsj.com/numbersguy/lottery-math-101-801/` you can read more about this idea:

> Low numbers are particularly popular, some of them because birthdays are a popular source of numbers to play. Research conducted by Tom Holtgraves showed that bettors also avoid numbers with repeated digits, though these are just as likely to turn up in lotteries as numbers without. [R239]

For still more information, see "Q3.4: Can RANDOM.ORG help me win the lottery?" at `www.random.org/faq/#Q3.4`.

Exercise 11.9.12. [R][Section 11.5][Goal 11.3] It's always 5.26%.

Compute the house advantage with an American wheel for each of the roulette bets in Section 11.5 to show that it's the same for each bet.

Exercise 11.9.13. [S][Section 11.5][Goal 11.3] Single zero roulette.

(a) Compute the house advantage for a roulette wheel with one extra wedge.
(b) Show that the house advantage in single zero roulette is approximately but not exactly half the house advantage in double zero roulette.

Exercise 11.9.14. [S][Section 11.5][Goal 11.3] Help this fellow out, please.

We found this roulette strategy on the web. The author says you should pick 19 numbers and bet on them. Since there are 38 spaces, you win half the time, with a 35 to 1 payoff. So when you win you're ahead $36 - 18 = 18$ dollars.

> Assuming you bet equally on all 19 numbers, this should result in a 50/50 shot of going 17 to 1 (unlike a 50/50 shot of going 1 to 1 if you bet on red/black or even/odd or 1-18/19-36).
>
> Am I missing something, or is it really that simple? [R240]

Answer his question.

Exercise 11.9.15. [U] What you're counting counts.

(a) What is the probability that a random word in English begins with the letter "t"?

 This is a question with several answers, which depend on how you select your "random word". You might count the words that begin with "t" in the dictionary. You might count those words in a newspaper, or on a website. There may be answers to the question on the web.

 Estimate the answer in several ways. Do the various assumptions lead to approximately equal answers?

(b) "e" is the most commonly used letter in English. What is the probability that a random letter in an English text is "e"?

 Attack this question as you did the previous one.

(c) What is "etaion shrdlu" and where does it come from?

Exercise 11.9.16. [S][Section 11.1][Goal 11.1] What is wrong with this estimate?

The following report appeared in the Offline column of *The New York Times* business section on March 8, 2008, where "Cubicle Coach" Marie Claire, says "take a chance" when considering whether to hire the ordinary candidate or one

> . . . who has the potential to be great, but has an equal chance of being awful?
>
> "You have a 66.7 percent chance of a positive result," the coach writes. "Yes, the unknown could flop, but she could also a) do as well as the known, or b) actually be a star." [R241]

(a) What assumption is Claire making that leads her to her estimate of 66.7%?

(b) Suppose Claire is correct when she assumes that the probabilities of great and awful are equal. Show that the chance of a positive result (great, or just OK) is somewhere between 50% and 100%.

12

Break the Bank—Independent Events

> I've thought that there may be more collisions ... in life than in books. Maybe the element of coincidence is played down in literature because it seems like cheating or can't be made believable. Whereas life itself doesn't have to be fair, or convincing.
>
> Shirley Hazzard
>
> *The Transit of Venus* [R242]

Unlikely things happen—just rarely! Here we calculate probabilities for combinations like runs of heads and tails. Then we think about luck and coincidences.

Chapter goals

Goal 12.1. Experiment with multiple independent events.

Goal 12.2. Understand that rare events do happen!

Goal 12.3. Understand common cultural references to events and their probabilities.

12.1 A coin and a die

We know that if we flip a fair coin then the probability of getting heads is $\frac{1}{2}$—that's the very definition of "fair".

Suppose you have a coin in one hand and a die in the other. What's the probability that when you flip the coin and toss the die you see a head and a 4?

In order to calculate

$$\frac{\text{number of desired outcomes}}{\text{number of possible outcomes}}$$

we first list all 12 possible outcomes:

Coin	Die	Coin	Die
head	1	head	4
tail	1	tail	4
head	2	head	5
tail	2	tail	5
head	3	head	6
tail	3	tail	6

Tossing the coin and rolling the die are completely unrelated. If you find out that the coin landed heads up you know nothing about what the die shows. The tosses are *independent* events. Each of the twelve equally likely outcomes occurs with probability 1/12. In particular, that's the probability of a head and a 4.

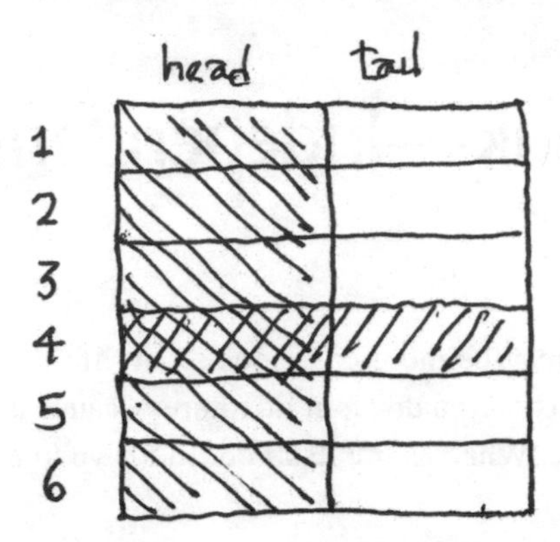

Figure 12.1. A coin and a die

Figure 12.1 displays those twelve possible outcomes in a square so that you can see the various probabilities as areas. The shaded column for heads takes up half the area, corresponding to a probability of 1/2. The shaded row for 4 takes up one sixth, corresponding to a probability of 1/6. The overlap represents a head and a 4: that doubly shaded rectangle has area $\frac{1}{2} \times \frac{1}{6} = \frac{1}{12}$.

The general principle is:

> When two events are independent the probability that both happen is the product of the probabilities for each one separately.

While we have this example in mind, we can think about a question that sounds similar: what is the probability of seeing a head or a four? Changing "and" to "or" in the question changes the answer dramatically. Of the 12 possible outcomes, in 6 the coin shows heads; in 2 the die shows four. So a first guess is that in $6 + 2 = 8$ of the 12 possible outcomes we see a head or a four. That first guess is wrong—it counts the combination "head and four" twice. The correct count is 7, so the probability is 7/12, not 8/12. That's clear in Figure 12.1, where 7 of the 12 small rectangles are shaded.

Suppose the question were asked slightly differently: what is the probability of seeing a head or a four *but not both*? The answer would be slightly different. We'd want to count just the singly shaded areas in the figure, for a probability of $(5 + 1)/12 = 50\%$.

This computation may help you remember several ideas that can help you navigate probability computations for independent events.

- "And" is straightforward. Multiply.
- "Or" is tricky. You can't just add. The sum is too big. To adjust it you have to think about whether in the particular problem you face the "or" means "but not both".

12.2 Repeated coin flips

What is the probability that you will get two heads when you flip two coins? Since neither coin cares about the other one, the two flips are independent. The same is true if you flip one coin twice: the second flip is independent of the first since the coin can't remember how it landed the first time.

That means we should multiply the two probabilities: $\frac{1}{2} \times \frac{1}{2} = \frac{1}{4}$ in order to count just one of the four equally likely outcomes for (first coin, second coin), or, if we have just one coin to play with, outcomes for (first flip, second flip).

H H
H T
T H
T T

The probability of two tails is also $\frac{1}{4}$.

In the other two cases there's one head and one tail, with probability $\frac{2}{4} = \frac{1}{2}$.

If you flip three coins (or one coin three times) there are eight possible outcomes, which we can list by putting first four heads and then four tails in front of the four possibilities for two coins:

HHH THH
HHT THT
HTH TTH
HTT TTT

The probability of three heads is clearly 1/8. You can see that by looking at the list of outcomes, or by computing

$$\frac{1}{2} \times \frac{1}{2} \times \frac{1}{2} = \frac{1}{8}.$$

The same is true for three tails.

To find the probability for two heads and a tail, or for two tails and a head, you can count the cases. The answer is 3/8.

If you flip 10 coins there are

$$2 \times 2 \times 2 \times 2 \times 2 \times 2 \times 2 \times 2 \times 2 \times 2 = 2^{10} = 1024$$

possible outcomes. Just one of those is all heads, with probability $(1/2)^{10} = 1/1024$, which is just about one tenth of one percent. (We've seen the approximation $1024 \approx 1000$ before, in Chapter 1 where we discussed the different meanings of "kilo" in kilogram and kilobyte.)

What is the probability of just one head in ten tosses? That head could come from any one of the ten flips, so there are ten ways to get one head:

HTTTTTTTTT
T**H**TTTTTTTT
...
TTTTTTTTT**H**

so the probability is $10/1024 \approx 1/100 = 1\%$.

Any combination of heads and tails can happen. Counting the number of ways for each is harder when there are a few heads and a few tails. You can study that mathematics in a statistics course, not here. The probability for exactly half heads and half tails (5 of each) turns out to be $252/1024 \approx 25\%$. That means that about a quarter of all tosses of 10 coins will show a 50 : 50 split.

To develop your intuition about what happens in practice we conducted an experiment flipping ten coins at a time, many times, counting the number of heads (between none and ten) each time. Figure 12.2 shows the results. (We didn't actually flip coins—we did the experiment in Excel, with the spreadsheet `flipcoins.xlsx`. You can repeat it if you wish.)

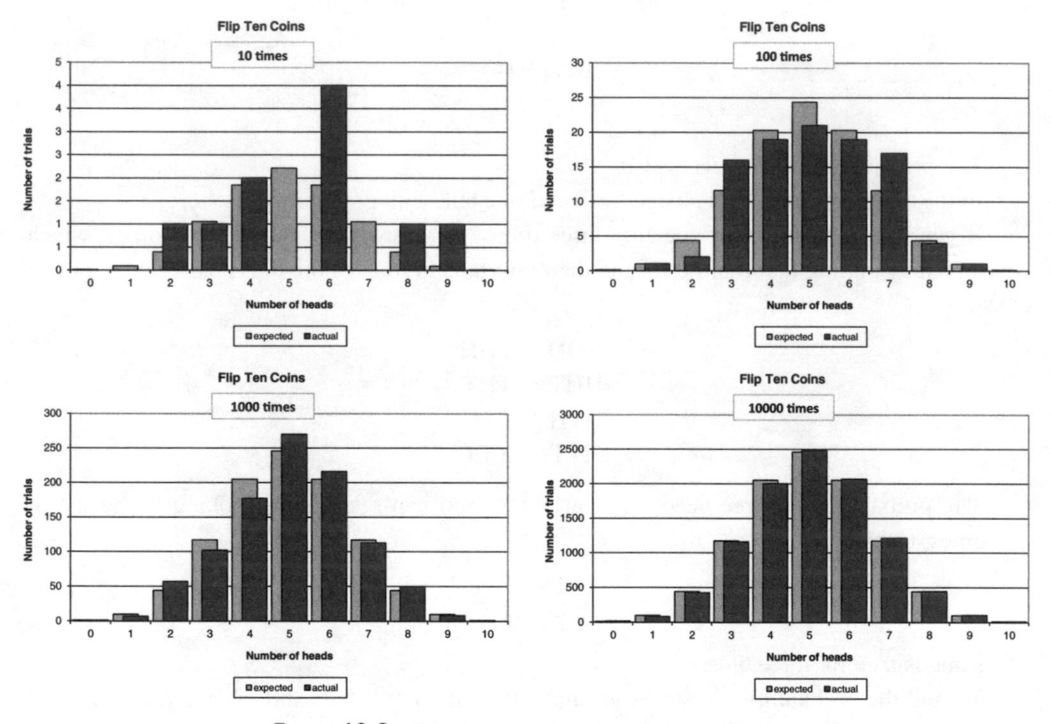

Figure 12.2. Flipping ten coins at once, many times

The categories on the x-axis are the number of heads (0 to 10).
The black bars record how many times each number appeared.
The gray bars are what the mathematics predicts.

The figure shows that when we flipped our ten coins just ten times we never saw exactly 5 heads, but did see 8 and 9 heads once each. Continuing to 100 ten-coin flips the observed values look like a better match to the gray bars, which show the predicted counts, calculated by looking at all the possibilities. But it was still true that five heads out of ten coins did not occur quite as often as the mathematics predicts. At 1,000 tries there were noticeably more occurrences of six heads out of ten than of four.

At 10,000 tries the observed values are close to the expected ones, but even there the number of times four heads occurred is visibly less than what it "should be". The outline of both column charts is the common bell curve we studied in Section 6.11. The mean, median and mode are all at the same place: half heads and half tails, with probability about 25%—odds of 1 : 3. The standard deviation is about 1. The probability of four or five or six heads is about 66%. Two thirds of the time that's what you'll see.

The gray bars look the same in the last three charts. That's because Excel chooses the scale on the *y*-axis so that it conveniently fits the values it needs to graph. The number of tries is ten times as large as you move from one experiment to the next. The scale on the *y*-axis covers ten times as much as well.

This experiment shows that when you flip several fair coins over and over again then in the long run you will see approximately half heads and half tails most of the time and that the distribution will approach the standard bell curve. Mathematicians have been working for about three hundred years to say precisely what "in the long run", "approximately half heads", "most of the time" and "approach" mean. One result is a formula for the standard deviation of the approximated bell curve. That formula says that if you toss 100 coins rather than just 10 the standard deviation is 5—about two thirds of the time you will see between 45 and 55 heads in one hundred coin tosses. Some day you may learn the formula in a statistics course. You don't need it now to begin to think sensibly about the probabilities for multiple coin flips.

12.3 Double your bet?

Here's a strategy that's bound to win money tossing fair coins.

Bet a dollar. If you lose, you're out $1, so bet $2 next time. If you win, you collect $4: your $2 bet and the $2 your opponent put up. You've spent $3 altogether, so you're ahead a dollar. If you lose that second bet, try a third at $4. If you win, you get $8, which is still a dollar more than the $4 + $2 + $1 = $7 you've spent so far. Next time bet $8 if you have to.

It seems clear that you will always win your dollar—eventually. But . . . you should always be suspicious of any "system" that guarantees a gambling win. This doubling scheme is one of the most common. Here are two reasons why it won't work, both related to the fact that long runs, though rare, do occur.

In Section 12.2 we found that the probability of 10 heads in a row is

$$\left(\frac{1}{2}\right)^{10} = \frac{1}{1024}.$$

That's close to 1000 : 1 against. It can happen, but it's very unlikely. Ten heads in a row will occur about once every thousand times, so if thousands of people try the system this will happen to some of them. You can't know that it won't happen to you.

That doesn't yet explain why the system must fail.

One reason is that you will run out of money to bet. Losing ten times in a row will cost you $1,023. Your next bet must be $2,048. If you don't have that much left, you're out more than $1,000, which you just risked in hopes of making $1.

You might feel safer with a million dollar bankroll. But if you lose twenty times in a row (which will happen from time to time) then you're out your million dollars and the next bet is more than a million. So if you didn't have two million to begin with you're stuck.

It doesn't matter how much you start with. You will lose it all sooner or later.

You can see the second flaw in the system when you think about the person you're betting against. If it's a casino, they will have a *house limit*—the maximum bet they're willing to accept.

If that's set at $1,000 you won't be able to make your tenth bet after losing nine times in a row. If it's $1,000,000 you won't be able to make your twentieth. Whatever it is, sooner or later you'll reach that limit. The higher it is, the more likely it is that you will think the system is working—but the more you'll lose when you encounter it.

Someone using this system will win small amounts frequently and lose more than her total winnings every once in a long while. What makes the system tempting is people's false belief that long runs never happen. We know better.

12.4 Cancer clusters

In an online post titled "Cancer Clusters: Findings Vs Feelings" you can read that

> A variety of factors often work together to create the appearance of a cluster where nothing abnormal is occurring. Looking for clusters is analogous to drawing a bull's eye after you have thrown darts at the wall at random. In this situation, there is possibly a place in which a bull's eye can be drawn that will leave multiple darts in close proximity to some common center. According to the American Cancer Society, cancer was diagnosed in an estimated 1,268,000 Americans in 2001. Finding clusters in cancer data is, thus, something like looking for patterns in the location of more than a million darts thrown at a dartboard the size of the United States. [R243]

To illustrate this phenomenon, we downloaded the populations of the 3,134 counties in the United States in the 2000 Census. The total population then was 279,517,404. The 1,268,000 cancer cases diagnosed in 2001 mean the national rate was $1,268,000/279,517,404 \approx 0.004536$, or just under one half of one percent—about one person in every 200. Then we did an experiment. We randomly assigned the cancer cases to the counties in proportion to their population. Then we sorted the list to see where cancer seemed to be concentrated. Table 12.3 shows the beginning and the end of that list.

Table 12.3. Simulated cancer distribution

State	County	Population	Cancer cases	Incidence rate
Nebraska	McPherson County	533	5	0.938 %
Texas	Roberts County	887	8	0.902 %
Idaho	Adams County	3,476	30	0.863 %
Montana	Petroleum County	493	4	0.811 %
Montana	Powder River County	1,858	15	0.807 %
		...		
Texas	Sherman County	3,186	5	0.157 %
Texas	Shackelford County	3,302	5	0.151 %
Colorado	Mineral County	831	1	0.120 %
Nebraska	Arthur County	444	0	0.000 %
Texas	Loving County	67	0	0.000 %

You can just imagine the editor of the local paper in MacPherson County, Nebraska or Roberts County, Texas writing an angry editorial wondering why the cancer rate there was twice the national average and demanding a federal investigation, while the Chambers of Commerce in Arthur County, Nebraska and Loving County, Texas wrote press releases bragging about what healthy places they were to live in.

Table 12.4 shows what happened when we repeated the experiment. Note that Petroleum County, Montana moved from fourth most cancerous to cancer free!

Table 12.4. Simulated cancer distribution (again)

State	County	Population	Cancer cases	Incidence rate
Texas	Kenedy County	414	5	1.208 %
Colorado	Hinsdale County	790	8	1.013 %
Montana	Garfield County	1,279	11	0.860 %
Texas	King County	356	3	0.842,7%
South Dakota	Hyde County	1,671	14	0.837,8%
		...		
Nebraska	Banner County	819	1	0.122 %
Texas	Roberts County	887	1	0.113 %
Montana	Petroleum County	493	0	0.000 %
Nebraska	Hooker County	783	0	0.000 %
Texas	Loving County	67	0	0.000 %

You can run the experiment for yourself: the data and the Excel commands that drive the simulation are in the spreadsheet `cancerclusters.xlsm`.

12.5 The hundred year storm

After heavy rains in March, 2010 *The Boston Globe* reported that

> FEMA [the Federal Emergency Management Agency] designates a property as being in a high-risk area if historical data show it has a 1 percent or greater chance of flooding during any given year. Although the rain of last weekend and Monday in most areas did not meet the federal standard of what is considered a 100-year storm, flooding was significant enough in many places to compare with federal projections, said Mike Goetz, FEMA's New England risk assessment branch chief. [R244]

Does "historical data" mean that FEMA actually located a century of flood records and checked off the places with at least one flood? The designation "100-year storm" is even trickier.

If you search the internet for "hundred year storm" the first few hits are to a rock band. A little further on you come to `www.livescience.com/environment/090922-100-year-storm.html`, where you can read an answer to the question "does a 100-year storm happen just once a century?" The answer is "no". Those big storms

> ... have a lower statistical likelihood of happening than your average thunderstorm, they can and do happen, sometimes within just a few years of each other. [R245]

We know enough to make quantitative sense of this. The probability of a "100-year storm" in any one year is just 1%. If the weather this year is independent of the weather last year then predicting a storm in any particular year is like flipping a coin with a only a 1% chance of heads (wet) and a 99% chance of tails (dry). Then we know that the probability of a run of 100 dry years is

$$0.99^{100} = 0.366032341 \approx 37\%.$$

That means that about 1/3 of the time you will go a whole century without seeing a hundred year storm and 2/3 of the time you will see at least one. You may see two less than 100 years apart. You can't properly argue that if there has been no storm in a long time you are "due for one" or that if there's a horrendous storm in 2010 you are "safe for another century". In fact on April 3 of that year *The Boston Globe* reported on "the two so-called 50-year storms that pounded Massachusetts over the last few weeks." [R246]

12.6 Improbable things happen all the time

Although it's unlikely that any particular person will win the state lottery, it's certain that someone will win.

Although it's unlikely that the Doonesbury cartoon characters Toggle and Mike's daughter will meet (as they did in that comic strip on March 21, 2009), that's no less likely than any other particular pairing. [R247]

When you encounter an unlikely event you're likely to look for a reason. Sometimes there is none—some rare things happen with probability one (someone wins the lottery), some unlikely things happen because there are lots of tries—heads ten or even twenty times in a row, cancer clusters.

On September 22, 2009 Carl Biakil, "The Numbers Guy," blogged in *The Wall Street Journal* about a Bulgarian lottery that produced the same six number winning combination twice in a row. Since each particular combination has a winning probability less than one in 5.2 million, the government investigated to check for cheating. "It just happened," a spokeswoman for the Bulgarian embassy in Washington, D.C., said. [R248]

Yet again, this time from Andrew Gelman's blog on May 26, 2011:

> It was reported last year that the national lottery of Israel featured the exact same 6 numbers (out of 45) twice in the same month, and statistics professor Isaac Meilijson of Tel Aviv University was quoted as saying that "the incident of six numbers repeating themselves within a month is an event of once in 10,000 years."
>
> . . .
>
> But wait a second . . . How many lotteries are there out there? A quick Wikipedia search yields the following:—62 international lotteries. I think I'm undercounting here because it looks like several countries have multiple "Pick *m* out of *n*" lotteries but I'm counting each country only once. 46 states or jurisdictions of the United States have lotteries. Some of these appear to be joint between states, however. I think a safe approximate guess is 100 major lotteries worldwide.
>
> These lotteries have different rules—some are more frequent than twice a week, some less frequent, some are easier to win than "pick 6 out of 45", some are harder to win. But a quick calculation is that if the Israeli lottery will have a repeat in a single month, once in 10,000 years, that if there are 100 lotteries out there, you'll

see "the incident of six numbers repeating themselves within a month" roughly once in 100 years. [R249]

12.7 Exercises

Exercise 12.7.1. [S][R][Section 12.1][Goal 12.1] Craps.

(a) Write down the 36 ways in which two dice can fall.
(b) Find the probability of each of the possible totals 2, 3, . . ., 12.
(c) Check that the probabilities sum to 1.
(d) What is the probability of throwing doubles?

Exercise 12.7.2. [S][Section 12.1][Goal 12.1] Sicherman dice.

George Sicherman found a strange way to number the faces of a pair of dice that leads to the same probability distribution of totals as an ordinary pair. One die is marked (1, 3, 4, 5, 6, 8) and the other (1, 2, 2, 3, 3, 4). [R250]

(a) Write down the 36 ways in which two Sicherman dice can fall.
(b) Check that the totals from a pair of Sicherman dice have the same probabilities as the totals from ordinary dice.
(c) Compare the odds of throwing doubles with Sicherman dice to those of throwing doubles with ordinary dice.

Exercise 12.7.3. [S][Section 12.1][Goal 12.1] Nontransitive dice.

Here are three strangely numbered dice:

A — 3,3,3,3,3,6
B — 2,2,2,5,5,5
C — 1,4,4,4,4,4

(a) Suppose you roll die A and your friend rolls die B. Find the probability that your die shows a higher number.

[See the back of the book for a hint.]

(b) Answer the previous question if you roll B and she rolls C.
(c) Answer the previous question if you roll C and she rolls A.
(d) Which is the most powerful die?

Exercise 12.7.4. [S][Section 12.1][Goal 12.1][Goal 12.3] Rain rain go away.

(a) What is the probability that it will rain tomorrow where you live?
(b) What is the probability that it will rain tomorrow in London?
(c) What is the probability that it will rain tomorrow in both places?
(d) What is the probability that it will rain tomorrow in neither place?
(e) The web page `www.srh.noaa.gov/ffc/?n=pop` explains how the National Weather Service calculates the "probability of precipitation". Read it and write about what you read.

Exercise 12.7.5. [S][Section 12.1][Goal 12.1][Goal 12.3] Who gave the money?

On April 24, 2009 an Associated Press story in *The Boston Globe* wondered why a dozen colleges with female presidents got large gifts from a single anonymous donor. The writer asked

> Coincidence? Unlikely. With about 23 percent of US college presidents women, the odds of a dozen randomly selected institutions all having female leaders are 1 in 50 million. [R251]

Verify the computation in the quoted paragraph.

Exercise 12.7.6. [S][R][Section 12.1][Goal 12.1] Blackjack.

Unlike craps and roulette, where each throw or spin is independent of the one before it, in blackjack the probabilities for the next card depend on the cards that have been dealt so far.

Since the dealer never changes his strategy, the player can overcome the house advantage by carefully counting the cards that have been dealt and betting more when his odds of winning are better. (For more on this, see `casinogambling.about.com/cs/blackjack/a/whyblackjacj.htm`.)

(a) What is the probability that the first card dealt will be a face card (a jack, queen or king)?
(b) At what points in the deal can the probability of a face card be the same as it was at the start?
(c) What is the probability that the next card dealt will be a face card if the 12 cards dealt so far have been J829K6Q9A3K4?
(d) When might the probability of a face card be the same as it was at the start as the deal in the previous question continues?

Exercise 12.7.7. [S][Section 12.2][Goal 12.1][Goal 12.2] Impossible?

What is the probability that a fair coin will come up heads 100 times in a row?
[See the back of the book for a hint.]

Exercise 12.7.8. [S][Section 12.2][Goal 12.1] Would you bet?

(a) What is the probability that a fair coin lands tails up eight times in a row?
(b) Suppose you're offered a bet at 250:1 odds against a fair coin landing tails up eight times in a row. What would you do—take one side of the bet or the other, or decide not to play at all? (Your answer should depend both on your answer to the previous part of the problem and on your own personal ideas about risk and money.)
(c) Answer the previous question if you are operating a casino. Compute the house advantage and explain how much money you would make in the long run.
(d) Which is more likely, that eight flips will be TTTTTTTT or that they will be THTHTHTH?

Exercise 12.7.9. [U][Section 12.2][Goal 12.1] Simulating dice.

Use the spreadsheet at `flipcoins.xlsx` to roll dice by setting the probability of a head to be $1/6$ instead of $1/2$. Then the experiment is the same as rolling ten dice and counting the number of one's that come up.

(a) Describe what happens as you increase the number of tries.
(b) Explain why the probability that you see all ten dice showing a one is $(1/6)^{10}$. Did that ever happen in your experiment? Did you try long enough to expect it to happen?

(c) Explain why the probability that none of the ten dice shows a one is $(5/6)^{10}$. Did that ever happen in your experiment? Did you try long enough to expect it to happen?

Exercise 12.7.10. [U][Section 12.2][Goal 12.2] Online experiments.

Play with some of the applications at `www.random.org/`. Write about what you discover. What was interesting? What was surprising?

Exercise 12.7.11. [S][R][Section 12.2][Goal 12.1] Runs of twenty.

If everyone in the United States flipped a coin twenty times about how many people would see twenty heads? Twenty tails? Heads and tails alternating perfectly?

Exercise 12.7.12. [U][Section 12.4][Goal 12.2][Goal 12.3] Cancer clusters.

(a) Why do you think the counties with the highest and lowest (simulated) cancer incidence are all rural?
(b) Redo the experiment when the population is subdivided into fifty states rather than three thousand counties. Why is the result less dramatic?

[See the back of the book for a hint.]

Exercise 12.7.13. [U][Section 12.5][Goal 12.1][Goal 12.3] Misinformation on the web.

Criticize this argument from `www.numbersplanet.com/`:

> Would you play "14 22 38 49 59 16" on your Powerball ticket if you knew they had already hit on Oct 10th 2009? I would think not! While it's entirely possible that the same six numbers can draw more than once, it simply doesn't happen, or if it does, not very often. [R252]

Exercise 12.7.14. [S][Section 12.5][Goal 12.1][Goal 12.3] More misinformation on the web.

Here's an online discussion about using Powerball statistics to beat the system. The website starts with the data in this table summarizing the result of drawing of six numbers multiple times.

Evens	Odds	Drawings
6	0	17
5	1	127
4	2	331
3	3	431
2	4	353
1	5	118
0	6	16

and then says

> The important statistic is the "3 Even / 3 Odd: 431" statistic . . . [t]elling you that you should be playing six numbers that has a "3 Even and 3 Odd! number combination". [R253]

(a) Plot the data and show that it approximates the normal bell curve, as it should.

(b) Criticize the argument suggesting that you should always bet on a three even three odd combination.
(c) Fix the author's grammar.

Exercise 12.7.15. [S][Section 12.6][Goal 12.2][Goal 12.3] The Tour de France.

On August 6, 2008, London's *Telegraph* featured an article headlined "Olympic Games drug testing means 'cheaters escape and innocents tarnished'". That article featured a comment from Professor Donald Berry discussing doping charges against the cyclist Floyd Landis.

> With 126 samples taken in the Tour de France, assuming 99 per cent specificity, the false-positive rate is 72 per cent. So, an apparently unusual test result may not be unusual at all when viewed from the perspective of multiple tests. ... This is well understood by statisticians, who routinely adjust for multiple testing. [R254]

Professor Berry was then head of the Division of Quantitative Sciences and chair of the Department of Biostatistics and Frank T. McGraw Memorial Chair of Cancer Research, MD Anderson Cancer Center, so presumably he knows what he is talking about.

Check his conclusion that a drug test that reports a false positive only 1% of the time will return a false positive with 72% probability when it's repeated 126 times.

(News reports in 2010 suggested that Landis was in fact guilty, even though you could not reliably conclude that if all you had were the results of those particular tests.)

Exercise 12.7.16. [U][Section 12.6][Goal 12.2][Goal 12.3] Old Friends Farm.

An email message from Old Friends Farm (`www.oldfriendsfarm.com/`) in Amherst, MA notes that

> Two members of our [six person] crew realized that they had met 10 years ago on a farm in France, and not seen each other since. What are the chances of that!? [R255]

Discuss the coincidence.

Exercise 12.7.17. [U][Section 12.6][Goal 12.2][Goal 12.3] The law of averages.

Mma Makutsi warns Mma Ramotswe:

In *The Double Comfort Safari Club* Alexander McCall Smith puts these words in Mma Makutsi's mouth.

> There is something called the law of averages—you may have heard of it. It says that if you haven't trodden on a snake yet, then you may tread on one soon-soon. [R256]

Discuss how this opinion fits with the ideas in this chapter.

Review exercises

Exercise 12.7.18. [A] Find the probability of each event.

(a) Tossing two fair coins and getting H T as the result.
(b) Tossing a fair coin twice and getting H T as the result.
(c) Drawing a red card from a deck of playing cards.

(d) Drawing a red ace from a deck of playing cards.
(e) Drawing an ace from a deck of cards and then drawing a king.
(f) Rolling two fair dice and getting snake eyes (1 and 1).
(g) Rolling two fair dice and getting boxcars (6 and 6).
(h) Rolling two fair dice and getting a total of 7.

13

How Good Is That Test?

In Chapter 12 we looked at probabilities of independent events—things that had nothing to do with one another. Here we think about probabilities in situations where we expect to see connections, such as in screening tests for diseases or DNA evidence for guilt in a criminal trial.

Chapter goals

Goal 13.1. Interpret and build two way contingency tables.

Goal 13.2. Understand how to compute probabilities for dependent events.

Goal 13.3. Understand the implications of false positives.

13.1 UMass Boston enrollment

Table 13.1 summarizes student enrollment at UMass Boston in 2006 by category two ways: graduate/undergraduate and male/female. We can use the data to answer some probability questions about a random student.

Table 13.1. UMass Boston enrollment, 2007

	Undergraduate	Graduate	Total
Female	5,680	2,388	8,068
Male	4,328	1,037	5,365
Total	10,008	3,425	13,433

- What is the probability that a student chosen at random is an undergraduate? The last row of the table has the numbers we need:

$$\frac{\text{number of undergraduates}}{\text{number of students}} = \frac{10,008}{13,433}$$
$$= 0.745$$
$$\approx 75\%.$$

Three quarters of the students are undergraduates.

- What is the probability that a student is female?
  For that computation we use the totals in the last column:

$$\frac{\text{number of females}}{\text{number of students}} = \frac{8{,}068}{13{,}433} = 0.600610437 \approx 60\%.$$

- What is the probability that a student is a female undergraduate?
  Use the count in the first column of the first row:

$$\frac{\text{number of female undergraduates}}{\text{number of students}} = \frac{5{,}680}{13{,}433} = 0.4228392764 \approx 42\%.$$

In each of these probability calculations we used the total number of students (13,433) in the denominator.

Continuing . . .

- What is the probability that a female student is an undergraduate?
  Since this is a question about the female students, we need a different denominator:

$$\frac{\text{number of female undergraduates}}{\text{number of female students}} = \frac{5{,}680}{8{,}068} = 0.70401586514 \approx 70\%.$$

- What is the probability that an undergraduate is female?
  That's a different question. This time we know the student is an undergraduate. That calls for a different denominator:

$$\frac{\text{number of female undergraduates}}{\text{number of undergraduates}} = \frac{5{,}680}{10{,}008} = 0.56754596322 \approx 57\%.$$

The last two questions sound similar, but have different answers, because each begins with a different assumption. In the first we know the student is female and wonder whether she's an undergraduate. In the second, we know that the student is an undergraduate and wonder whether it's a she.

We're not finished thinking about these probabilities. We found that there's a 60% probability that a student is female. But if we know the student is an undergraduate then that probability drops to 57%, because the proportion of women is different for undergraduates than for the student body as a whole. This is not what happened when we thought about a coin and a die in Section 12.1. The probability that the die shows a four is the same whether the coin comes up heads or tails. Those events are *independent*. The facts "is female" and "is an undergraduate" are *dependent*. When you know one of them you know something about the probability of the other.

We learned in Section 12.1 that when events are independent you multiply to compute the probability that both happen:

$$\begin{aligned}\text{probability(coin H and die 4)} &= \text{probability(coin H)} \times \text{probability(die 4)}\\ &= \frac{1}{2} \times \frac{1}{6}\\ &= \frac{1}{12}.\end{aligned}$$

For dependent events that won't work. We found that

$$\text{probability(female and undergraduate)} = 42\%$$

but

$$\begin{aligned}\text{probability(female)} \times \text{probability(undergraduate)} &= 60\% \times 75\%\\ &= 45\%.\end{aligned}$$

Those answers are close, but not the same, because the proportion of females among the undergraduates is close to but not the same as the proportion among the graduate students.

In the rest of this chapter we will look at the probabilities for dependent events, working with displays like Table 13.1 in examples where the consequences matter much more than they do here.

13.2 False positives and false negatives

Many women have periodic mammograms to look for breast cancer. Many men have periodic PSA tests to look for prostate cancer. In each there are four possibilities. We'll spell them out for breast cancer.

- *True positive*: a woman has breast cancer and the mammogram says so.
- *True negative*: a woman does not have breast cancer and the mammogram says she doesn't.
- *False positive*: a woman doesn't have breast cancer but the mammogram mistakenly says she does.
- *False negative*: a woman does have breast cancer but the mammogram doesn't detect it.

If the test were perfect there would be no false positives and no false negatives—but there are very few perfect tests. In order to understand what the test results mean you can build a table like the one in the first section of this chapter.

We'll do that with a real example. Figure 13.2 appeared in the article "False positives, false negatives, and the validity of the diagnosis of major depression in primary care" in the September 1998 *Archives of Family Medicine*.

It summarizes the results of a study of 372 patients who were screened by family physicians for clinical depression.

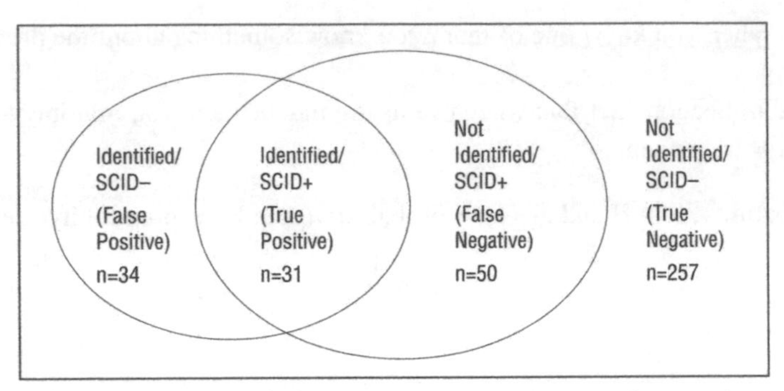

Proportions of patients identified as depressed by physicians, Structured Clinical Interview for Diagnostic and Statistical Manual of Mental Disorders, Revised Third Edition *(SCID), both methods, or neither method (N = 372).*

Figure 13.2. Diagnosing depression [R257]

The numbers in the four categories in the figure are easier to understand when we put them in Table 13.3.

Table 13.3. Diagnosing depression

		depressed		
		yes	no	total
diagnosed	yes	31	34	68
	no	50	257	307
	total	81	291	372

Two by two tables like this are called *contingency tables*. Figure 13.4 shows the standard names for the four cells with raw data: true positive, false positive, false negative and true negative. In this example they have values 31, 34, 50 and 257.

		condition present	
		yes	no
screened positive	yes	true positive	false positive
	no	false negative	true negative

Figure 13.4. A two way contingency table

The totals tell us that 65 people in the population of 372 (17%) were diagnosed as depressed, and that 81 (22%) were depressed. Those numbers are pretty close. But does that make it a good test? To answer that question we need to look at the columns separately.

- The first column tells us there were 31 true positives and 50 false negatives from the total of 81 subjects who were in fact depressed. So if a subject was depressed the probability that he or she was diagnosed correctly is only $31/81 \approx 38\%$. That is the true positive rate. There's a 62% chance the condition was missed. That 62% is the false negative rate.

- The second column says that even when a subject was not depressed the chance of a diagnosis of depression was $34/291 = 0.117 \approx 12\%$. That is the false positive rate. It means about 12% of the people diagnosed as depressed didn't in fact suffer from that condition.

Whether this is a "good" test is a difficult decision.

Although the chance of misdiagnosis of depression when it doesn't exist is fairly low—about 12%—the 62% false negative rate says that test will identify fewer than half the depressed people.

13.3 Screening for a rare disease

A test with a small false positive rate looks like a good candidate for screening large populations for a nasty disease. However, if the disease is rare, the test may not be as good as it looks. In this section we'll study two examples, one made up and one real.

Suppose a drug company has developed a test for the rare disease X. Clinical trials show that the test is 90% accurate at detection, so the false negative rate is 10%. Those trials also show that the false positive rate is only 1%.

These are the important questions:

1. What is the probability that a person who suffers from X tests positive?
2. What is the probability that a person who tests positive suffers from X?

If the test were perfect—no false positives, no false negatives—each question would have the same answer: 100%. But the two facts "suffers from X" and "tests positive for X" are not exactly the same. Knowing either one makes the other more likely, but not certain. We want to find out how much more likely in each case.

Question 1 is easy: the drug company's clinical trials found that there is a 90% probability that a person who suffers from X tests positive for X.

Whether that test is as good as it sounds depends in part on the answer to the second question. That answer depends on two things: the false positive rate and the number of people who actually have X. Suppose just one person in every 1,000 suffers from X (one tenth of one percent of the population). Then even though the false positive rate is only 1%, most of the positive results will come from healthy people. We can use a contingency table to find the actual value for "most of".

Since percentages (particularly small percentages) are often confusing, we'll build our table for an imaginary population of 100,000 people that just matches the statistical profile for this test. In a population of 100,000, one out of every 1,000 will have the disease. That's 100 people. Of those 100, 90% (so 90 people) will test positive. The other 10 will be the false negatives. Of the 99,900 healthy people, one percent (999) will test positive. The other 98,901 will be the true negatives. Table 13.5 shows the contingency table.

Now we can answer the second question. The probability that someone who tests positive is actually ill with X is only $90/1089 = 8.26\%$.

Is this acceptable? Maybe, maybe not. If the test is inexpensive and there's a second test (perhaps more expensive) that can weed out the false positives, and the disease can be treated successfully if detected, perhaps the screening is a good idea. If all the people who test positive must undergo expensive painful unreliable treatment, which would be unnecessary for

Table 13.5. Screening for disease X

		suffers from X yes	no	total
test + for X	yes	90	999	1,089
	no	10	98,901	98,911
	total	100	99,900	100,000

more than 90% of them, then the screening is probably a bad investment of scarce health care resources.

For a real application of this technique to the statistics of screening for breast cancer, work Exercise 13.7.7 .

13.4 Trisomy 18

In this section we'll work through the numbers when considering whether to call for routine prenatal screening for the rare birth defect trisomy 18. The story that follows is from a July 2008 web posting T. Davies wrote at `askville.amazon.com/understand-False-Positive-test-Trisomy-18/AnswerViewer.do?requestId=12714458` in response to an online query. His prose is dense and complicated. You have to read it carefully over and over again to extract the meaning. Our first job is to understand what he's trying to say. [R258]

The trisomy 18 test finds half the cases with only 1% false positives. Davies compares that with screening for Down's syndrome which detects 70% of the cases with 5% false positives. "Unfortunately," he says, "the authors [of the screening study] fail to emphasise the importance of the relative incidence of the two conditions at birth before concluding that screening for trisomy 18 should be introduced."

Davies notes that there are about 12.6 instances of Down's syndrome per 10,000 births. The incidence for trisomy 18 is just 1.3 per 10,000 births.

A positive test leads to a second procedure, an amniocentesis to check whether the positive is true or false. Davies calculates that screening 10,000 pregnant women for Down's syndrome "would result in 8.8 cases being detected at the cost of 500 amniocenteses (5% of 10,000). This means that one case of Down's syndrome is detected for every 57 amniocenteses performed." For trisomy 18 his figures are 0.65 cases detected, 100 amniocenteses, so 154 amniocenteses to detect one true case.

Let's check his arithmetic for trisomy 18. To build the contingency table we need three numbers:

- The false negative rate. Since the test detects 50% of cases the other 50% are the false negatives.
- The false positive rate. It's only 1%.
- The incidence rate. The second paragraph tells us it's 1.3 per 10,000 births.

Contingency Table Calculations
Ethan Bolker and Maura Mast, for *Common Sense Mathematics*

Set population to 1 to see percentages in the table.
Set population to a large round number like 1000 to see meaningful numbers of cases.

The incidence, false positive and false negative rates are formatted as percentages.
You can use actual values in Excel formulas to set them - for example, =1/1000 for 0.1%.

population:	10000
incidence:	0.0130%
false+:	1.00%
false-:	50.00%

	actual+	actual-	total	
test+	0.65	99.987	100.637	0.65% prob(actual+ when test+)
test-	0.65	9898.713	9899.363	99.99% prob(actual- when test-)
total	1.3	9998.7		

Figure 13.6. Screening for Trisomy 18

Figure 13.6 shows a screenshot of the spreadsheet `ContingencyTable.xlsx` with entries for this problem. Cell `B12` is named `INCIDENCE`; it contains the formula `=1.3/10000`, formatted as a percent. Cell `B17` for the number of true positive results contains the formula

```
=POPULATION*INCIDENCE*(1-FALSENEG)
```

which in this example is

$$10{,}000 \times \frac{1.3}{10{,}000} \times (1 - 0.5) = 0.65,$$

confirming Davies' "0.65 cases per 10,000 women tested." The spreadsheet shows 100.637 positive tests, which matches Davies' estimate of 100. So it would take 100 amniocenteses to find 0.65 cases of trisomy 18. That works out to $100/0.65 = 154$ amniocenteses to find each case. (Exercise 13.7.3 asks you to check Davies's arithmetic for Down's syndrome.)

You don't have to know what amniocentesis is, but you do have to know that it has some risks: there is a small chance that it will lead to a miscarriage. Davies claims that "a screening programme would cause the abortion of at least as many normal fetuses as it would detect cases of trisomy 18." ("Abortion" his synonym for "miscarriage," not the politically charged "abortion" so much in the news.)

That would be true if the risk of miscarriage from amniocentesis was about one in 150. It's probably smaller. Several web sources provide statistics like these:

> Miscarriage is the primary risk related to amniocentesis. The risk of miscarriage ranges from 1 in 400 to 1 in 200. In facilities where amniocentesis is performed regularly, the rates are closer to 1 in 400. [R259]

If we use one in 300 instead of one in 150 as the probability of miscarriage from an amniocentesis then it costs about one unnecessary miscarriage to detect two cases of trisomy 18. That's still a pretty high risk.

Davies compares his risk estimate to the much lower estimate for similar screening for Down's syndrome, noting that "In many places it is still undecided whether screening for Down's syndrome is worth the disbenefits for the prospective parents."

13.5 The prosecutor's fallacy

The Cornell University Legal Information Institute posted a discussion of *McDaniel v. Brown* when that case was on the docket of the Supreme Court. They wrote

> Following a state conviction for sexual assault, Troy Brown filed a petition for writ of habeas corpus in the United States District Court for the District of Nevada. The District Court allowed Brown to present new evidence: a report from Dr. Lawrence Mueller. This report detailed a statistical error ("prosecutor's fallacy") made by the prosecution during the presentation of DNA evidence. Based on Dr. Mueller's report, the District Court dismissed the DNA evidence from consideration, found insufficient evidence to convict Brown, and ordered a retrial.
>
> . . .
>
> At trial, Renee Romero, a forensic scientist at the Washoe County Crime Lab, testified that the DNA found in the victim's underwear matched Brown's DNA; only one in three million people would match the DNA tested. The prosecutor asked Romero to express this statistic as "the likelihood that the DNA found . . . is the same as the DNA found in [Brown's] blood." Romero concluded that the likelihood was 99.999967 percent. Based on this statistic, the prosecutor then asked Romero if it would be fair to conclude that there was a 0.000033 percent chance that the DNA did not belong to Brown. Romero agreed with the prosecutor, stating that that this was "not inaccurate." [R260]

Romero's arithmetic is right: one in three million is 0.000033 percent. But her thinking is wrong.

The prosecutor's fallacy is the claim that the one in three million probability of a random match is the same as the probability that the defendant is innocent. We can use a contingency table to show why those probabilities are different.

First we need an estimate of the population in which a possible DNA match might be found. To make the arithmetic easier, we'll take that to be 9 million people (Los Angeles is near enough to Nevada). Then the "one in three million" statistic says we should expect three DNA matches from the innocent people in that population. This contingency table summarizes the data:

		truth		
		guilty	innocent	total
DNA	match	1	3	4
	nonmatch	0	8,999,996	8,999,996
	total	1	8,999,999	9,000,000

Make sure you understand the first row of the table. One person is guilty and is a DNA match. The three matches from innocent people are false positives. In other words, the first row of that table tells us that if the only evidence in the case is the DNA match the odds are 3 : 1 that the suspect is innocent! The probability that he's guilty is only 25%. That's a far cry from the "99.999967% guilty" that the prosecutor asked the jury to believe.

The defense didn't make this argument using a hypothetical 9,000,000 population of potential suspects. Instead they questioned the "one in three million" chance of a match. The defendant had near relatives in the area which increased the chances of a match to about one in 6,500, according to a defense specialist. That would reduce the chance of an accidental match to $6499/6500 = 0.999846154 \approx 99.98\%$. We're not surprised that the change from 99.999967 percent to 99.98% did not convince the jury to acquit. 99.98% still sounds very much like a sure thing.

But it's not, because of the prosecutor's fallacy. That was the basis for the appeal. Suppose we reduce the population from which the match might come to just 100,000—the nearby area where there may be close relatives. Then the 1 in 6,500 chance of a match means there will be about 15 matches in that population in addition to the one match for the guilty party. The numbers in the revised contingency table below show there is now a 15 : 1 chance that the DNA match fingers an innocent person rather than the true criminal.

		truth		
		guilty	innocent	total
DNA	match	1	15	16
	nonmatch	0	99,984	99,984
	total	1	99,999	100,000

Nevertheless, the story did not end well for Brown.

> The Supreme Court [overturning the appeals court order for a retrial] said in a per curiam opinion that overstated estimates of a DNA match at trial did not warrant reversal of a conviction when there is still "convincing evidence of guilt." [R261]

13.6 The boy who cried "Wolf"

After unusual disasters like terrorist attacks, earthquakes, severe storms or airplane crashes you often hear finger-pointing discussions about the incompetence of the agencies charged with predicting (perhaps even preventing) what happened. Those discussions may start with a search that discovers warning signs that were ignored.

Sometimes there were real lapses, and policies and practices must be designed to prevent a recurrence. But often blame is unjustified. Table 13.7 explains why, even without numbers. You might call this *qualitative reasoning*.

With numbers in the first column you can compute the probability that a disaster occurs with no warning at all. That's not good. To guard against it, there should be more warnings. Then with numbers in the first row you can compute the probability that a particular warning actually corresponds to a disaster about to happen. But more warning don't lead to more disasters, just to more false positives.

Table 13.7. Should it have been predicted?

		what happens		
		disaster	nothing	total
warning?	yes	rare	usually	infrequent
	no	rare	almost always	almost always
	total	rare	almost always	always

That means there are often good reasons for ignoring a warning. State and governmental agencies have to balance the severity of the warning with the cost and inconvenience of asking the public to respond. For example, an earthquake warning may lead to an order to evacuate an entire city. The expense and disruption from repeated evacuations that are not followed by an earthquake may be worse than the consequences in the rare instance when the earthquake happens. Just because after the fact you look back and find clues in the seismic record that suggested an earthquake might be imminent doesn't mean evacuation was the right call.

13.7 Exercises

Exercise 13.7.1. [S][Section 13.2][Goal 13.1][Goal 13.3] Chronic fatigue syndrome.

On August 24, 2010 a headline in *The Boston Globe* read "Researchers link chronic fatigue syndrome to class of virus". The story reported on a study of 37 patients with the disease. 32 tested positive for a particular suspicious virus. Only 3 of 44 healthy people tested positive. [R262]

A 2003 study in the *Archives of Internal Medicine* reported that "The overall . . . prevalence of CFS . . . was 235 per 100,000 persons." [R263]

(a) Construct the contingency table for this diagnostic tool. You may do this by hand, or with the spreadsheet `ContingencyTable.xlsx`.
[See the back of the book for a hint.]
(b) Explain why this test is potentially important for research on chronic fatigue syndrome but might not be a good screening test.

Exercise 13.7.2. [S][W][Section 13.2][Goal 13.1][Goal 13.3] Pregnancy tests.

A home pregnancy test kit website says

> Home tests are usually 97% accurate when all instructions are followed correctly and the results are read on time.
>
> A false positive pregnancy test is when the test says that you are pregnant but actually you are not. This is a one off case and a positive pregnancy test is a pretty good indication that you are pregnant. False positive pregnancy tests are rare—though there are instances and conditions where they can occur. [R264]

Assume that "97% accurate" means a false positive rate and a false negative rate of 3%. Since a woman is unlikely to use a home pregancy test unless she thinks she's probably pregnant, assume that 80% of the women who try one are in fact pregnant.

Explain why a positive test indicates a pregnancy more than 99% of the time even though the false positive rate is 3%.

Exercise 13.7.3. [U][Section 13.3][Goal 13.1][Goal 13.3] Prenatal screening.

Check the calculations for Down's syndrome testing using the data in the quotation in Section 13.3.

Exercise 13.7.4. [U][C][Section 13.2][Goal 13.1][Goal 13.3] Spam.

Spam is junk email. Most mail systems have a spam filter that tries to decide whether each piece of email you get is spam. When the spam filter finds something it thinks is spam, it may throw it away, or put it in a junk mail folder so that you can decide whether to throw it away without reading it.

Before my university department set up a spam filter I ran my own. (The "I" here is Ethan Bolker, one of the authors, not the generic authorial "we" we use in most of the book.)

I got about 250 emails each day. My spam filter trapped about 175 of them. Of those about five were legitimate, and should have been delivered directly to me. My inbox, which should have contained just the legitimate messages was usually about half spam. So (in words) my spam filter is pretty good (but not perfect) at recognizing legitimate email but not very good at calling spam spam.

(a) Build a two way contingency table with row categories "marked spam" and "not marked spam", column categories "spam" and "legitimate".
(b) Compute and interpret the false positive and false negative rates.
(c) Explain why both the false positives and the false negatives make dealing with my email harder.
(d) I can adjust the settings in my spam filter to reduce the false positive rate. Explain why that would increase the false negative rate.
(e) Is the number of spam emails I received consistent with the claim in the August 6, 2008 issue of *The New Yorker* that there are more than a hundred billion spam emails every day? [R265]
(f) What is the original meaning of the word "spam"? Does the company that sells (the real) spam object to the new meaning?
(g) How do you deal with spam? (If your email provider does all the filtering for you, you may not even know it's throwing things away before you see them, so you may need to do some research on your email provider's web site to find the answers to these questions.)
    - Who provides your email service (your university, your internet service provider, Google, Yahoo, ...)?
    - Do you have any say in how your email provider filters spam for you? If so, what do you tell it?
    - Estimate the data you need to build the two way table for your spam statistics and compute the false negative and false positive rates.

Here are some web sites to look at if you want to find out more about spam.

- `www.imediaconnection.com/content/3649.asp`. There are some useful tips here about how to keep other people's spam filters from thinking mail from you is spam.
- Tools your system administrator might use: `www.spamcop.net/`, `www.spamhaus.org/`

Exercise 13.7.5. [S][Section 13.2][Goal 13.1][Goal 13.3] Plagiarism.

In 2006 UMass Boston experimented with the plagiarism detection software described at `www.turnitin.com` that claims it can identify plagiarism in essays students write. UMass did not purchase the software after the experiment. Perhaps the possibility of false positives contributed to that decision.

Suppose that the software can actually detect every cheater and that it's 99% accurate in declaring honest students honest. (We made up these numbers since the company does not advertise them.) Sounds like a pretty good test.

(a) Estimate how many papers are submitted by students at your school each semester.
(b) Suppose that most students are honest. Estimate how many students will be falsely accused of cheating.
(c) What are the advantages and disadvantages of using the software? (There are several arguments on both sides of the question. Think of as many as you can.)
(d) Read and write about this article from *The New York Times*: `www.nytimes.com/2010/07/06/education/06cheat.html`

Exercise 13.7.6. [U][Section 13.2][Goal 13.3] Airport screening.

In response to the article "Screening programme evaluation applied to airport security" in the December 10, 2007 issue of the *British Medical Journal*, Ganesan Karthikeyan wrote

> It is probably true that airport security in its present form is not an efficient screening measure. However, one important difference exists between screening for disease in individual patients and screening for, say, explosives in airports. While one missed cancer on screening can cause the loss of at the most, one life, the number of potential lives lost per missed screening at airports can be substantially larger. This has to be factored into any attempts at evaluation of the process. [R266]

It's clear that a false negative is a disaster. Discuss the consequences of a high false positive rate.

Exercise 13.7.7. [S][W][Section 13.2][Goal 13.1][Goal 13.3] Breast cancer screening.

In his "Chances Are" column in *The New York Times* on April 25, 2010 Steven Strogatz wrote about a diagnostic puzzle presented to several doctors:

> The probability that [a woman in this cohort] has breast cancer is 0.8 percent. If a woman has breast cancer, the probability is 90 percent that she will have a positive mammogram. If a woman does not have breast cancer, the probability is 7 percent that she will still have a positive mammogram. Imagine a woman who has a positive mammogram. What is the probability that she actually has breast cancer?
>
> ...
>
> [When 24 doctors were asked this question], their estimates whipsawed from 1 percent to 90 percent. Eight of them thought the chances were 10 percent or less, 8 more said 90 percent, and the remaining 8 guessed somewhere between 50 and 80 percent. Imagine how upsetting it would be as a patient to hear such divergent opinions. [R267]

(a) What is the correct answer?

[See the back of the book for a hint.]

(b) What percentage of the 24 doctors got the correct answer?

Exercise 13.7.8. [U][Section 13.2][Goal 13.1][Goal 13.3] Identity fraud.

On July 24, 2011 Jane Allen wrote in a letter to the editor of *The Boston Globe* in response to an article headlined "Identity fraud dragnet hardly seems worth the expense or trouble".

> [T]he state Registry of Motor Vehicles sends out 1,500 suspension letters a day. Last year, as a result of the software, State Police said there were 100 arrests for fraudulent identity and 1,860 licenses were revoked.
>
> That means that about 390,000 people were questioned for the sake of finding fewer than 2,000 transgressors. This hardly seems worth the $1.5 million grant for the software, let alone the investment of personnel. [R268]

(a) Check Allen's arithmetic in the second paragraph.
(b) Construct the contingency table for this screening. Identify the true and false positives and negatives. Explain the costs and benefits.

Exercise 13.7.9. [U][C][Section 13.5][Goal 13.1] Candy leads to crime.

An article headlined "Happy Halloween! Kids who eat candy every day grow up to be violent criminals" in the October 2, 2009 *Daily Finance*, begins

> Quick, hide the candy jar! Feeding your child candy every day could help turn Junior into a violent criminal, according to a large study in Britain, which found that 69 percent of the participants who had committed violence by 34 had eaten sweets or chocolate nearly every day during childhood. [R269]

You can find the full text at `www.dailyfinance.com/2009/10/02/happy-halloween-kids-who-eat-candy-every-day-grow-up-to-be-viol/`.

(a) Read the rest of the article. Build the contingency table with columns for whether or not someone ate candy as a child, rows for whether or not they committed violence as an adult.
(b) Explain why this is an example of the prosecutor's fallacy.
(c) Some of the online comments on that article recognize the fallacy—for example

> 10-03-2009 @ 10:21PM
> Bski said . . .
> I bet you, 99% of criminals ate bread daily by the time they were 10 years old!!!!

Write your own blog entry, using your understanding of two way contingency tables to enlighten any readers. If you like what you've written you may still be able to post your comment on the article's blog.

Exercise 13.7.10. [U][Section 13.2][Goal 13.3] Domestic violence.

In Andrew Gelman's blog on "Statistical Modeling, Causal Inference, and Social Science" commenter Mike Spagat writes that

> Even within exceptionally violent environments most households will still not have a violent death. So a very small false positive rate in a household survey will cause substantial upward bias in violence estimates. [R270]

Write a paragraph or two explaining this to someone who is interested and smart enough to understand this but has not studied the material in this chapter. Consider making up some numbers to illustrate your argument.

Exercise 13.7.11. [U][Section 6.12] Surgery for prostate cancer?

An article in *The Boston Globe* headlined "Surgery offers no advantage for early prostate cancer, study finds" reported on a clinical trial involving 731 men diagnosed with prostate cancer. About half had surgery; the rest were monitored.

> After 12 years, nearly 6 percent of men who had immediate surgery died of the cancer, compared with slightly more than 8 percent of those patients who were observed, which was not a great enough difference to reach statistical significance. [R271]

(a) About how many men were in each category?
(b) About how many deaths were there in each category?
(c) Construct the contingency table for this study.

Exercise 13.7.12. [S][A][W] Teenage drug use.

Here's a made up story.

The dean at a fancy private high school is very worried. She suspects that about 20% of the 1000 students on campus are using drugs. She has asked all the parents to administer a home drug test to their kids (since it's a private school she can actually require them to do it). She has read on the web that

> With home drug testing methods believed to produce reliable and accurate results, many of us overlook the cases of false positives and draw conclusions on the suspect before reconfirming the result. But, researchers from the Boston University have found out that drug tests may produce false positives in 5–10% of cases and false negatives in 10–15% of cases. [R272]

We found several blogs that seem to report on this same study. None gives a link or a precise reference. We haven't been able to locate the original.

Answer the following questions, assuming the worst cases (10% false positive rate, 15% false negative rate).

(a) Build the contingency table for this drug screening scenario. To do that you will have to figure out

    How many students are drug users?
    How many of the drug users test positive? How many test negative?
    How many students are drug free?
    How many of the drug free students test positive? How many test negative?

You may do the arithmetic with by hand or with the spreadsheet at `ContingencyTable.xlsx`.

(b) What is the true positive rate?
(c) Student John Smith tested positive. What is the probability that he is really on drugs?
(d) Student Jane Doe tested negative. What is the probability that she is really drug free?
(e) Answer the previous two questions if you assume the best cases for reported false values in the Boston University study.

Exercise 13.7.13. [U][Section 13.6] The boy who cried "wolf".

Use Table 13.7 to analyze the children's story with that title.

Exercise 13.7.14. [S] Playing the lottery.

Table 13.8 illustrates the ultimate example of the error you can make reading a column instead of a row.

Table 13.8. Playing the lottery

		bought a ticket		
		yes	no	total
won the lottery	yes	1	0	1
	no	many	very many	very many
	total	many	very many	very many

(a) Suppose you won the lottery. What is the probability that you bought a ticket?
(b) Suppose you bought a ticket. What is the probability that you won the lottery?

Hints

Exercise 1.8.4 To answer the question you'll have to think about the size of the restaurant, the number of years it's been open, the number of customers served in a day, and the fraction of those customers who order soup with matzoh balls (probably not a big fraction).

You may need to do some research if you don't know what a matzoh ball is.

Please don't contact the restaurant and ask them to answer the question for you.

Exercise 1.8.12 For the second question you have to think about whether people trying in one minute are the same as people trying the next minute, or trying an hour later.

Exercise 1.8.20 By now you shouldn't need this hint, but here it is: you will need to know the population of Peru to answer some of the questions.

Exercise 1.8.32 Make and justify estimates for

- The redemption value of a soda can or bottle.
- The 2006 cost of years at MIT and at California state schools.

Then estimate how many cans or bottles you need to collect for one year's tuition. How many is that per day or per hour? Do your answers seem reasonable? Do you have to think about how many days per year or hours per day you could (reasonably) spend collecting cans?

Exercise 2.9.1 From: Laura M Keegan

To: "eb@cs.umb.edu"
Subject: Homework Help
Date: Mon, 16 Sep 2013 03:21:07 +0000

Hello Professor Bolker,

I have a questions about number 2 on the homework, I have not been able to figure out exactly what I am supposed to do, I am not sure of the type of article I am supposed to be searching for or just how to get started looking for one? Sorry, this is kind of a silly question but this part of the homework has taken me a really long time and I still can't find/don't know exactly what to look for.

Thank you,
Laura Keegan

From: Ethan Bolker

To: Laura M Keegan
Subject: Homework Help
Date: Mon, 16 Sep 2013 09:21:15 -0400 (EDT)

Laura

There really is no "exactly what [you are] supposed to do". Just read the newspaper or a magazine, find a story where numbers with units appear in several versions (miles per gallon and gallons per mile, dollars per hour and dollars per year, barrels of oil and liters of gasoline, accidents per week and accidents per person, . . .) and show me that the arithmetic is right and the numbers make sense. You don't need to discuss the whole article—just extract a paragraph or a sentence two with the numbers.

The exercises in the book suggest the kind of thing I have in mind. If you find a good one I can put it in the book.

There's no particular "type of article" to look for. If you find one on a subject you're interested in rather than just thinking of this as a homework problem you have to do you might even enjoy it.

Although I'd like to encourage you to read the newspaper, you can look on the net for a story.

Ethan

Exercise 2.9.2 Warning: it's not 45 miles/hour.

The problem does not say how far it is from Here to There. You're free to pick some convenient distance to work with if you like.

Exercise 2.9.24 This is a hint about what not to do. A web search might lead you to a graphic from *The New York Times* at `www.nytimes.com/imagepages/2011/03/15/science/15food\_graphic.html` that goes as far as 2008. It's no real help. Nor will it help to spend (waste?) time scrolling through web postings that refer to it.

Exercise 2.9.28 `www.census.gov/compendia/statab/2012/tables/12s1103.pdf`

Exercise 2.9.41 The metric system will help: a cubic centimeter of water weighs one gram. You can convert pints to ounces and then to cubic centimeters; pounds to kilograms and then to grams.

Exercise 2.9.44 To make a number larger than what the calculator can display you can't just enter it from the keypad. Try multiplying together two numbers each of which is nearly as large as possible.

Exercise 2.9.52 You can find an inflation calculator on the web, or read ahead in Chapter 4.

Exercise 3.10.7 Look up the Massachusetts budget for 2013, and the percentage of revenue that comes from this personal income tax.

Exercise 3.10.11 The answers to some of these questions will be very small numbers. Keep careful track of the decimal points.

For the last question, which is more informative, the absolute or the relative comparison?

Exercise 3.10.23 Leap years are a little more complicated than you think.

Exercise 3.10.24 You can do this problem with algebra, but you may find it easier by working with some definite numbers that aren't given in the problem. So pretend that fuel economy before the increase is, say, 20 miles per gallon and you will drive 100 miles. Then work out how much gas you save (in percentage terms) if that 20 miles per gallon increases 20% to 24 miles per gallon. For the second part of the problem, convert 20 miles per gallon to gallons per mile, or to gallons per hundred miles. Then a 20% improvement will decrease that number by 20%, since "improvement" means fewer gallons per mile.

Exercise 3.10.30 You can't really get started on this problem if you don't know what "wholesale price" and "markup" mean. If that's the case, look them up.

You may find it easier to work with a book whose wholesale price is $1 or $100.

Exercise 3.10.35 Read carefully. The $6.6 billion figure is what the banks will lose in revenue, not what they make now or will make after the cut takes effect.

Exercise 3.10.43 For the last part of the exercise you will need estimates for the student populations in each of those years. That number was growing, but not nearly as fast as the total debt.

Exercise 4.7.3 You won't find any help visiting the State House web site.

The Bureau of Labor Statistics inflation calculator doesn't go far enough back. What should you do?

Exercise 4.7.5 The inflation calculator can tell you the year when a nickel would buy what you need a dime for in 2011.

Exercise 4.7.9 Be careful when you move the decimal point to find the percent, which will be negative and pretty small.

Exercise 4.7.12 One of the raises in this 2010 article hadn't taken place when the article was written, so the inflation rate for that year wasn't known then. But it is known now.

When you use the inflation calculator, remember that the base pay for these raises was the 2005 value, not the one in 2006. That was after the first raise.

Exercise 4.7.13 Consider computing with a particular salary and inflation rate as examples. Choose numbers that are easy to work with.

Exercise 4.7.14 You may want to use the internet for part of your answer to the last question. You should not need it before that.

Exercise 4.7.23 (a) Careful. Why isn't the answer not 30%?

(b) Use trial-and-error. Guess an answer, test it using the argument from the first part of the exercise, adjust your answer until it's right.

Exercise 5.7.4 • You don't need to fuss with how many courses make up the credits. Just imagine one course for 12 (or 9 or 6) credits in the new semester, with a numerical grade. For the past work, imagine one 55 credit course with a grade of 1.80. It doesn't matter what particular courses led to that GPA. Then figure out the grade for the current course that makes the overall GPA at least 2.0.

- Another hint: you can do this with algebra, or by guess-and-check. Guess a grade for the new course, compute the new GPA, and see whether it's too high, too low or just right. Adjust your guess accordingly.

Exercise 5.7.8 • You don't have to know the numerical values that go with each letter grade.

- A good semester GPA with very few credits won't change the year GPA much if there are lots of credits in the other semester.

Exercise 5.7.9 There's a fair amount of tedious arithmetic in parts (a) and (c). If you know how to use Excel (which *Common Sense Mathematics* takes up in the next chapter) you can save yourself some time and grief by entering the columns in Table 5.6 in a spreadsheet to answer (a), and then changing just the weights to answer (c).

Exercise 5.7.10 Estimate the percentage of the U.S. population that's urban in the northeast by treating it as a Fermi problem. Combine reliable data you find on the net with some common sense.

When you have that percentage (call it W for the moment) you can express the overall average inflation rate of 2.85% as a weighted average of the urban northeast rate (2.58%) with the weight W and the unknown inflation rate for the rest of the country with weight $100\% - W$.

Exercise 5.7.13 If you find an "average" statistic for the last question you will need to think about which kind of average it is.

Exercise 6.14.5 You may draw your bar chart neatly by hand, or use Excel.

Exercise 6.14.14 Start with a copy of `WingAeroHistogram.xlsx`. Create new data in columns `D` and `E` below the existing data there for the two new histograms. Fill in column `F` appropriately. Then you can copy and paste from what's there to fill in columns `G` and `H` and the sums and the mean.

If Excel wants to treat your salary ranges as dates, try formatting the cells as text.

To make the charts, copy the histogram that's there, then find the place in Excel where you can change the source data to the new rows in columns `D` and `E`.

Exercise 6.14.16 When you enter the data in two columns in Excel put the categories (usability scores) on the left since they are the labels for the x-axis. Put the numbers of websites on the right since they are the values that go with the categories, and should plot vertically, on the y-axis.

If your Excel is anything like ours, it may well think that you want to display both columns on the y-axis, since both columns are numbers. If it does that it will label the x axis with the numbers 1 to 10.

If that happens, delete the data series corresponding to the bars you don't want (the percentages). Then right click on the chart and explore until you find the place that allows you to enter the fields you want to use as x-axis category labels.

Exercise 6.14.20 The mean amount spent for Fire/EMS services per person is not the Excel `AVERAGE` of the amounts spent per person by each city. It's wrong to average those numbers since they are already averages. You must weight them by the city populations in order to compute the total amount spent by all the people in all the cities. Then divide by the total population.

You can't compute the median with Excel's `MEDIAN` function for the same reason. Further hint: almost everyone lives in New York.

You can't answer the question "What do firefighters earn?" by finding the mean of the twelve numbers in column C. Compute the mean correctly as a weighted average. You will probably want to start by creating a column labeled

```
total Fire/EMS expenses
```

and fill in the value for each city.

Exercise 6.14.27 Warning: 2:30 is 2 hours and 30 minutes. That's 2.5 hours, not 2.3 hours.

Exercise 6.14.28 If you don't know what "per capita" means, look it up.

Exercise 6.14.33 For part (d), use guess-and-check in Excel. Set up the computation so that when you change the percentage of meat the percentage of cheese and the total change automatically.

Exercise 7.8.9 Finding the places where the lines in your Excel chart cross is the key to the last part of the problem.

Exercise 7.8.13 For part (a) you will have to look up the conversion factors among various forms of energy in order to convert watt-hours to gallons of gasoline. Part (d) is about psychology, not quantitative reasoning. Your answer might begin "The display is valuable because ... On the other hand ..."

Exercise b It should not be hard to find a website that helps with the second question.

Exercise 7.8.22 What did electricity cost in the spring of 2011?

Exercise 7.8.30 The answer depends on the relative lengths of the lines, not on the absolute difference in the lengths.

Exercise 8.5.3 Look at the data starting in about 1980.

Exercise 8.5.5 Try a Google search for

Per capita consumption of selected beverages in gallons .

Exercise 8.5.16 For the second question, all you can really look for is the order of magnitude. If that doesn't match, try to explain why.

Exercise 9.6.2 (a) Inflation is usually reported as a percent increase.

(b) No hint needed.

(c) No hint needed.

(d) Read this one carefully to think about what depends on what. Don't just jump at the word "percent".

(e) The washing happens over and over again on the same day—that's how dirty they were.

(f) No hint needed.

(g) Interest on unpaid balances accumulates.

(h) What units would the weatherman use to report the rate at which snow was accumulating?

(i) Think about how the number of people exposed to germs depends on the number of people sick.
(j) (Electronic) word of mouth generates new subscribers from old ones.

Exercise 9.6.15 Use the `ExponentialGrowth.xlsx` spreadsheet, or find the answer by trying different values for T (# of years) in the formula until you find one that gets you close to $3000.)

Exercise 9.6.16 Your computations should suggest that it won't be infinite, which might have been your first guess.

Exercise 9.6.17 You can answer the first question using guess-and-check until you're close enough. You'll need a web search for the second and a very good web search for the third, which is open-ended.

Exercise 9.6.19 Guess a number of hours, try your guess in the two exponential equations, then adjust your guess up or down until the answers match.

Exercise 9.6.22 Compare the relative change in the number of email accounts for the two time periods (early 1900s to end of 1999, end of 1999 to the time of the blog post).

Exercise 10.6.8 Start with the calculation $(1 + 0.0143)^{365}$. The answer is hard to believe.

Exercise 10.6.9 What annual interest rate would you need to turn $4,000 into $1.5 million in 94 years? Compare that rate to the increase due just to inflation.

Exercise 11.9.4 You might want to do this exercise in Excel. Then you can see the odds for all the horses, and see how the payoffs change when you change the track's take.

Exercise 11.9.8 Consider starting at `www.census.gov/govs/state/10lottery.html`. There may be a better site that gives you totals, or lets you download directly into a spreadsheet.

Exercise 11.9.11 You might find it easiest to answer this question by imagining that you played the lottery 100 days in a row.

Exercise 12.7.3 There are 36 equally likely outcomes. One possibility is that you roll a 3 with die A (5 ways) and your friend rolls a 2 with die B (3 ways). So there are 15 ways that can happen, for a probability of 15/36 that you win that way. Now figure out the probabilities for the other possibilities.

Exercise 12.7.7 The answer is very small, but it's not zero. Use the Google calculator to find it.

Exercise 12.7.12 You can get the state populations from the data in `cancerclusters.xlsm` or find it on the web and download it. We might do it for you in a later edition of this book.

Exercise 13.7.1 The first quote tells you the false positive and false negative rates. The second tells you the incidence.

Exercise 13.7.7 Build the contingency table, based on a population of 1,000 women tested. You may do this by hand or with the spreadsheet `ContingencyTable.xlsx`.

References

The several hundred entries here identify the sources for data and quotations in the text and exercises. Citing sources is a necessary part of good academic work. That does not mean you need to follow these links: you should be able to read the text and work the exercises without having to consult the original sources.

R1 Lewis Carroll, *A Tangled Tale*, Answers to Knot 4. www.gutenberg.org/files/29042/29042-8.txt (last visited July 14, 2015).

R2 Henry Wadsworth Longfellow, *Kavanagh*, 1849. www.archive.org/stream/talekavanagh00longrich/talekavanagh00longrich_djvu.txt (last visited July 14, 2015).

Chapter 1

R3 Joint House and Senate Hearing, 110 Congress, FISA hearing, www.gpo.gov/fdsys/pkg/CHRG-110jhrg38878/html/CHRG-110jhrg38878.htm (last visited July 14, 2015).

R4 D. Priest and W. Arkin, Top Secret America, *The Washington Post* (July 2010), projects.washingtonpost.com/top-secret-america/articles/a-hidden-world-growing-beyond-control/ (last visited August 25, 2015).

R5 J. Beltrane, Artificial Mini-Hearts Developed, Historic Canada Blog, *The Canadian Encyclopedia*, thecanadianencyclopedia.ca/en/article/artificial-mini-hearts-developed/ (updated December 16, 2013) (last visited August 16, 2015).

R6 O. Judson, Darwin Got It Going On, *The New York Times* (May 4, 2010), opinionator.blogs.nytimes.com/2010/05/04/darwin-got-it-going-on/ (last visited December 6, 2015).

R7 Baba Brinkman bio, www.bababrinkman.com/bio/ (last visited July 14, 2015), quoted with permission.

R8 Redrawn from data accompanying D. C. Denison, Taking a different measure, *The Boston Globe* (October 14, 2010), www.boston.com/news/science/articles/2010/10/14/akamai_keeping_an_eye_on_its_greenhouse_emissions/ (last visited August 1, 2015). The original graphic is no longer available.

R9 E. Horowitz Want to join Forbes 400? It's $1.7b minimum, *The Boston Globe* (September 29, 2015), www.bostonglobe.com/business/2015/09/29/how-rich-are-forbes-really/Zz8CaFo6iYMyamIKf6MFsJ/story.html (last visited September 30, 2015).

R10 B. Herbert, A Very Bright Idea, *The New York Times* (May 17, 2010), www.nytimes.com/2010/05/18/opinion/18herbert.html (last visited July 20, 2015).

R11 Donnelly/Colt, search for "every minute" at donnellycolt.com (last visited October 11, 2015).

R12 H. Bray, Smartphone apps may help retail scanning catch on, *The Boston Globe* (March 12, 2012), bostonglobe.com/business/2012/03/11/modiv-and-aislebuyer-apps-turn-smartphones-into-retail-scanners/DYgfE8JmimVWDjblpbKDzI/story.html (last visited July 29, 2015).

R13 L. Greenhouse, Never Before, *The New York Times* (March 21, 2012), opinionator.blogs.nytimes.com/2012/03/21/never-before/ (last visited July 20, 2015).

R14 C. Shea, Our blog vocabulary, our selves, *The Boston Globe* (July 25, 2010), www.boston.com/bostonglobe/ideas/articles/2010/07/25/our_blog_vocabulary_our_selves/ (last visited July 12, 2015).

R15 F. Cook, *To the Top of the Continent*, 1908. Reissued by The Mountaineers Books, Seattle, 2001, pp. 107–108.

R16 Millions jam street-level crime map website, BBC News (February 1, 2011), www.bbc.co.uk/news/uk-12336381 (last visited July 20, 2015).

R17 C. C. Miller, Another Try by Google to Take On Facebook *The New York Times* (June 28, 2011), www.nytimes.com/2011/06/29/technology/29google.html (last visited July 20, 2015).

R18 Comment on J. Calmes, Obama Draws New Hard Line on Long-Term Debt Reduction, *The New York Times* (September 19, 2011), www.nytimes.com/2011/09/20/us/politics/obama-vows-veto-if-deficit-plan-has-no-tax-increases.html (last visited February 21, 2016).

R19 M. Kakutani, Touring the Ruins of the Old Economy, *The New York Times* (September 26, 2011), www.nytimes.com/2011/09/27/books/boomerang-by-michael-lewis-review.html (last visited July 0, 2015).

R20 A. Waters and K. Heron, No Lunch Left Behind, *The New York Times* (February 19, 2009), www.nytimes.com/2009/02/20/opinion/20waters.html (last visited July 20, 2015).

R21 Water Sense Kids, turn off the tap!, Environmental Protection Agency, www.epa.gov/WaterSense/kids/tap-off.html (last visited July 16, 2015).

R22 Associated Press, Lady Liberty's crown to reopen July 4, *The Boston Globe* (May 9, 2009), www.boston.com/news/nation/articles/2009/05/09/lady_libertys_crown_to_reopen_july_4/ (last visited July 20, 2015).

R23 L. Collins, Just a Minute, *The New Yorker* (December 7, 2009), www.newyorker.com/magazine/2009/12/07/just-a-minute (last visited July 26, 2015).

R24 R. Vecchio, Global Psyche: On Their Own Time, *Psychology Today* (July 1, 2007), www.psychologytoday.com/articles/200707/global-psyche-their-own-time (last visited August 16, 2015).

R25 John McPhee, "Season on the Chalk", *The New Yorker* (March 12 2008), reprinted in *Silk Parachute*, Farrar Straus and Giroux, New York, 2010, page 23.

R26 A. Singer, The Future Buzz, Social Media, Web 2.0 and Internet Stats (2009), thefuturebuzz.com/2009/01/12/social-media-web-20-internet-numbers-stats/ (last visited July 16, 2015). Quoted with permission.

R27 D. Abel, State panel OKs expansion of nickel deposit to bottled water, *The Boston Globe* (July 15, 2010), www.boston.com/news/local/massachusetts/articles/2010/07/15/state_panel_oks_expansion_of_nickel_deposit_to_bottled_water/ (last visited July 20, 2015).

R28 E. Ailworth, Drivers curb habits as cost of gas soars (April 21, 2011), www.boston.com/business/personalfinance/articles/2011/04/21/drivers_curb_habits_as_cost_of_gas_soars/ (last visited July 20, 2015).

R29 Graphic on a T-shirt from the Homemade Cafe, www.homemade-cafe.com/index.php (last visited July 12, 2015). They're delighted by the publicity.

R30 A. Doerr, So many books, so little time, *The Boston Globe* (Mar 4, 2012), www.bostonglobe.com/arts/books/2012/03/04/many-books-little-time/znyJ5MhKbMweJNNMk3yIYM/story.html (last visited August 17, 2015).

R31 Annual Estimation of Cell Phone Crashes 2013, National Safety Council, www.nsc.org/DistractedDrivingDocuments/CPK/Attributable-Risk-Summary.pdf (last visited September 8, 2015).

R32 S. J. Dubner, Should the U.S. Really Try to Host Another World Cup?, *The New York Times* (July 19, 2010), freakonomics.blogs.nytimes.com/2010/07/19/should-the-u-s-really-try-to-host-another-world-cup/ (last visited July 20, 2015).

R33 Putting Three Kids Through College by Redeeming Cans and Bottles, ABC News (March 30, 2006), abcnews.go.com/2020/story?id=1787254 (last visited October 1, 2015).

R34 E. Kolbert, Flesh of Your Flesh, *The New Yorker* (November 29, 2009), www.newyorker.com/arts/critics/books/2009/11/09/091109crbo_books_kolbert (last visited July 20, 2015).

R35 Giga Quotes, www.giga-usa.com/ (last visited July 20, 2015).

R36 J. Dodge, Data Glut as Gene Research Yields Information Counted in Terabytes, Researchers Struggle to Visualize and Process It, *The Boston Globe* (February 24, 2003), pqasb.pqarchiver.com/boston/access/293479831.html (last visited July 20, 2015).

R37 Editorial, Don't forward those photos, *The Boston Globe* (September 7, 2010) www.boston.com/bostonglobe/editorial_opinion/editorials/articles/2010/09/07/dont_forward_those_photos/ (last visited July 20, 2015).

R38 K. Kraus, When Data Disappears *The New York Times* (August 6, 2011), www.nytimes.com/2011/08/07/opinion/sunday/when-data-disappears.html (last visited July 20, 2015).

R39 S. R. Loss, T. Will, P. P. Marra, The impact of free-ranging domestic cats on wildlife of the United States, *Nature Communications* 4 (2013) 1396-1403, doi:10.1038/ncomms2380. www.nature.com/ncomms/journal/v4/n1/full/ncomms2380.html (last visited July 16, 2015). Quoted with permission.

R40 www.census.gov/population/international/data/idb/informationGateway.php (last visited July 6, 2015).

R41 S. Mayerowitz, United lures top fliers with promise of a hot meal, Associated Press, reported in *The Boston Globe* (August 22, 2014), www.bostonglobe.com/business/2014/08/21/united-lures-top-fliers-with-promise-hot-meal/uBUbYBhinOZvUptp37MbYM/story.html (last visited July 20, 2015).

Chapter 2

R42 NASA's metric confusion caused Mars orbiter loss, CNN (September 30, 1999), www.cnn.com/TECH/space/9909/30/mars.metric/ (last visited July 20, 2015).

R43 R. P. Larrick and J. B. Soll, The MPG Illusion, *Science* (June 20, 2008), Vol. 320 no. 5883 pp. 1593–1594.

R44 Gasoline Vehicle Label, U. S. Environmental Protection Agency, www.epa.gov/carlabel/gaslabel.htm (last visited July 31, 2015).

R45 Federal Bureau of Investigation, Uniform Crime Reports, Crime in the United States 2011, www.fbi.gov/about-us/cjis/ucr/crime-in-the-u.s/2011/crime-in-the-u.s.-2011/tables/table8statecuts/table_8_offenses_known_to_law_enforcement_california_by_city_2011.xls (last visited July 20, 2015).

R46 E. Moskowitz, Urgent fixes will disrupt rail lines *The Boston Globe* (April 28, 2010), www.boston.com/news/local/massachusetts/articles/2010/04/28/urgent_fixes_will_disrupt_rail_lines/ (last visited July 20, 2015).

R47 Screen captures from www.online-calculator.com/ (last visited September 12, 2015). That website says "Want to use Online-Calculator commercially or for business, time your employees (nasty!), use it in power-point documents, in presentations.—Go for it. Anything else—I'm sure it will be Fine." An email from them makes permission even clearer: "Yes—please feel free to use the images shown, or any other images/text/calculators in the textbook."

R48 Christopher Marlowe, allpoetry.com/The-Face-That-Launch'd-A-Thousand-Ships (last visited August 23, 2015).

R49 FAA shutdown would cost gov't $200 million a week, Associated Press story reported by *USA Today*, (2011). usatoday30.usatoday.com/news/washington/2011-07-22-faa-aviation-shutdown_n.htm (last visited July 20, 2015).

R50 J. Scott Orr, Does keeping the penny still make sense?, Newhouse News Service, *The Seattle Times*, (2006). community.seattletimes.nwsource.com/archive/?date=20060604&slug=penny04 (last visited July 17, 2015).

R51 Government Accounting Office, Benefits and Considerations for Replacing the $1 Note with a $1 Coin, (2012). www.gao.gov/products/GAO-13-164T (last visited July 17, 2015).

R52 Data from U.S. Mint annual reports, www.usmint.gov/about_the_mint/?action=annual_report. (last visited July 17, 2015).

R53 B. Baker, How hot was it? It was so hot that . . . , *The Boston Globe* (June 22. 2012). www.bostonglobe.com/2012/06/21/howhot/hwZQeTGjrRLU8v1XYY0PyJ/story.html (last visited October 1, 2015).

R54 M. Gromov, Crystals, Proteins, Stability and Isoperimetry, *Bulletin (New Series) of the American Mathematical Society*, Volume 48, Number 2, April 2011, p. 240, © 2011, American Mathematical Society.

R55 D. Joling, Global warming opens Arctic for Tokyo-London undersea cable, *The Seattle Times* (January 21, 2010), www.seattletimes.com/nation-world/global-warming-opens-arctic-for-tokyo-london-undersea-cable/ (last visited July 17, 2015).

R56 Sticker Shock, Editorial, *The New York Times* (June 5, 2011), www.nytimes.com/2011/06/05/opinion/05sun1.html (last visited July 17, 2015).

R57 M. L. Wald, Hybrids vs. Nonhybrids: The 5-Year Equation, *The New York Times* Green Blog (February 23, 2011), green.blogs.nytimes.com/2011/02/23/hybrids-vs-nonhybrids-the-5-year-equation/ (last visited August 22, 2015).

R58 Clean Energy, calculations and references, Environmental Protection Agency, www.epa.gov/cleanenergy/energy-resources/refs.html (last visited July 16, 2015).

R59 Comments on M. D. Shear, Rising Gas Prices Give G.O.P. Issue to Attack Obama, *The New York Times* (February 19 2012), www.nytimes.com/2012/02/19/us/politics/high-gas-prices-give-gop-issue-to-attack-obama.html (last visited July 17, 2015).

R60 About Scholastic, Scholastic and *Harry Potter and the Deathly Hallows* make publishing history with 8.3 million copies sold in first 24 hours, (2007). www.scholastic.com/aboutscholastic/news/press_07222007_RT.htm (last visited July 17, 2015).

R61 Senior-citizen volunteers fight Medicare fraud, Associated Press, reported in *The Star Democrat* (Easton, MD) 2010), www.stardem.com/article_c3eb003d-d240-59ba-99bd-ac0f458c5cfe.html (last visited July 17, 2015).

R62 T. Haupert, Carrots 'N' Cake, Grocery Shopping 101: Unit Price (2011), carrotsncake.com/2011/01/grocery-shopping-101-unit-price.html (last visited July 16, 2015). Quoted with permission.

R63 K. Chang, With shuttle's end, X Prize race to the moon begins, *The Seattle Times* (2011). seattletimes.com/text/2015689643.html (last visited July 18, 2015).

R64 Astrobiotic Technology, www.astrobotic.com/ (last visited August 22, 2015).

R65 J. Buzby and H. Farah, Chicken Consumption Continues Longrun Rise, USDA Economic Research Service, Web Archive, webarchives.cdlib.org/sw1wp9v27r/ers.usda.gov/AmberWaves/April06/Findings/Chicken.htm (last visited August 1, 2015).

R66 Apple Press release, Apple's App Store Downloads Top 25 Billion, images.apple.com/pr/library/2012/03/05Apples-App-Store-Downloads-Top-25-Billion.html (last visited July 27, 2015).

R67 Comment on G. Collins, Send in the Clowns, and Cheese *The New York Times* (April 4, 2012), www.nytimes.com/2012/04/05/opinion/collins-send-in-the-clowns-and-cheese.html (last visited July 27, 2015).

R68 TPC at Sawgrass, wikipedia. en.wikipedia.org/wiki/TPC_at_Sawgrass (last visited July 18, 2015).

R69 E. Hazen, USA Coverage, How Many Driving Accidents Occur Each Year?, www.usacoverage.com/auto-insurance/how-many-driving-accidents-occur-each-year.html (last visited July 16, 2015). Quoted with permission.

R70 Car-Accidents.com, Car Accident Statistics, www.car-accidents.com/pages/stats.html (last visited July 18, 2015).

R71 B.P. Phillips, An Impartial Price Survey of Various Household Liquids as Compared to a Gallon of Gasoline, Annals of Improbable Research (14.4), www.improbable.com/2008/09/17/an-impartial-price-survey-of-various-household-liquids-as-compared-to-a-gallon-of-gasoline/ (last visited July 18, 2015). Reproduced with permission.

R72 Wikipedia, from Federal Highway Administration–MUTCD or somewhat more than one and a half blocks three feet deep (mutcd.fhwa.dot.gov/shsm_interim/index.htm. [Public domain], commons.wikimedia.org/wiki/File\%3ASpeed_Limit_55_sign.svg (last visited August 1, 2015).

R73 A. L. Myers and J. Billeaud, Arizona using donations for border fence, *The Washington Times* (November 24, 2011), www.washingtontimes.com/news/2011/nov/24/arizona-using-donations-for-border-fence (last visited July 18, 2015).

R74 K. Chang, Rosetta Spacecraft Set for Unprecedented Close Study of a Comet, *The New York Times* (August 5, 2014), www.nytimes.com/2014/08/06/science/space/rosetta-spacecraft-set-for-unprecedented-close-study-of-a-comet.html (last visited July 18, 2015).

R75 D. Streitfeldmarch, In a Flood Tide of Digital Data, an Ark Full of Books, *The New York Times* (March 3, 2012), www.nytimes.com/2012/03/04/technology/internet-archives-repository-collects-thousands-of-books.html (last visited July 18, 2015).

R76 C. Li and W. Qian, Floating trash threatens Three Gorges Dam, *China Daily* (August 2, 2010), www.chinadaily.com.cn/m/hubei/2010-08/02/content_11083052.htm (last visited July 17, 2015).

R77 P. Krugman, Bits and Barbarism, *The New York Times* (December 22, 2013), www.nytimes.com/2013/12/23/opinion/krugman-bits-and-barbarism.html (last visited July 18, 2015).

R78 J. Fitzgerald, Solar use will push energy costs up in Mass., *The Boston Globe* (February 12, 2014) (last visited July 18, 2015).

R79 Smoot, en.wikipedia.org/wiki/Smoot (last visited October 11, 2015).

R80 S. Nordenstam and B. Hirschler, Nobel Prize for seeing how life works at molecular level, Reuters (October 8, 2014), uk.reuters.com/article/2014/10/08/us-nobel-prize-chemistry-idUKKCN0HX0V220141008 (last visited July 27, 2015).

R81 C. Lyons, A Sticky Tragedy: The Boston Molasses Disaster, *History Today* (Volume 59 Issue 1 January 2009), www.historytoday.com/chuck-lyons/sticky-tragedy-boston-molasses-disaster (last visited October 21, 2015).

R82 Great Molasses Flood, Wikipedia, en.wikipedia.org/wiki/Great_Molasses_Flood (last visited October 21, 2015).

Chapter 3

R83 A. Halsey III, 28 percent of accidents involve talking, texting on cellphones, *The Washington Post* (January 12, 2010), www.washingtonpost.com/wp-dyn/content/article/2010/01/12/AR2010011202218.html (last visited July 17, 2015).

R84 Graphic redrawn from data in *The Boston Globe* (September 9, 2008).

R85 George Pólya, *How to Solve It*, Princeton University Press (Reprint edition 2014).

R86 J. Miller, State's rainy day fund has dwindled over past decade, *The Boston Globe* (October 4, 2015), www.bostonglobe.com/metro/2015/10/04/state-rainy-day-fund-depleted/gX3v8T9erHrlyBvGFpOefL/story.html (last visited October 5, 2015).

R87 J. Surowicki, The Financial Page, *The New Yorker* (March 29, 2010) p. 45, archives.newyorker.com/?i=2010-03-29#folio=044 (last visited July 27, 2015).

R88 B. Reddall and E. Scheyder, Obama seeks new drill ban as oil still spews, Reuters (June 23, 2010), in.reuters.com/article/2010/06/23/idINIndia-49558320100623 (last visited July 19, 2015).

R89 M. Levenson and E. Moskowitz, Patrick expected to seek increased income tax, *The Boston Globe* (January 16, 2013), bostonglobe.com/metro/2013/01/16/patrick-set-propose-income-tax-increase/V2PY3WnBElmKrmi3aHptkN/story.html (last visited July 18, 2015).

R90 J. McGuire, Data on Brain Injury in Massachusetts: A Snapshot , Massachusetts Executive Office of Health and Human Services, Presentation to the Brain Injury Commission (February 7 2011), www.mass.gov/eohhs/docs/eohhs/braininjury/201102-presentation-mcguire.rtf (last visited July 19, 2015).

R91 Federal Funding for Public Broadcasting: Q&A, Western Reserve Public Media (2015), westernreservepublicmedia.org/federal-funding.htm (last visited October 1, 2015).

R92 D. Gonzalez, Births by U.S. visitors: A real issue?, *The Arizona Republic* (August 17, 2011), www.azcentral.com/news/articles/2011/08/17/20110817births-by-us-visitors-smaller-issue.html (last visited July 19, 2015).

R93 Mercury in the Fog, *City on a Hill Press* (April 21, 2012), www.cityonahillpress.com/2012/04/21/mercury-in-the-fog/ (last visited July 19, 2015).

R94 J. Garthwaite, Coastal California Fog Carries Toxic Mercury, Study Finds, *The New York Times* (March 28, 2012), green.blogs.nytimes.com/2012/03/28/coastal-california-fog-carries-toxic-mercury-study-finds/ (last visited July 19, 2015).

R95 Harkin, Hagan Re-Introduce Bill to Promote Responsible Use of Taxpayer Dollars in Higher Ed, U.S. Senate Committee on Health, Education, Labor and Pensions (March 12, 2013), www.help.senate.gov/ranking/newsroom/press/harkin-hagan-re-introduce-bill-to-promote-responsible-use-of-taxpayer-dollars-in-higher-ed (last visited July 19, 2015).

R96 Affiliated Managers *The Boston Globe* (October 6, 2010), www.boston.com/yourtown/beverly/articles/2010/10/06/affiliated_managers/ (last visited August 12, 2015). Figure redrawn from data in the original graphic.

R97 A. Kingswell, five nines: chasing the dream?, Continuity Central (2010), www.continuitycentral.com/feature0267.htm (last visited July 20, 2015). Quoted with permission.

R98 Digest of Educational Statistics, National Center for Educational Statistics, nces.ed.gov/programs/digest/d10/tables/dt10_285.asp (last visited July 19, 2015).

R99 J. Eilperin, New Nature Conservancy atlas aims to show the state of the world's ecosystems, *The Washington Post* (April 12, 2010), www.washingtonpost.com/wp-dyn/content/article/2010/04/11/AR2010041103556.html (last visited July 19, 2015).

R100 M. Iwata, Gulf oil spill could lead to drop in global output, Dow Jones Newswires reported in *The Denver Post*, www.denverpost.com/nacchio/ci_15330002 (last visited September 21, 2015).

R101 Goldman Sachs limits pay, earns $4.79 billion in fourth quarter, Associated Press, reported in the *Tampa Bay Times* (January 21, 2010), www.tampabay.com/news/business/banking/goldman-sachs-limits-pay-earns-479-billion-in-fourth-quarter/1067187 (last visited September 12, 2015).

R102 A. Bjerga, Food Stamps Went to Record 41.8 Million in July, Bloomberg (October 5, 2010), www.bloomberg.com/news/articles/2010-10-05/food-stamp-recipients-at-record-41-8-million-americans-in-july-u-s-says (last visited August 23, 2015).

R103 W. Parry, Gamblers spending less time, money in AC casinos, Associated Press, reported in on NBC news (December 7, 2010), www.nbcnews.com/id/40547628/ns/business-us_business/t/gamblers-spending-less-atlantic-city-casinos (last visited September 12, 2015).

R104 S. Syre, Harvard reports big gains on investments, *The Boston Globe* (September 23, 2011), www.boston.com/business/markets/articles/2011/09/23/harvard_endowment_posts_big_investment_gain/ (last visited August 1, 2015). Graphic showing the historical data no longer available.

R105 T. S. Bernard and B. Protess, Banks to Make Customers Pay Fee for Using Debit Cards, *The New York Times* (September 29, 2011), www.nytimes.com/2011/09/30/business/banks-to-make-customers-pay-debit-card-fee.html (last visited July 19, 2015).

R106 W. Yardley, Oregon town weighs future with fossil fuel, New York Times News Service, reported in *The Bulletin* (Bend, Oregon) (April 22, 2012), www.bendbulletin.com/article/20120422/NEWS0107/204220325/ (last visited July 19, 2015).

R107 J. Abelson, Seeking savings, some ditch brand loyalty, *The Boston Globe* (January 29, 2010), www.boston.com/business/articles/2010/01/29/shoppers_are_ditching_name_brands_for_store_brands/ (last visited August 2, 2015).

R108 D.C. Denison, UMass research funding reaches record, *The Boston Globe* (February 10, 2011). www.boston.com/business/articles/2011/02/10/umass_research_funding_reaches_record/ (last visited July 19, 2015).

R109 A. Beam, Long gone?, *The Boston Globe* (March 4, 2011), www.boston.com/lifestyle/articles/2011/03/04/from_landlines_to_e_mail_to_movies_to_print_their_demise_has_been_greatly_exaggerated/ (last visited July 19, 2015).

R110 J. Sununu, Creature double feature, *The Boston Globe* (March 21, 2011), www.boston.com/bostonglobe/editorial_opinion/oped/articles/2011/03/21/creature_double_feature/ (last visited July 19, 2015).

R111 H. Bray, Facebook vs. Google+, *The Boston Globe* (September 29, 2011), www.bostonglobe.com/business/2011/09/28/facebook-google/yWa8anwKHgelcklsasoDkP/story.html (last visited August 27, 2015).

R112 G. Washburn, Game time, *The Boston Globe* (June 30, 2010), www.boston.com/sports/basketball/celtics/articles/2010/06/30/game_time_when_james_makes_decision_dominoes_to_fall/ (last visited August 27, 2015).

R113 S. Ohlemacher, 13 million get unexpected tax bill from Obama tax credit, Associated Press reported in Wilmington NC *StarNewsOnline* (December 17, 2010), www.starnewsonline.com/article/20101217/ARTICLES/101219738 (last visited July 20, 2015).

R114 Direct Debit, Fedloan Servicing, U.S. Department of Education, www.myfedloan.org/make-a-payment/ways-to-pay/direct-debit.shtml (last visited September 30, 2015).

R115 M. Viser, Benefits take hit in Patrick budget, *The Boston Globe* (January 13, 2008), www.boston.com/news/local/articles/2008/01/13/benefits_take_hit_in_patrick_budget/ (last visited July 20, 2015).

R116 Driving More Efficiently, U. S. Department of Energy, www.fueleconomy.gov/feg/driveHabits.jsp#speed-limit (last visited July 20, 2015).

R117 Labels redrawn from www.fueleconomy.gov/feg/images/speed_vs_mpg_2012_sm.jpg (last visited August 2. 2015).

R118 G. Edgers, Holiday shows bringing box office joy, *The Boston Globe* (December 27, 2012), www.bostonglobe.com/arts/2012/12/27/boston-ballet-holiday-pops-leading-wave-holiday-arts-show-boom/FELXeihGF2IT2SiooJ7JPI/story.html (last visited August 2, 2015). Graphic redrawn from data in the article.

R119 T. Luna, Amazon to begin collecting Mass. sales tax Friday, *The Boston Globe* (October 29, 2013), www.bostonglobe.com/business/2013/10/28/amazon-begin-collecting-massachusetts-sales-tax-friday/acyUMexp4ChUk2I6JtdJyI/story.html (last visited July 20, 2015).

R120 M. Yglesias, Last year, 25 hedge fund managers earned more than double every kindergarten teacher combined, Vox.com (May 6, 2014), www.vox.com/2014/5/6/5687788/last-year-25-hedge-fund-managers-earned-more-than-double-every (last visited July 20, 2015).

R121 S. Carrell and J. Zinman, In Harm's Way? Payday Loan Access and Military Personnel Performance, January 2013, *Rev. Financ. Stud.* (2014) 27 (9): 2805–2840 first published online May 28, 2014 doi:10.1093/rfs/hhu034, *The Review of Financial Studies*, rfs.oxfordjournals.org/.

R122 J. Boak, School spending by affluent is widening wealth gap, Associated Press (September 30, 2014), bigstory.ap.org/article/e59ee3ddaef64e2ebc75e081bf9f1d40/school-spending-affluent-widening-wealth-gap (last visited July 20, 2015).

R123 R. Burns and L. C. Baldor, U.S. nuclear woes: Pentagon chief orders a shakeup, Associated Press report in the *Vancouver Columbian* (November 14, 2014), www.columbian.com/news/2014/nov/14/us-nuclear-woes-pentagon-chief-orders-a-shakeup/ (last visited August 28, 2015).

R124 M. O'Brien, The top 400 households got 16 percent of all capital gains in 2010, *The Washington Post* (November 25, 2014), www.washingtonpost.com/blogs/wonkblog/wp/2014/11/25/the-top-400-households-got-16-percent-of-all-capital-gains-in-2010/ (last visited August 28, 2015).

R125 Ted Williams, Wikipedia, en.wikipedia.org/wiki/Ted_Williams (last visited July 21, 2015).

R126 A. Ryan and M. E. Irons, Boston city councilors vote for $20,000-a-year raise, *The Boston Globe* (October 8, 2014), www.bostonglobe.com/metro/2014/10/08/boston-city-councilors-set-consider-raise/FqmrkRlm5wNDkvTBTEv8uI/story.html (last visited July 27, 2015).

R127 P. Whittle, Dogfish remain abundant off of Maine, East Coast, Associated Press reported in *The Washington Post* (September 28, 2014), www.washingtontimes.com/news/2014/sep/28/dogfish-remain-abundant-off-of-maine-east-coast/ (last visited July 21, 2015).

R128 New Zealand Permanent Warning, Wikipedia, commons.wikimedia.org/wiki/File:New_Zealand_Permanent_Warning_-_Steep_Up_Grade.svg (last visited August 13, 2015). The copyright holder of this work allows anyone to use it for any purpose including unrestricted redistribution, commercial use, and modification.

Chapter 4

R129 Screenshot from www.bls.gov/data/inflation_calculator.htm.

R130 Health Care Cost Institute, Changes in Health Care Spending in 2011 (2012), www.healthcostinstitute.org/files/HCCI_IB3_Spending.pdf (last visited July 17, 2015).

R131 M. M. Crow, Growing a better NIH, *The Boston Globe* (June 19, 2011), www.boston.com/bostonglobe/ideas/articles/2011/06/19/growing_a_better_nih (last visited July 21, 2015).

R132 Historical Consumer Price Index (CPI-U) Data, inflationdata.com/inflation/consumer_price_index/historicalcpi.aspx (last visited August 29, 2015).

R133 Excel chart from data at www.dol.gov/minwage/chart1.htm (last visited September 24, 2015).

R134 History of Federal Minimum Wage Rates Under the Fair Labor Standards Act, 1938–2009, Wage and Hour Division (WHD), United States Department of Labor, www.dol.gov/whd/minwage/chart.htm (last visited August 2, 2015).

R135 N. Bilton, For Sale: A $160,000 Apple Computer, *The New York Times* (November 11, 2010), bits.blogs.nytimes.com/2010/11/11/for-sale-a-16000-apple-computer/ (last visited July 21, 2015).

R136 Secretary of the Commonwealth of Massachusetts, The Massachusetts State House Model, www.sec.state.ma.us/trs/trsbok/mod.htm (last visited July 21, 2015).

R137 C. Lohmann, Raise bottle deposit to 10 cents, Letter to the Editor, *The Boston Globe* (December 29, 2011), www.bostonglobe.com/opinion/letters/2011/12/29/raise-bottle-deposit-cents/0x0nLO7DKi10wE69Ru5GiN/story.html (last visited July 21, 2015).

R138 M. Cieply, Hollywood Math: Bad to Worse. *The New York Times* (January 10, 2011), query.nytimes.com/gst/fullpage.html?res=9E0CE1DE173CF933A25752C0A9679D8B63 (last visited October 2, 2015).

R139 A. J. Liebling, The Jollity Building, *The New Yorker* (April 26, 1941), www.newyorker.com/magazine/1941/04/26/the-jollity-building (last visited July 27, 2015).

R140 A. J. Liebling; Introduction by David Remnick, *Just Enough Liebling*, North Point Press, us.macmillan.com/justenoughliebling/ajliebling (last visited July 27, 2015).

R141 S. Bauer, Committee approves 1 percent pay raise for state, University of Wisconsin workers, Associated Press reported in the Minneapolis-St.Paul *Star Tribune* (June 26, 2013), www.startribune.com/nation/213097941.html (last visited July 22, 2015).

R142 Private colleges vastly outspent public peers, Bloomberg News reported in *The Boston Globe* (July 10, 2010), www.boston.com/news/education/higher/articles/2010/07/10/private_colleges_vastly_outspent_public_peers/ (last visited July 22, 2015).

R143 N. Mitchell, DPS, teachers' union reach accord, Chalkbeat Colorado (June 19, 2012), co.chalkbeat.org/2012/06/19/dps-teachers-union-reach-accord/ (last visited July 22, 2015).

R144 S. E. Harger, County wages dropped nearly 14 percent in last decade, *The Portland Tribune* (June 6, 2012), portlandtribune.com/scs/83-news/110775-report-county-wages-dropped-nearly-14-percent-in-last-decade (last visited July 22, 2015).

R145 Newspaper sales slid to 1984 level in 2011, Reflections of a Newsosaur, newsosaur.blogspot.com/2012/03/newspaper-sales-slid-to-1984-level-in.html (last visited July 22, 2015).

R146 M. J. Perry, Free-fall: Adjusted for Inflation, Print Newspaper Advertising Will be Lower This Year Than in 1950, *Carpe Diem* (September 6, 2012), mjperry.blogspot.com/2012/09/freefall-adjusted-for-inflation-print.html (last visited July 22, 2015).

R147 D. Owen, Penny Dreadful, *The New Yorker* (March 31, 2008), www.newyorker.com/magazine/2008/03/31/penny-dreadful (last visited July 28, 2015).

Chapter 5

R148 Grading System, Office of the Registrar, UMass Boston, www.umb.edu/registrar/grades_transcripts/grading_system (last visited July 28, 2015).

R149 Grading System, Office of the Registrar, UMass Boston, www.umb.edu/registrar/grades_transcripts/grading_system (last visited July 28, 2015).

R150 United States Department of Labor, Bureau of Labor Statistics, Consumer Price Index Frequently Asked Questions. www.bls.gov/cpi/cpifaq.htm (last visited July 17, 2015).

R151 D. Strumpf, New-vehicle prices plunge, report says, Associated Press reported in *The Columbus Dispatch* (September 5, 2008), www.dispatch.com/content/stories/business/2008/09/05/new_vehicle_prices_0905.ART_ART_09-05-08_C10_AQB7T2V.html (last visited July 22, 2015).

R152 Data from M. Arsenault, Out-of-state donations filling Warren's campaign coffers, *The Boston Globe* (October 20, 2011), www.bostonglobe.com/metro/2011/10/19/out-state-donations-filling-elizabeth-warren-campaign-coffers/HColE63GGanoV1NZ1M6YtM/story.html (last visited July 22, 2015).

R153 H. Dondis and P. Wolff, Chess Notes *The Boston Globe* (March 10, 2008), secure.pqarchiver.com/boston-sub/doc/405109540.html?FMT=FT&FMTS=CITE:FT&type=current&date=Mar+10\%2C+2008&author=DONDIS\%2C+HAROLD\%3B+Wolff\%2C+Patrick&pub=Boston+Globe&edition=&startpage=&desc=CHESS+NOTES (last visited July 22, 2015).

R154 A. Ryan, Month may become dimmest on record, *The Boston Globe* (June 23, 2009) www.boston.com/news/local/massachusetts/articles/2009/06/23/so_far_june_sunlight_in_boston_is_lowest_in_past_century/ (last visited July 22, 2015).

R155 *The Hightower Lowdown*, May 2010, Volume 12, Number 5. www.hightowerlowdown.org/node/2330 (last visited July 28, 2015).

R156 Rising food prices mean a more costly Thanksgiving, *The Boston Globe* (November 19, 2011), www.bostonglobe.com/business/2011/11/19/rising-food-prices-mean-more-costly-thanksgiving/nlGoSN1DsnVRfn1M0hTXdI/igraphic.html (last visited July 22, 2015).

R157 From wire reports, No fill-ups at Kyrgyzstan base for U.S., *USA Today* (June 2, 2010), usatoday30.usatoday.com/printedition/news/20100602/capcol02_st.art.htm (last visited July 22, 2015).

R158 D. Hemenway, Why your classes are larger than "average", *Mathematics Magazine*, Vol. 55, No. 3 (May, 1982), pp. 162–164, Mathematical Association of America, Article DOI: 10.2307/2690083, www.jstor.org/stable/2690083 (last visited August 3, 2015).

Chapter 6

R159 Baby Infant Growth Chart Calculator, www.infantchart.com/ (last visited August 3, 2015).

R160 en.wikipedia.org/wiki/File:Standard_deviation_diagram.svg (last visited August 3, 2015). Licensed under the Creative Commons Attribution 2.5 Generic License.

R161 Raising Taxes on Rich Seen as Good for Economy, Fairness, Pew Research Center (July 16, 2012), www.people-press.org/2012/07/16/raising-taxes-on-rich-seen-as-good-for-economy-fairness (last visited July 28, 2015). Quoted with permission.

R162 Data source: Surveillance, Epidemiology, and End Results (SEER) Program (www.seer.cancer.gov) SEER*Stat Database: Incidence—SEER 9 Regs Limited-Use, Nov 2008 Sub (1973–2006), National Cancer Institute, DCCPS, Surveillance Research Program, Cancer Statistics Branch, released April 2009, based on the November 2008 submission. Graphic drawn by Ben Bolker.

R163 E. Moskowitz, Cash-strapped T proposes 23 percent fare increase, *The Boston Globe* (March 28, 2012), bostonglobe.com/metro/2012/03/28/mbta-unveils-percent-fare-hike-limited-service-cuts-also-proposed/moCl42rwr0Nf5xyx20ZQGP/story.html (last visited July 22, 2015).

R164 M. C. Fisk and J. Lawrence, Walmart to Settle Massachusetts Suit for $40 Million (Update2), *Bloomberg News* (December 2, 2009), www.bloomberg.com/apps/news?pid=newsarchive&sid=a2AClc9J8WwE (last visited October 2, 2015).

R165 Graphic redrawn from data from data J. P. Kahn, Missed connections in our digital lives, *The Boston Globe* (April 15, 2012), www.bostonglobe.com/metro/2012/04/14/missed-connections-our-digital-lives/bPHauWdvUl5XAd1ol7SOQL/igraphic.html (last visited August 2, 2015).

R166 Jakob Nielsen, Aspects of Design Quality Nielsen Norman Group (November 3, 2008), www.nngroup.com/articles/aspects-of-design-quality/ (last visited July 22, 2015), image © Neilsen Norman Group, reproduced with permission.

R167 K. Geldis, The Richest Counties in America, *TheStreet* (February 13, 2012), www.thestreet.com/story/11415107/3/the-richest-counties-in-america.html (last visited July 22, 2015).

R168 Wikipedia, commons.wikimedia.org/wiki/File:Distribution_of_Annual_Household_Income_in_the_United_States_2010.png (last visited August 11, 2015). Creative Commons Attribution-Share Alike 3.0 Unported license.

R169 Graphic redrawn from data scraped from a Nate Silver *New York Times* graphic published October 31, 2012. The original seems not to be available.

R170 Data from D. Slack, Boston spends most on firefighters in US, *The Boston Globe* (March 30, 2009), www.boston.com/news/local/massachusetts/articles/2009/03/30/boston_spends_most_on_firefighters_in_us/, data for graphic at www.boston.com/news/local/massachusetts/articles/2009/03/30/fire_spending/ (last visited August 11, 2015).

R171 D. Slack and J. C. Drake, Error made in fire dept. report, *The Boston Globe* (March 31, 2009) www.boston.com/news/local/massachusetts/articles/2009/03/31/error_made_in_fire_dept_report/ (last visited July 22, 2015).

R172 Data from J. Stripling and A. Fuller, Presidents Defend Their Pay as Public Colleges Slash Budgets, *The Chronicle of Higher Education* (April 3, 2011), chronicle.com/article/Presidents-Defend-Their/126971 (last visited August 11, 2015).

R173 Paul Erdős, en.wikipedia.org/wiki/Paul_Erdos (last visited July 22, 2015).

R174 The distribution of Erdős numbers, wwwp.oakland.edu/enp/trivia/ (last visited October 11, 2015).

R175 M. Schlueb and D. Damron, Activists press officials to put sick-leave proposal to voters, *Orlando Sentinel* (August 6, 2012), articles.orlandosentinel.com/2012-08-06/news/os-sick-leave-ballot-race-20120806_1_ballot-language-signatures-sick-time (last visited July 22, 2015).

R176 user kmbunday, Two examples of innumeracy in books for parents about gifted children, Davidson Institute (April 8, 2013), giftedissues.davidsongifted.org/BB/ubbthreads.php/topics/152941/Re_Innumeracy_in_Gifted_Educat.html (last visited July 29, 2015).

R177 M. Woolhouse, A Boston taco tells the tale of far-reaching food cost woe, *The Boston Globe* (February 06, 2015). www.bostonglobe.com/business/2015/02/05/food-prices-spike-increasing-cost-taco/vU3c42L99X9fBt25opSKkO/story.html (last visited December 16, 2015).

Chapter 7

R178 Document from www.squashedfrogs.co.uk/ (last visited August 31, 1015), no longer available.

R179 The Cook Nuclear Plant. www.cookinfo.com/cookplant.htm (last visited July 17, 2015).

R180 K. P. Erb, IRS Announces 2014 Tax Brackets, Forbes.com (October 31, 2013), www.forbes.com/sites/kellyphillipserb/2013/10/31/irs-announces-2014-tax-brackets-standard-deduction-amounts-and-more/ (last visited August 12, 2015).

R181 Data from G. Anrig, 10 Reasons to Eliminate the Tax Break for Capital Gains, The Century Foundation (October 20, 2011), tcf.org/blog/detail/10-reasons-to-eliminate-the-tax-break-for-capital-gains (last visited August 4, 2015).

R182 M. Kanellos, From Edison's Trunk, Direct Current Gets Another Look, *The New York Times* (November 17, 2011), www.nytimes.com/2011/11/18/business/energy-environment/direct-current-technology-gets-another-look.html (last visited July 22, 2015).

R183 J. DiMiceli, Public Street Trees—A Choice, Newton Conservators (April 2012), www.newtonconservators.org/newsletters/apr12.pdf (last visited July 22, 2015). Quoted with permission.

R184 Data from H. Bray, Pay full price for iPhone, avoid contract, *The Boston Globe* (June 14, 2012), bostonglobe.com/business/2012/06/14/bgcom-techlab/AskcWIPBv1qccvmqIx7DmK/story.html (last visited July 22, 2015).

R185 H. McGee, How Much Water Does Pasta Really Need?, *The New York Times* (February 24, 2009), www.nytimes.com/2009/02/25/dining/25curi.html (last visited July 22, 2015).

R186 R. Randazzo, New Arlington Valley solar site packs power, *The Arizona Republic* (May 1, 2013). www.azcentral.com/business/arizonaeconomy/articles/20130501new-arlington-valley-solar-site-packs-power.html (last visited July 22, 2015).

R187 M. Dickerson, Wind-power industry seeks trained workforce, *Los Angeles Times* (March 1, 2009), articles.latimes.com/2009/mar/01/business/fi-wind-bootcamp1/ (last visited July 22, 2015).

R188 J. M. Roney, World Solar Power Topped 100,000 Megawatts in 2012, Earth Policy Institute (July 31, 2013), www.earth-policy.org/indicators/C47/solar_power_2013 (last visited July 23, 2015).

R189 Average electricity consumption per electrified household, www.wec-indicators.enerdata.eu/household-electricity-use.html (last visited November 14 2015).

R190 E. Ailworth, Chilling out by the quarry, *The Boston Globe* (August 16, 2010), www.boston.com/business/technology/articles/2010/08/16/chilling_out_by_the_quarry/ (last visited July 23, 2015).

R191 J. Coifman, An energy program too efficient for its own good, *The Boston Globe* (March 26, 2011), www.boston.com/bostonglobe/editorial_opinion/oped/articles/2011/03/26/an_energy_program_too_efficient_for_its_own_good/ (last visited July 23, 2015).

R192 Attorney General: Connecticut electricity tax could cost Mass. $26M, *The Norwich Bulletin* (June 7, 2011), www.norwichbulletin.com/x832282338/Attorney-General-Connecticut-electricity-tax-could-cost-Mass-26M (last visited July 23, 2015).

R193 J. Carney, President Obama and Vice President Biden's 2013 Tax Returns, The White House Blog (April 11, 2014), www.whitehouse.gov/blog/2014/04/11/president-obama-and-vice-president-biden-s-2013-tax-returns (last visited July 23, 2015).

R194 Adjusted gross income, Wikipedia, en.wikipedia.org/wiki/Adjusted_gross_income (last visited November 14, 2015).

R195 B. Rooney, Pandora raises IPO target to $200 million, CNNMoneyTech (June 10, 2011), money.cnn.com/2011/06/10/technology/pandora_ipo/index.htm (last visited July 23, 2015).

R196 Sotomayor will help usher out 2013 in NYC, Associated Press report in *The Boston Globe* (December 30, 2013), www.bostonglobe.com/news/nation/2013/12/30/justice-sotomayor-lead-times-square-ball-drop/NacOVg3INmXYcLcLIYSz4J/story.html (last visited July 29, 2015).

R197 sifxtreme, Explain travel times and distances on flight, Travel Stack Exchange (April 15, 2014), travel.stackexchange.com/questions/26083/explain-travel-times-and-distances-on-flight (last visited July 23, 2015).

R198 B. Teitell, 17 holiday blunders (and how to avoid them), *The Boston Globe* November 27, 2015, www.bostonglobe.com/lifestyle/2015/11/27/the-mistakes-you-make-this-holiday-season-and-how-avoid-them-maybe/cvAPjtOkhi8QoHcZDTig1L/story.html (last visited December 16, 2015).

Chapter 8

R199 J. Keohane, Imaginary fiends, *The Boston Globe* (2010). www.boston.com/bostonglobe/ideas/articles/2010/02/14/imaginary_fiends/ (last visited July 17, 2015).

R200 Data from www2.fbi.gov/ucr/cius2008/data/table_01.html (last visited August 21, 2015) and www.gallup.com/poll/123644/Americans-Perceive-Increased-Crime.aspx (last visited August 21, 2015).

R201 Wikipedia, Anscombe's quartet, en.wikipedia.org/wiki/Anscombe's_quartet (last visited July 23, 2015).

R202 Photo: "The Leaning Tower of Pisa SB" by Saffron Blaze—Own work. Licensed under CC BY-SA 3.0 via Wikimedia Commons commons.wikimedia.org/wiki/File:The_Leaning_Tower_of_Pisa_SB.jpeg#/media/File:The_Leaning_Tower_of_Pisa_SB.jpeg (last visited August 13, 2015). The data can be found in D. S. Moore and G. McCabe, Introduction to the Practice of Statistics (last visited August 31, 2015).

R203 United States Postal Service, Postage Rates and Historical Statistics, about.usps.com/who-we-are/postal-history/rates-historical-statistics.htm (last visited August 14, 2015).

R204 D. Dineen, Despite its many benefits, corporate use of aircraft still vilified, *The Boston Globe* (May 26, 2012), www.bostonglobe.com/opinion/letters/2012/05/25/despite-its-many-benefits-corporate-use-aircraft-still-vilified/mbQ6mINMQXbAayzWvFn6NI/story.html (last visited July 23, 2015).

R205 Mark Twain, Life on the Mississippi, www.gutenberg.org/files/245/245.txt (last visited July 23, 2015).

R206 R. Munroe, xkcd, xkcd.com/552/ (last visited August 14, 2015). From xkcd.com/about/: You are welcome to reprint occasional comics pretty much anywhere (presentations, papers, blogs with ads, etc). If you're not outright merchandizing, you're probably fine. Just be sure to attribute the comic to xkcd.com.

Chapter 9

R207 High-Level Radioactive Waste, Nuclear Information and Resource Service, www.nirs.org/factsheets/hlwfcst.htm (last visited October 14, 2015). Quoted with permission.

R208 Data from V. Cooper, University of New Hampshire, used with permission.

R209 T. R. Malthus, An Essay on the Principle of Population. www.gutenberg.org/etext/4239 (last visited July 17, 2015).

R210 S. Clifford, Other Retailers Find Ex-Blockbuster Stores Just Right, *The New York Times* (April 8, 2011), www.nytimes.com/2011/04/09/business/09blockbuster.html (last visited July 23, 2015).

R211 E. Osnos, Green Giant, *The New Yorker* (December 21, 2009), www.newyorker.com/magazine/2009/12/21 (last visited July 29, 2015).

R212 S, Graham and M. Hebert, Writing to Read, Carnegie Corporation (2010), all4ed.org/wp-content/uploads/2010/04/WritingToRead.pdf (last visited July 23, 2015).

R213 Educating mothers saves lives, study says, Associated Press reported in *The Boston Globe* (September 17, 2010), www.boston.com/news/world/europe/articles/2010/09/17/educating_mothers_saves_lives_study_says/ (last visited July 23, 2015).

R214 V. Heffernan, The Trouble With E-Mail, *The New York Times* (May 29, 2011), opinionator.blogs.nytimes.com/2011/05/29/the-trouble-with-e-mail/ (last visited July 23, 2015).

R215 M. Rosenberg, India's Population, About.com (April 1, 2011), geography.about.com/od/obtainpopulationdata/a/indiapopulation.htm (last visited July 23, 2015).

R216 M. B. Farrell, MIT grad led team that built faster YouTube player, *The Boston Globe* (September 24, 2012), www.bostonglobe.com/business/2012/09/23/building-faster-youtube/JqbVsEFUJfa5tpQmgbujkL/story.html (last visited July 23, 2015).

R217 N. Silver, *The Signal and the Noise*, page 32, Penguin Press (September 27, 2012).

R218 L. Neyfakh, Cuba, you owe us $7 billion, *The Boston Globe* (April 18, 2014), www.bostonglobe.com/ideas/2014/04/18/cuba-you-owe-billion/jHAufRfQJ9Bx24TuzQyBNO/story.html (last visited July 23, 2015).

R219 J. Barron, As Time Goes By, What's This Piano Worth?, *The New York Times* (December 13, 2012), cityroom.blogs.nytimes.com/2012/12/13/as-time-goes-by-whats-this-piano-worth/ (last visited July 23, 2015).

Chapter 10

R220 Banking Tutor, PracticalMoneySkills.com, www.practicalmoneyskills.com/flash/bank_tutor/index.html (last visited August 14, 2015). From www.practicalmoneyskills.com/foreducators/roadmap.php: All resources are free and can be used by educators, parents and consumers of all ages.

R221 United States Federal Trade Commission, Consumer Information, Free Credit Reports (2013). www.consumer.ftc.gov/articles/0155-free-credit-reports (last visited July 17, 2015).

R222 Credit Reports and Scores, USA.gov, www.usa.gov/topics/money/credit/credit-reports/bureaus-scoring.shtml (last visited July 23, 2015).

R223 J. B. McKim, Rates for big loans tumble, *The Boston Globe* (November 20, 2010), www.boston.com/realestate/news/articles/2010/11/20/rates_for_big_loans_tumble/ (last visited July 23, 2015).

R224 B. Applebaum, Without Loan Giants, 30-Year Mortgage May Fade Away, *The New York Times* (March 3, 2011), www.nytimes.com/2011/03/04/business/04housing.html (last visited July 23, 2015).

R225 By P. McMorrow, The end of 30-year fixed- rate mortgage?, *The Boston Globe* (March 4, 2011) www.boston.com/bostonglobe/editorial_opinion/oped/articles/2011/03/04/the_end_of_30_year_fixed__rate_mortgage/ (last visited July 23, 2015).

R226 N. H. caps rates on payday loans, Associated Press reported in *The Boston Globe* (January 1, 2009), www.boston.com/business/articles/2009/01/01/nh_caps_rates_on_payday_loans/ (last visited July 23, 2015).

R227 J, Saltzman, Charity sues R.I. hospital over donation in 1912, *The Boston Globe* (February 23, 2008), www.boston.com/news/local/articles/2008/02/23/charity_sues_ri_hospital_over_donation_in_1912/ (last visited July 24, 2015).

R228 The Nilson Report (May 19, 2014), www.nilsonreport.com. Quoted with permission.

Chapter 11

R229 National Gambling Impact Study Commission, Lotteries. govinfo.library.unt.edu/ngisc/research/lotteries.html (last visited July 17, 2015).

R230 upload.wikimedia.org/wikipedia/commons/5/5d/13-02-27-spielbank-wiesbaden-by-RalfR-094.jpg (last visited September 18, 2015). Licensed under the Creative Commons Attribution-Share Alike 3.0 Unported, 2.5 Generic, 2.0 Generic and 1.0 Generic license.

R231 L. P. Weston, Dump the Insurance on your Clunker, *MSN Money* (March 2007), reposted at www.insurancemommy.com/Images/dumpyourclunker.pdf (last visited October 4, 2015).

R232 M. Novak, 9 Albert Einstein Quotes That Are Totally Fake, paleofuture.gizmodo.com/9-albert-einstein-quotes-that-are-totally-fake-1543806477 (last visited March 13, 2014).

R233 D. Fears, Possible cut to beach testing a health threat, critics say, Washington Post report in *The Boston Globe* (March 4, 2012), www.bostonglobe.com/news/nation/2012/03/04/elimination-funding-for-beach-contamination-monitoring-could-health-hazard-environmentalists-say/9sGB4Sz1U2m3CvM6jaINhN/story.html (last visited July 24, 2015).

R234 M. Torres, What Is a TV Extended Warranty?, About.com, tv.about.com/od/warranties/a/buyexwarranty.htm (last visited July 29, 2015).

R235 CyberCemetery, govinfo.library.unt.edu/ (last visited July 24, 2015).

R236 www.masslottery.com/winners/pf_faqs.html (last visited July 30, 2015).

R237 S. Bishop, Lottery suspense builds in Mass. for $355m prize, *The Boston Globe* (January 5, 2011), www.boston.com/news/local/massachusetts/articles/2011/01/05/lottery_suspense_builds_in_mass_for_355m_prize/ (last visited July 24, 2015).

R238 M. Levenson, Megabucks plan rankles lottery players, *The Boston Globe* (March 21, 2009), www.boston.com/news/local/massachusetts/articles/2009/03/21/megabucks_plan_rankles_lottery_players/ (last visited July 24, 2015).

R239 C. Bialik, Lottery Math 101, *The Wall Street Journal* (September 22, 2009), blogs.wsj.com/numbersguy/lottery-math-101-801/ (last visited July 30, 2015).

R240 CrzRsn, Roulette strategy . . . , forums.finalgear.com/off-topic/roulette-strategy-26295/ (March 27 2008), (last visited September 3, 2015).

R241 P. B. Brown, Avoiding a Problem C.E.O., *The New York Times* (March 8, 2008), www.nytimes.com/2008/03/08/business/08offline.html (last visited July 29, 2015).

Chapter 12

R242 S. Hazzard, *The Transit of Venus*, Viking Press, New York, 1980, p. 62.

R243 D. Robinson, Cancer Clusters: Findings Vs Feelings, Medscape. www.medscape.com/viewarticle/442554_5 (last visited July 15, 2015).

R244 J. B. McKim, Flooded with evidence, *The Boston Globe* (March 20, 2010), www.boston.com/business/articles/2010/03/20/flooded_with_evidence/ (last visited July 24, 2015).

R245 A. Thompson, What is a 100-Year Storm?, LiveScience (September 22, 2009), www.livescience.com/environment/090922-100-year-storm.html (last visited July 24, 2015).

R246 M. Valencia, After the rains, the reckoning, *The Boston Globe* (April 3, 2010)), www.boston.com/news/local/massachusetts/articles/2010/04/03/after_record_rainfall_secondary_effects_begin_to_emerge/ (last visited October 4, 2015).

R247 G. Trudeau, Doonesbury (March 21, 2009), www.gocomics.com/doonesbury/2009/03/21 (last visited August 11, 2015).

R248 C. Bialik, Lottery Math 101, *The Wall Street Journal* (September 22, 2009), blogs.wsj.com/numbersguy/lottery-math-101-801/ (last visited July 24, 2015).

R249 A. Gelman, Lottery probability update, Statistical Modeling, Causal Inference, and Social Science (May 26 2011), andrewgelman.com/2011/05/26/lottery_probabi/ (last visited July 24, 2015). Quoted with permission.

R250 Sicherman dice, Wikipedia, en.wikipedia.org/wiki/Sicherman_dice (last visited August 11, 2015).

R251 J. Pope, Colleges bewildered by anonymous major gifts, Associated Press report in *The Boston Globe* (April 24, 2009), www.boston.com/news/nation/articles/2009/04/24/colleges_bewildered_by_anonymous_major_gifts/ (last visited October 4, 2015).

R252 Ever wonder if the lottery numbers you play everyday have actually already hit, before you started playing them?, *Numbers Planet*, www.numbersplanet.com/ (last visited July 24, 2015).

R253 Odd vs. Even statistics, *Numbers Planet*, www.numbersplanet.com/ (last visited July 24, 2015).

R254 R. Highfield, Olympic Games drug testing means 'cheaters escape and innocents tarnished', *The Telegraph* (August 6 2008), www.telegraph.co.uk/science/science-news/3348936/Olympic-Games-drug-testing-means-cheaters-escape-and-innocents-tarnished.html (last visited July 24, 2015).

R255 From an email from Old Friends Farm, www.oldfriendsfarm.com/ (last visited August 3, 2015). "Go ahead and use the quote—thanks for asking!"

R256 Alexander McCall Smith, *The Double Comfort Safari Club*, Pantheon Books, 2010, page 116.

Chapter 13

R257 M. S. Klinkman, J. C. Coyne, S. Gallo and T. L. Schwenk, False Positives, False Negatives, and the Validity of the Diagnosis of Major Depression in Primary Care, *Arch Fam Med.* 1998;7(5): 451–461, www.ncbi.nlm.nih.gov/pubmed/9755738 (last visited October 4, 2015). Licensed under a Creative Commons Attribution-Noncommercial-No Derivative Works 3.0 United States License. (creativecommons.org/licenses/by-nc-nd/3.0/

R258 T. Davies, askville.amazon.com/understand-False-Positive-test-Trisomy-18/AnswerViewer.do?requestId=12714458 (last visited July 25, 2015).

R259 Amniocentesis, American Pregnancy Association, americanpregnancy.org/prenataltesting/amniocentesis.html (last visited July 15, 2015).

R260 M. Lynn and C. Maier, McDaniel v. Brown (08-559), Legal Information Institute, LII Supreme Court Bulletin, Cornell University Law School, www.law.cornell.edu/supct/cert/08-559 (last visited September 18, 2015). This material is covered by a Creative Commons license, viewable at creativecommons.org/licenses/by-nc-sa/2.5/.

R261 D. Badertscher, U.S. Supreme Court Update: McDaniel v. Brown, Criminal Law Library Blog (January 26, 2010), www.criminallawlibraryblog.com/2010/01/us_supreme_court_update_mcdani.html (last visited July 25, 2015).

R262 R. Stein, Researchers link chronic fatigue syndrome to class of virus, Washington Post report in *The Boston Globe* (August 24, 2010), www.boston.com/news/nation/articles/2010/08/24/researchers_link_chronic_fa\tigue_syndrome_to_class_of_virus (last visited July 25, 2015).

R263 M. Reyes *et. al.*, Prevalence and incidence of chronic fatigue syndrome in Wichita, Kansas. *Arch Intern Med.* 2003 Jul 14;163(13):1530–6, www.ncbi.nlm.nih.gov/pubmed/12860574 (last visited December 15, 2015).

R264 How Common Is A False Positive Pregnancy Test And What Causes It?, BabyHopes.com, www.babyhopes.com/articles/falsepositive.html (last visited July 25, 2015).

R265 M. Specter, Damn Spam, Annals of Technology, *The New Yorker* (August 6, 2007), www.newyorker.com/reporting/2007/08/06/070806fa_fact_specter (last visited July 25, 2015).

R266 G. Karthikeyan, The cost of a “negative test”, response to Screening programme evaluation applied to airport security, *British Medical Journal* (December 27 2007), www.bmj.com/rapid-response/2011/11/01/cost-negative-test (last visited September 4, 2015). Quoted with permission.

R267 S. Strogatz, Chances Are, *The New York Times* (April 25, 2010), opinionator.blogs.nytimes.com/2010/04/25/chances-are/ (last visited March 2, 2016).

R268 J. Allen, Identity fraud dragnet hardly seems worth the expense or trouble, *The Boston Globe* (July 24, 2011), www.boston.com/bostonglobe/editorial_opinion/letters/articles/2011/07/24/identity_fraud_dragnet_hardly_seems_worth_the_expense_or_trouble/ (last visited July 25, 2015). Quoted with permission.

R269 E. Wahlgren, Happy Halloween! Kids who eat candy every day grow up to be violent criminals, originally published on DailyFinance.com (October 2, 2009), www.dailyfinance.com/2009/10/02/happy-halloween-kids-who-eat-candy-every-day-grow-up-to-be-viol/ (last visited July 30, 2015). Quoted with permission.

R270 A. Gelman, The Reliability of Cluster Surveys of Conflict Mortality: Violent Deaths and Non-Violent Deaths, Statistical Modeling, Causal Inference, and Social Science (August 11 2011), andrewgelman.com/2011/08/the_reliability/ (last visited July 25, 2015). Quoted with permission.

R271 D. Kotz, Surgery offers no advantage for early prostate cancer, study finds, *The Boston Globe* (July 18, 2012), bostonglobe.com/lifestyle/health-wellness/2012/07/18/surgery-offers-survival-advantage-for-older-men-with-early-stage-prostate-cancer-study-finds/T5XM7APIuoZuav6PbJzYuI/story.html (last visited July 25, 2015).

R272 How to Avoid False Positives While Conducting a Home Drug Test, lapoliticaesotracosa.blogspot.com/2012/05/how-to-avoid-false-positives-while.html (last visited July 25, 2015).

Index

About the Authors

Ethan Bolker was born in Brooklyn, New York in 1938. In his first year at Erasmus Hall High School he was hooked by Hugo Steinhaus's *Mathematical Snapshots*. In his senior year he captained the Math Team, then went on to major in mathematics at Harvard College, where he earned his degree *summa cum laude*. He turned down medical school to continue in mathematics at Harvard, where Andy Gleason supervised his PhD.

He was an instructor at Princeton University and an Assistant and Associate Professor at Bryn Mawr College before coming to UMass Boston as a Full Professor in 1972. Ethan retired in 2014 and awaits official endorsement of his Emeritus status. While at UMass he chaired the joint Mathematics and Computer Science Departments for 12 years. He earned the Chancellor's Award for Scholarship in 1979 and for Teaching in 2003.

Ethan has taught more than 30 different courses at all levels (graduate and undergraduate) in both mathematics and computer science. For 15 years he has spent a day a week in an elementary school, working with teachers and small groups of students in grades K–7.

He has held visiting research associate positions at the University of California at Berkeley, at Harvard in both the Mathematics Department and the Graduate School of Design, and at MIT. For 25 years he consulted at BGS Systems and BMC Software on computer capacity planning and queueing theory.

His research interests have included measure theory, convex geometry, integral geometry, the geometry of polyhedra, combinatorics, utility theory, rigidity of frameworks and queueing theory. He has published more than 50 refereed papers.

Ethan is the author of four books: *Java Outside In* (with Bill Campbell); *Using Algebra*; *First Year Calculus*, (with Joseph Kitchen, Jr.); and *Elementary Number Theory, an Algebraic Approach*.

Maura Mast became Dean of Fordham College at Rose Hill in August 2015. She is the first woman to be dean of the college and the first dean with a background in science and mathematics. Prior to coming to Fordham, she was Vice Provost for Undergraduate Studies and Associate Professor of Mathematics at the University of Massachusetts Boston.

Maura earned her PhD in mathematics from the University of North Carolina and her Bachelor of Arts degree from the University of Notre Dame with a double major in mathematics and anthropology. Maura is an active researcher in the field of differential geometry, primary focusing on understanding geodesic behavior as a means of exploring the relationship between geometric properties and analytic properties of a manifold. Before coming to UMass

Boston, she was associate professor of mathematics at the University of Northern Iowa; she has also held visiting positions at Northeastern University, Wellesley College and the University of Notre Dame. Maura has been recognized repeatedly for her teaching abilities: she received the Chancellor's Award for Excellence in Undergraduate Teaching from the University of North Carolina, the College of Natural Sciences Dean's Award for Teaching Excellence from the University of Northern Iowa, and the Science Dean's Award for Outstanding Achievement in Teaching, Research and Service from the University of Massachusetts Boston. Maura is a successful grant writer and has received National Science Foundation and other funding for her research and teaching projects.

In 2012, Maura was selected to be a Fellow of the American Council on Education (ACE). This program, the premier higher education leadership development programs in the United States, condenses years of learning about higher education and leadership into a single year. As a Fellow, Maura held a placement with the President of Merrimack College; participated in extended research, discussions and conferences on topics such as diversity, change management, student success, strategic planning, and leadership; and visited with campus leaders at over 40 higher education institutions in the United States, the United Arab Emirates, the Costa Rica.

Maura's other professional activities include mathematics course development and participation in projects that provide access to mathematics to diverse audiences. In addition to co-authoring the *Common Sense Mathematics* textbook, she was a contributing author to the University of Texas Charles A. Dana Center's New Mathways Project textbook series. Maura served as chair of the MAA's Special Interest Group on Quantitative Literacy and speaks nationally about contemporary approaches to quantitative literacy teaching and learning; mathematics and social justice; and access to mathematics for all students. Maura is a strong advocate for the participation of women and girls in mathematics at all levels. She has served on the Executive Committee for the Association for Women in Mathematics as Clerk and as board member and she co-chaired the Joint Committee on Women in the Mathematical Sciences. Maura regularly organizes panels and discussions on issues relevant to women mathematicians at mathematics meetings.